Foundations of ELECTRIC CIRCUITS

J.R. COGDELL

University of Texas at Austin
ECE Department
Austin, Texas

PRENTICE HALL
Upper Saddle River, New Jersey 07458

Library of Congress Cataloging-in Publication Data

Cogdell, J. R.
  Foundations of electric circuits / J. R. Cogdell.
    p. cm.
  Includes bibliographical references and index.
  ISBN 0-13-907742-1
  1. Electric circuits I. Title.
TK454.C33 1999
621.3—dc21
                                     98—47447
                                        CIP

Acquisitions editor: **Alice Dworkin**
Editorial/production supervision: **Sharyn Vitrano**
Copy editor: **Barbara Danziger**
Managing editor: **Eileen Clark**
Editor-in-chief: **Marcia Horton**
Director of production and manufacturing: **David W. Riccardi**
Manufacturing buyer: **Pat Brown**
Editorial assistant: **Dan DePasquale**

 © 1999 by Prentice Hall, Inc.
Pearson Education
Upper Saddle River, New Jersey 07458

All rights reserved. No part of this book may be
reproduced, in any form or by any means,
without permission in writing by the publisher.

The author and publisher of this book have used their best efforts in preparing this book. These
efforts include the development, research, and testing of the theories and programs to determine
their effectiveness. The author and publisher make no warranty of any kind, expressed or implied,
with regard to these programs or the documentation contained in this book. The author and
publisher shall not be liable in any event for incidental or consequential damages in connection
with, or arising out of, the furnishing, performance, or use of these programs.

Printed in the United States of America

15

ISBN 0-13-907742-1

Prentice-Hall International (UK) Limited, *London*
Prentice-Hall of Australia Pty. Limited, *Sydney*
Prentice-Hall Canada Inc., *Toronto*
Prentice-Hall Hispanoamericana, S.A., *Mexico*
Prentice-Hall of India Private Limited, *New Delhi*
Prentice-Hall of Japan, Inc., *Tokyo*
Prentice-Hall (Singapore) Asia Pte. Ltd., *Singapore*
Editora Prentice-Hall do Brasil, Ltda., *Rio de Janeiro*

Contents

Preface vii

Chapter 1 Basic Circuit Theory 2

Objectives 3

1.1 Introduction to Electrical Engineering 4
Place of Electrical Engineering in Modern Technology 4

1.2 Physical Basis of Circuit Theory 5
Energy and Charge 5
What Is Circuit Theory? 6

1.3 Current and Kirchhoff's Current Law 7
Definition of Current 7
Kirchhoff's Current Law 10
Check Your Understanding 12

1.4 Voltage and Kirchhoff's Voltage Law 12
Definition of Voltage 12
Kirchhoff's Voltage Law (KVL) 14
Check Your Understanding 19

1.5 Energy Flow in Electrical Circuits 19
Voltage, Current, and Power 19
Check Your Understanding 23

1.6 Circuit Elements: Resistances and Sources 23
Resistances and Switches 23
Voltage and Current Sources 26
Analysis of DC Circuits 28
Check Your Understanding 33

1.7 Series and Parallel Resistances, Voltage and Current Dividers 33
Series Resistances and Voltage Dividers 33
Parallel Resistances and Current Dividers 35
Check Your Understanding 40

Chapter Summary 41

Glossary 42

Problems 43
Section 1.3: Current and Kirchhoff's Current Law 43
Section 1.4: Voltage and Kirchhoff's Voltage Law 44
Section 1.5: Energy Flow in Electrical Circuits 45
Section 1.6: Circuit Elements: Resistors, Switches, and Sources 46
Section 1.7: Series and Parallel Resistances; Voltages and Current Dividers 48
General Problems 51
Answers to Odd-Numbered Problems 54

Chapter 2 The Analysis of DC Circuits 56

Objectives 57

Introduction 58

2.1 Superposition 58
Superposition Illustrated 58
Principle of Superposition 59
Check Your Understanding 63

2.2 Thévenin's and Norton's Equivalent Circuits 63
Example to Justify the Concept 63
Thévenin's Equivalent Circuit 66
Impedance Level 73
Source Transformations 74
Check Your Understanding 75

2.3 Node–Voltage Analysis 76
Basic Idea 76
Node–Voltage Technique 78
Some Refinements 80

Critique 83
Check Your Understanding 84

2.4 Loop-Current Analysis 85

Simple Method of Loop-Current Analysis 85
Some Extensions and Fine Points 88
Summary of Methods of Circuit Analysis 92
Check Your Understanding 93

2.5 Controlled Sources 93

What are Control Sources? 93
What is the Significance of Control Sources? 100
How Does the Present of Control Sources Affect Circuit Analysis? 101

Chapter Summary 101

Glossary 103

Problems 104

Section 2.1: Superposition 104
Section 2.2: Thévenin's and Norton's Equivalent Circuits 105
Section 2.3: Node-Voltage Analysis 107
Section 2.4: Loop-Current Analysis 109
General Problems 110
Section 2.5: Control Sources 113
Answers to Odd-Numbered Problems 113

Chapter 3 The Dynamics of Circuits 116

Objectives 117

3.1 Theory of Inductors and Capacitors 118

Time and Energy 118
Inductor Basics 118
Capacitor Basics 123
Check Your Understanding 128

3.2 First-Order Transient Response of *RL* and *RC* Circuits 128

Classical Differential Equation Solution 129
A Simpler Method 131
RC Circuits 135

3.3 Advanced Techniques 136

Circuits with Multiple Resistors 136
Initial and Final Values 137
Pulse Problem 142
Higher-Order Transients 144
An RLC 147
Check Your Understanding 149

Chapter Summary 150

Glossary 151

Problems 151

Section 3.1: Theory of Inductors and Capacitors 151
Section 3.2: First-Order Transient Response of RL and RC Circuits 153
Section 3.3: Advanced Techniques 156
General Problems 159
Answers to Odd-Numbered Problems 159

Chapter 4 The Analysis of AC Circuits 162

Objectives 163

4.1 Introduction to Alternating Current (AC) 164

Importance of AC 164
Sinusoids 164
AC Circuit Problem 168
Check Your Understanding 169

4.2 Representing Sinusoids with Phasors 169

Sinusoids and Linear Systems 169
Mathematics of the Complex Plane 170
Phasor Idea 177
Back to the Circuit Problem 179
Check Your Understanding 185

4.3 Impedance: Representing the Circuit in the Frequency Domain 186

The Final Shortcut 186

4.4 Phasor Diagrams for *RL*, *RC*, and *RLC* Circuits 191

RL Circuits 192
RC Circuits 196
RLC Circuits 199
Check Your Understanding 202

Chapter Summary 202

Glossary 204

Problems 205

    Section 4.1: Introduction to Alternating Current (AC) 205

    Section 4.2: Representing Sinusoids with Phasors 206

    Section 4.3: Impedance: Representing the Circuit in the Frequency Domain 207

    General Problems 211

    Answers to Odd-Numbered Problems 213

Chapter 5   Circuit and System Analysis Using Complex Frequency 214

Objectives 215

    Introduction to Complex Frequency Techniques 216

5.1 Complex Frequency 216

    Check Your Understanding 221

5.2 Impedance and the Transient Behavior of Linear Systems 221

    Generalized Impedance 221
    Transient Analysis 222
    Check Your Understanding 231

5.3 Transient Response of *RLC* Circuits 232

    Types of Natural Responses in RLC Circuits 232
    RLC Circuit Transient Behavior 234
    Check Your Understanding 238

5.4 Filters and Bode Plots 238

    Spectra 238
    Filter Functions 239
    Bode Plots 245
    Frequency Response 247
    High-Pass Filter 248
    Check Your Understanding 252

5.5 Systems 252

    System Notation 252
    Representing Differential Equations in System Notation 257

Chapter Summary 260

Glossary 261

Problems 261

    Section 5.1: Complex Frequency 261
    Section 5.2: Impedance and the Transient Behavior of Linear Systems 263
    Section 5.3: Transient Response of RLC Circuits 265
    Section 5.4: Filters 266
    Section 5.5: System Analysis 267
    Answers to Odd-Numbered Problems 267

Chapter 6   Power In AC Circuits 270

Objectives 271

6.1 AC Power and Energy Storage: The Time-Domain Picture 272

    Importance of Power and Energy 272
    Average Values of Electrical Signals 272
    Effective or Root-Mean-Square (RMS) Valve 275
    Power and Energy Relations for R, L, and C 279
    General Case for Power in an AC Circuit 283
    Check Your Understanding 287

6.2 Power and Energy in the Frequency Domain 288

    Real and Reactive Power from Phasors 288
    Complex Power 291
    Reactive Power in Power Systems 295
    Reactive Power in Electronics 297
    Check Your Understanding 301

6.3 Transformers 301

    Transformer Principles 301
    Transformer Applications in AC Power Systems 308
    Residential AC Power 312
    Electrical Safety 314
    Check Your Understanding 318

Chapter Summary 318

Glossary 320

Problems 321

    Section 6.1: AC Power and Energy Storage: The Time-Domain Picture 321
    Section 6.2: Power and Energy in the Frequency Domain 324
    Section 6.3: Transformers 326

General Problems *328*
Answers to Odd-Numbered
   Problems *329*

Chapter 7   Electric Power Systems   330

Objectives   331

7.1   Three-Phase Power   332

Importance of Electric Power Systems *332*
Introduction to Three–Phase Power
   Systems *332*
Three–Phase Power Sources *334*
Three-Phase Loads *339*
Per-Phase Equivalent Circuits *347*
Check Your Understanding *349*

7.2   Power Distribution Systems   349

Three-Phase Transformers *350*
Per-Unit Calculations *353*
Transmission Properties *355*
Check Your Understanding *358*

7.3   Introduction to Electric Motors   359

Terminology *359*
Motor Characterization in Steady-State
   Operation *360*
The Motor with a Load *361*
Dynamic Operation *363*
Three-Phase Induction Motor Nameplate
   Interpretation *364*
Single-Phase Induction Motor *370*
Check Your Understanding *370*

Chapter Summary   371

Glossary   372

Problems   372

Section 7.1: Three-Phase Power *372*
Section 7.2: Power Distribution Systems *376*
Section 7.3: Introduction to Electric
   Motors *377*
General Problems *381*
Answers to Odd-Numbered Problems *382*

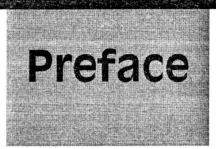

Preface

The need for this book.

About 10 years ago, I started teaching electrical engineering to nonmajors. We used a well-known text, but I found that my students had trouble with the book. The encyclopedic scope of this text of necessity forced it to be superficial. I wanted a text that presented the important ideas in depth and left many of the details for future learning in further study or professional practice. So I wrote *Foundations of Electrical Engineering,* choosing "Foundation" for the title to point to the few important principles that upheld the entire superstructure of electrical engineering. *Foundations of Electrical Engineering* was well received and went into a much-changed second edition in 1996.

The Foundations series.
The *Foundations of Electrical Engineering* has now been divided into three stand-alone books: *Foundations of Electric Circuits, Foundations of Electronics,* and *Foundations of Electric Power.* The purpose was to reduce the student's cost and to target specific one-semester (one- or two-quarter) courses as follows:

One-semester course	Foundation books required
Circuits for majors or nonmajors	*Foundations of Electric Circuits*
Circuits and electronics for nonmajors	*Foundations of Electric Circuits* (skipping Chaps. 5 and 7) and *Foundations of Electronics*
Electronics for nonmajors	*Foundations of Electronics*
Power for majors or nonmajors	*Foundations of Electric Power*

For a full-year survey of electrical engineering, including topics in circuits, electronics, and power, *Foundations of Electrical Engineering* is the best choice.

The present work corresponds roughly to one-third of the second edition of *Foundations of Electrical Engineering*. Specifically, we have separated out the work on circuits, adding an adapted version of the chapter on linear systems to give a fuller treatment of frequency domain techniques. Chapter 7 of the present volume, *Electric Power Systems*, goes beyond the material normally treated in a circuits book to apply the material to the distribution and transformation of electric power, mainly in electric motors. Most engineering students will have no other opportunity to learn about this important area of technology.

Prerequisites

We address this book to students who have completed one year of college calculus and physics. We work with linear differential equations and the algebra of complex numbers, but the techniques of solutions are explained in full. A few simple ideas from mechanics are introduced in discussions of dynamics and steady-state operations of motors, primarily Newton's second law.

Pedagogy of the book.

One way to view the structure of an engineering subject is shown in the following diagram:

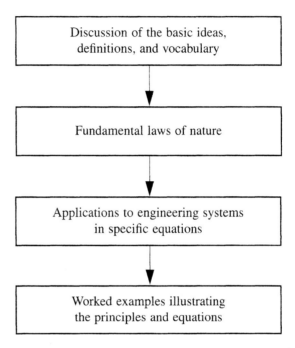

The instructor's emphasis goes from top to bottom. The structure in the instructor's (and author's) mind is primarily ideas and vocabulary, followed by laws, equations, then examples—from the general to the specific. But most students seem to learn in the opposite order: first examples, then equations, then laws, and finally ideas. One suspects some students never go beyond studying the examples, and certainly many believe the equations are what's important. When the instructor probes their understanding of the more general principles by giving a quiz problem that differs from previous examples, many students serve up memorized "solutions" or protest that "the quiz wasn't anything like the lectures and homework." The goals and needs of both instructors and students were paramount in the writing and design of this book.

Aids to learning

This text addresses the viewpoints of both students and instructors in a number of ways,

- **Ideas.** We have identified eight basic ideas of electrical engineering and identified them with a lightbulb icon in the margin, as shown at left. All but

Conservation of Energy

vocabulary for important words

LEARNING OBJECTIVE

This alerts students to important material

three of these ideas are used in the present volume. These icons hopefully will convince students that electrical engineering is based upon a few foundational ideas that come up again and again.

- **Key terms** are italicized and emphasized by a note in the margin, as shown, when they are first introduced and defined. *Vocabulary* cannot be overemphasized because we communicate with others and, indeed, do our own thinking with words. Each chapter ends with a Glossary that defines many key terms and refers to the context where the words first appear in the text.

- **Objectives.** Chapters begin with stated objectives and end with summaries that review how those objectives have been met. In between, we place marginal pointers that alert the student to where the material relates directly to one of the stated objectives. Our intention is to give road signs along the way to keep our travelers from losing their way.

EXAMPLE P.1 **This is the title of an example**

The numerous examples are boxed, titled, and numbered.

SOLUTION:
Solutions are differentiated from problem statements, as shown. Students are lured beyond passively studying the examples by a WHAT IF? challenge at the end.

WHAT IF? What if the student were asked to rework the example with a slight change?[1] The answer to the WHAT IF? challenge appears in a footnote for easy checking of results.

- **Check your understanding.** We have "Check Your Understanding" questions and problems, with answers, after major sections.

Problems

Three types of problems have different levels of difficulty and appear at three places in the development:

1. WHAT IF? challenges follow most examples. These problems present a slight variation on the examples and are intended to involve the student actively in the principles illustrated by the examples. The answers are given in a footnote.

2. Check Your Understanding problems follow most sections. These are intended as occasions for review and quick self-testing of the material in each section. The answers follow the problems.

3. The numerous end-of-chapter problems range from straightforward applications, similar to the examples, to quite challenging problems requiring insight and refined

---

[1] That would get them involved, wouldn't it?

problem-solving skills. Answers are given to the odd-numbered problems. We are convinced that the only path to becoming a good problem solver passes through a forest of nontrivial problems.

Acknowledgments

I gratefully acknowledge the assistance of many colleagues at the University of Texas at Austin: Lee Baker, David Bourell, David Brown, John Davis, Mircea Driga, "Dusty" Duesterhoeft, Bill Hamilton, Om Mandhana, Charles Roth, Irwin Sandberg, Ben Streetman, Jon Valvano, Bill Weldon, Paul Wildi, Quanghan Xu, and no doubt others. My warmest thanks go to my friend, Jian-Dong Zhu, who worked side by side with me in checking the answers to the end-of-chapter problems. Special thanks go to my son-in-law, David Brydon, who introduced me to Mathematica and to my son, Thomas Cogdell, who produced most of the figures in the text. My heartfelt thanks for numerous corrections, improvements, and wise advice go to the reviewers of *Foundations of Electrical Engineering*: William E. Bennett, U.S. Naval Academy; Richard S. Marleau, University of Wisconsin-Madison; Phil Noe, Texas A&M University; Ed O'Hair, Texas Tech University; and Terry Sculley and Carl Wells, Washington State University and to Dolon Williams for numerous corrections and suggestions.

I wish to thank my wife for her support and encouragement. Finally, I thank the Giver of all good gifts for the joy I have in teaching and writing about electrical engineering.

John R. Cogdell

Basic Circuit Theory

Introduction to Electrical Engineering

Physical Basis of Circuit Theory

Current and Kirchhoff's Current Law

Voltage and Kirchhoff's Voltage Law

Energy Flow in Electrical Circuits

Circuit Elements: Resistances and Sources

Series and Parallel Resistances; Voltage and Current Dividers

Chapter Summary

Glossary

Problems

objectives

1. To understand what a circuit is and why circuits are important in electrical engineering
2. To understand the definition of current and use Kirchhoff's current law (KCL) to express conservation of charge
3. To understand the definition of voltage and use Kirchhoff's voltage law (KVL) to express conservation of electric energy
4. To understand how to use the voltage and current to calculate the power into or out of a circuit element
5. To understand the relationship between voltage and current in a resistor as described by Ohm's law
6. To understand how to combine resistances connected in series and how voltage divides between series resistances
7. To understand how to combine resistances connected in parallel and how current divides between parallel resistances
8. Be understand how to analyze circuits containing one source and resistances in series and parallel

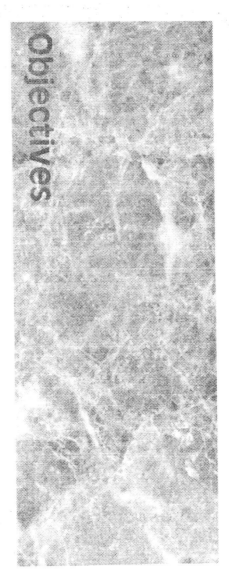

Electrical engineering begins with the subject of circuits for three reasons. The most important is that almost every electrical device, from a radio to an electric motor, is a circuit, or at least contains circuits. The second reason is that the study of circuits is neither as abstract or as mathematically sophisticated as other electrical subjects such as electromagnetic fields. Finally, this subject has produced the language of electrical engineering. Learn the vocabulary of circuits and you can break into the conversation about electrical engineering.

1.1 INTRODUCTION TO ELECTRICAL ENGINEERING

Place of Electrical Engineering in Modern Technology

electrical engineering

What is electrical engineering? Electrical engineering is, in one sense, the opposite of lightning. Lightning unleashes electrical energy unpredictably and destructively. *Electrical engineering* harnesses electrical energy for human good—for transporting energy and information, for lifting the burdens of toil and tedium. When electrical energy is important for its own sake, to turn motors or illuminate department stores, we think of the electrical power industry. When energy is important for its symbolic (information) content, we think of the electronics industry. Both use electrical energy for beneficial purposes.

This book explains the fundamental ideas and techniques of electric circuits. Our goal is to provide you with a strong foundation to solve basic and practical problems and to furnish you with a vocabulary of words and ideas for clear thinking and clear communication. All engineers need this background to contribute in a technology that is increasingly electrical.

Foundational ideas in electrical engineering. Our focus will be on the ideas upon which electrical engineering is built. The foundational ideas of electrical engineering are as follows:

- **Conservation of charge** (Kirchhoff's current law) is one of the fundamental principles used in writing circuit equations.
- **Conservation of energy** (Kirchhoff's voltage law) applies in two forms: Conservation of *electric* energy is one of the fundamental principles used in writing circuit equations; and conservation of energy generally is used in electrical/mechanical systems to develop equations and understanding.
- **The frequency domain** is a way of looking at the physical world in which frequency, not time, is the independent variable.
- **Equivalent circuits** model real devices by ideal electrical devices that have identical or similar characteristics.
- **Impedance level** determines how electrical devices interact.
- **Feedback** is a technique for bringing part of the output of an electronic device back to the input to improve performance.
- **Analog information** uses an electrical signal proportional to the information content.
- **Digital information** uses a two-valued code to represent the information content.

As we introduce, explain, and apply these ideas, we will remind you by the light bulb icon that a certain foundational idea is being used. In this way, we hope to convince you that electrical engineering is based on relatively few ideas; the rest is detail.

Why start with circuit theory? This book is about electric circuit theory. There are three reasons why the study of electrical engineering starts with circuit theory. It is an easy place to start, neither as abstract nor as mathematically sophisticated as other branches of electrical engineering. More important, practically every electrical device is a circuit. A radio is a circuit, as is the power distribution system that runs your lights and air conditioner. Hence, understanding of the methods of circuits opens the door to the study of all areas of electrical engineering.

Finally, the study of circuits provides a logical starting point because circuit theory has generated the language of electrical engineering. Even devices that are more sophisticated than an electrical circuit (an aircraft radar, for example) are described by electrical engineers in the language of circuits. Hence, our study of circuits introduces the ideas and language underlying much of electrical engineering.

1.2 PHYSICAL BASIS OF CIRCUIT THEORY

Energy and Charge

charge

Charge is a fundamental physical quantity. *Charge*, like mass, is a property of matter; indeed, charge joins mass, length, and time as one of the fundamental units from which all scientific units are derived. The unit of electric charge is the coulomb (abbreviated C), named in honor of Charles de Coulomb (1736–1806). There are two types of charge, *positive* and *negative*. The names fit because the two types of charges produce opposite effects. Thus, equations describing the effects of charges encompass both types of charges if we associate a positive number with one type and a negative number with the other. Traditionally, the electron has been assigned a negative sign and the proton a positive. The magnitude of the charge of the electron is the smallest possible charge; in the MKS system of units, this is[1]

$$e = -1.602 \times 10^{-19} \text{ coulombs (C)} \quad (1.1)$$

Because the mass of the electron is 9.11×10^{-31} kg, the charge-to-mass ratio of the electron is 1.76×10^{11} in the MKS system. The charge on the proton is positive in sign and equal in magnitude to that of the electron, but the proton mass is 1846 times greater. Because the charge-to-mass ratios of the fundamental bits of matter are so great, electrical effects usually dominate mechanical inertia effects. Hence, we usually talk about charge as if it were not tied to mass—as if it were massless.

electrostatic forces

Forces between charges. We know about electric charges because charges exert forces on other charges. There are two types of forces between charges. Charges attract or repel each other due to *electrostatic forces*, which are described by Coulomb's law. Electrostatic forces are responsible for lightning, because charges are separated in clouds by droplet separation; and electrostatic forces are used in photocopy machines to form images on glass drums with charged bits of dry ink.

magnetic forces

Magnetic forces depend on moving charges (that is, on currents) and are described by Ampere's force law. Magnetic forces turn motors, deflect electron beams in TV tubes, and effect energy conversion in generators. Electrical engineers thus have both electrostatic and magnetic effects to use in manipulating electrical energy.

energy

The importance of energy. *Energy* is the medium of exchange in a physical system, like money in an economic system. Energy is exchanged whenever one physical

---

[1] Important physical constants are tabulated on the back cover of this book.

thing affects another. In mechanics, it takes force and movement to do work (exchange energy), and in electricity, it takes electrical force and movement of charges to do work (exchange energy). The electrical force is represented by the voltage and the movement of charge by the current in an electrical circuit.

What Is Circuit Theory?

> **LEARNING OBJECTIVE 1.**
> To understand what is a circuit and why circuits are important in electrical engineering

A circuit problem. To see what circuit theory involves, consider the circuit consisting of the battery, light switch, two headlights, and connecting wire and chassis from a car, as shown in Fig. 1.1. When we pull the switch, we expect that the lights will glow and also get hot, which suggests that the battery is supplying energy to the headlights.[2] Figure 1.2 shows an electric circuit representing the physical situation depicted in Fig. 1.1, a 12.6-volt (V) battery symbol and two 5.25-ohm (Ω) resistances, modeling the headlights, with lines representing the wire and chassis return. We may use circuit theory to calculate the current in the wires, the power out of the battery, and the power into each headlight.

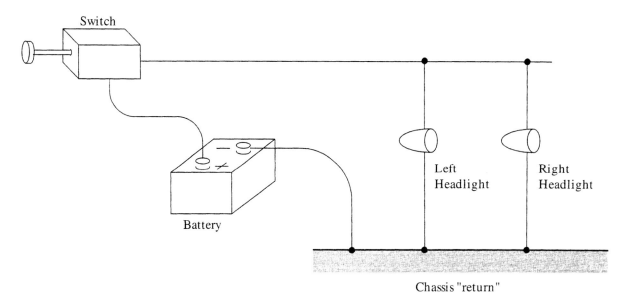

Figure 1.1 Automotive lighting system.

circuit theory

The nature of electrical circuit theory. Soon we will present the definitions and laws that will allow us to make such calculations. Here we show what circuit theory encompasses. The solution of an engineering problem normally proceeds through four stages: First, a real-world problem is identified; second, the problem is modeled; third, the model is analyzed; and fourth, the results are applied to the original physical problem. In the case of the battery and headlights, we skipped the first and last steps, but

---

[2] Most people assume that the energy flows through the wire, but it is more accurate to say that the wire guides the energy from the battery to the headlight.

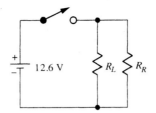

Figure 1.2 Circuit representing automotive lighting system. The headlights are represented by resistances, $R_L = R_R = 5.25\ \Omega$.

we did model a physical situation (battery, switch, and headlight) with common circuit symbols and we plan to analyze the circuit model soon. *Circuit theory* consists only of the third step: taking a given circuit model and solving for certain results through the application of known circuit laws. This third step is what you will learn to do in this book. We begin with the definitions of current and voltage.

1.3 CURRENT AND KIRCHHOFF'S CURRENT LAW

Definition of Current

LEARNING OBJECTIVE 2.
To understand the definition of current and use Kirchhoff's current law (KCL) to express conservation of charge

Current is charge in motion. In the experiment with the battery and headlights described before, we recognize that the headlights glow because of charges moving through the electrical conductors. An electrical conductor has mobile (conduction) electrons capable of moving in response to electric forces. A nonconductor has plenty of charges but its charges cannot move.

Consider a wire with a cross-section of A m² with charges moving with a velocity u from left to right, as pictured in Fig 1.3. If in a period of time Δt, ΔQ coulombs cross A in the indicated direction, we define the *current* to be

current

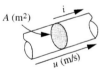

Figure 1.3 Wire with current.

$$i = \frac{\Delta Q}{\Delta t}\ \text{C/s or ampere, A} \qquad (1.2)$$

Note that the units of current are coulombs per second, but to honor André Ampère (1775–1836), we give this unit a special name, the ampere (A). A copper wire has a concentration of conduction electrons of $n_e = 1.13 \times 10^{29}$ electrons/m³. If the electrons are moving with a velocity u, the number of electrons crossing A in Δt would be $\Delta n = n_e A u\, \Delta t$. Hence, the current would be[3]

$$i = \frac{\Delta Q}{\Delta t} = \frac{e\,\Delta n}{\Delta t} = en_e A u\ \ \text{A} \qquad (1.3)$$

For example, consider electrons traveling downward in a No. 12 wire (0.081 in. in diameter) at a snail's pace of 0.1 mm/s. From Eq. (1.3), the charges constitute a current of $i = -5.8$ A downward, as shown in Fig. 1.4(a). We could also express this result by saying that the current is $i = +5.8$ A upward, Fig. 1.4(b).

---

[3] Note that e is negative, so the direction of numerically positive (physical) current is opposite to the direction of electron motion.

physical current

Reference directions. That we can express the current two ways reveals the importance of reference directions. We write "downward," to indicate the direction to which we are referring current flow. To specify a current, we require a reference direction plus a numerical value, which may be positive or negative. In this book, the current reference directions are indicated by arrowheads drawn on the lines representing the conductors, as in Fig. 1.4, or with arrows beside the wires, or with subscripts. The relationship between the reference direction and the numerical sign of the current is shown in Fig. 1.4, which expresses the same current in all three ways. The direction of the *physical current* is opposite to the direction of electron motion and hence the physical current is by definition numerically positive.

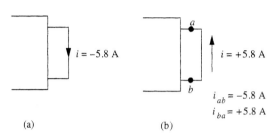

Figure 1.4 Four notations for expressing the same current. The physical current is from b to a.

Assigning reference directions. The engineer is responsible for assigning current reference directions as a beginning step in analyzing a circuit. These reference directions may be assigned without regard for the direction of the physical currents; they are assigned for bookkeeping purposes. This freedom in assigning reference directions will be clarified later when we state Kirchhoff's current law.

On the other hand, experienced engineers will usually define the current reference direction in the direction of the physical current if that direction is evident. They know that normally the physical current flows *out* of the $+$ terminal on the battery, through the circuit, and returns into the $-$ terminal. It is reasonable, therefore, to make your best guess as to which ways the physical currents go, but you must write and solve the equations to find out for sure.

Summary. Moving charges constitute a current. To specify the current in a conductor, we need both a reference direction and a numerical value, which can be positive or negative.

A mechanical analogy. For most students, an understanding of mechanics comes easier and earlier than an understanding of electrical phenomena. We have all experienced forces and motion, springs and inertia; and mechanics is taught early in most curricula. Because your intuition for mechanics is relatively well developed, we present mechanical analogs for many electrical quantities and phenomena.

Velocity is the simplest analog for electric current, and displacement therefore would be analogous to charge accumulation. By these analogies, we mean that the equations relating these quantities are similar.

For example, if we know the velocity of an object, $u(t)$, the change in displacement, $x_2 - x_1$, during the period $t_1 < t < t_2$ would be

$$x_2 - x_1 = \int_{t_1}^{t_2} u(t)\,dt \tag{1.4}$$

as shown in Fig. 1.5(a). Similarly, if we know the current in a wire, $i(t)$, the charge q passing a cross-section of the wire during the period $t_1 < t < t_2$ would be

$$q = \int_{t_1}^{t_2} i(t)\,dt \text{ coulombs} \tag{1.5}$$

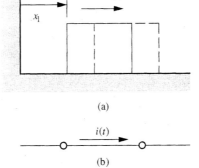

Figure 1.5 The mechanical analog for current (b) is velocity (a).

as shown in Fig. 1.5(b). Velocity and displacement in general are vector quantities. Charge motion is also in general a vector quantity; but in circuits, the wires channel the current in established directions, and a plus or minus sign is all that remains of the "vectorness" of the current. Thus, charge flow in circuit theory corresponds to linear motion in mechanics.

EXAMPLE 1.1 **Electron Motion**

A steady current of $+10^{-6}$ A flows toward the right in a copper wire of 0.001 in. diameter. Find the speed and direction of the electron motion. How many electrons pass a cross-section of the wire in 1 μs?

SOLUTION:
Because the current to the right is positive, the electrons must be traveling to the left. From Eq. (1.3), their velocity would be

$$u = \frac{I}{A n_e |e|} = \frac{10^{-6}}{\pi (0.0005 \times 0.0254)^2 \times 1.13 \times 10^{29} \times 1.60 \times 10^{-19}} \tag{1.6}$$

$$= 1.09 \times 10^{-7} \text{ m/s}$$

The total charge passing a cross-section in 10^{-6} s would be

$$q = \int_0^{10^{-6}} i\, dt = 10^{-6}\text{ A} \times 10^{-6}\text{ s} = 10^{-12}\text{ C} \tag{1.7}$$

The number of electrons would be

$$\frac{10^{-12}\text{ C}}{1.602 \times 10^{-19}\text{ C / electron}} = 6.24 \times 10^6 \text{ electrons} \tag{1.8}$$

WHAT IF? What if the speed is the same, but the diameter of the wire is doubled? What then would be the current?[4]

Kirchhoff's Current Law

Conservation of Charge

Conservation of charge and charge neutrality. All the evidence suggests that the universe was created charge-neutral, that is, there exists somewhere a positive charge for every negative charge. As was implied in our discussion on electrostatic forces, positive and negative charges can be separated by natural causes (lightning) or man-made causes (TV tubes), but most matter does not contain surplus charge, that is, most matter is charge-neutral. Furthermore, charge is neither created nor destroyed in electrical circuits. This conservation principle leads directly to a constraint on the currents at a junction of wires.

node

Kirchhoff's current law (KCL) at a junction of wires. The junction of two or more wires is called a *node*. The constraint imposed by conservation of charge and charge neutrality is known as *Kirchhoff's current law* (KCL) and can be stated as follows: Because charge is conserved, the sum of the currents leaving a node is zero at all times. We could have written "currents entering a node" just as well. Another equivalent statement would be that the sum of the currents entering a node is equal to the sum of the currents leaving that node.

Three forms of Kirchhoff's current law

1. The sum of the currents leaving a node is zero at all times.
2. The sum of the currents entering a node is zero at all times.
3. The sum of the currents referenced into a node is equal at all times to the sum of the currents referenced out of the node.

Figure 1.6 illustrates KCL at node *a*. Kirchhoff's current law (KCL) in the first form requires a + sign for currents referenced departing from the node and a − sign for currents referenced toward the node. In general form, KCL is

---

[4] The current would be 4 µA.

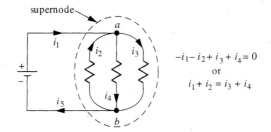

Figure 1.6 Conservation of charge at nodes a and b is expressed by Kirchhoff's current law (KCL). KCL also applies to the region encircled by the dashed line, a "supernode."

$$\sum_{\text{node}} \pm i_n = 0 \qquad (1.9)$$

The signs (+ or −) come from the reference directions; the i's are counted + if referenced departing from the node and − if referenced entering the node.

supernode

KCL applied to groups of nodes. Kirchhoff's current law is one of the fundamental laws of electric circuits because it expresses the conservation of charge. Not only does it apply to a simple node, but it applies also to groups of nodes, which are often called *supernodes*. For example, if in Fig. 1.6 we add KCL for node a to KCL for node b, the i_2 term must appear with opposite signs in the two equations; thus the two i_2 terms will cancel. For the same reason, the i_3 and i_4 terms will also cancel. Clearly, this cancellation will always occur for the internal currents in a group of nodes. Thus in Fig. 1.6, we can easily show that i_1 and i_5 are equal by applying KCL to the supernode enclosed with the dashed line.

EXAMPLE 1.2 KCL

Two sources are connected to the same load, as shown in Fig. 1.7. The load current is 12 A, and the first source supplies 8 A. How much current is supplied by the second source (i_2)?

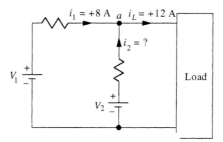

Figure 1.7 The load is supplied by two sources. Two currents are known.

SOLUTION:
We may apply Kirchhoff's current law to node a. Summing currents leaving the node, we have

$$-i_1 - i_2 + i_L = 0 \Rightarrow -(+8) - i_2 + (+12) = 0 \qquad (1.10)$$

In Eq. (1.10), the $-(+8)$ term, for example, distinguishes between the $-$ due to the reference direction (outside the parentheses) and the $+$ associated with the numerical sign of the current (inside the parentheses). Thus,

$$i_2 = -8 + 12 = 4 \text{ A} \tag{1.11}$$

WHAT IF? What if the reference direction for i_2 were down? What would i_2 be in that case?[5] Is the physical current different than before?[6]

Check Your Understanding

1. The charge on the electron is $e = -1.60 \times 10^{-19}$. What is the current in a wire in which 10^8 electrons pass a cross-section in 15 µs?
2. How many electrons pass a cross-section in 15 µs in a wire carrying 55 µA?
3. The physical current comes out of the plus or minus terminal of a battery. Which?

Answers. (1) 1.07 µA, sign not required; (2) 5.15×10^9 electrons; (3) plus, unless the battery is being charged.

1.4 VOLTAGE AND KIRCHHOFF'S VOLTAGE LAW

LEARNING OBJECTIVE 3.

To understand the definition of voltage and use Kirchhoff's voltage law (KVL) to express conservation of electric energy

voltage

Definition of Voltage

Voltage and energy exchanges. Voltage expresses the potential of an electrical system for doing work. *Voltage* is defined to be the work done by the electrical system in moving a charge from one point to another in a circuit, divided by the charge. In Fig. 1.8, we show the circuit modeling the battery–headlight experiment described earlier. We have identified points in the circuit with the letters a, b, c, d, e, and f. After the switch is closed, the motion of charges around the circuit effects the transfer of energy from the battery to the headlights, which are represented by the resistances. The work done by the electrical system in moving a charge from a to b is indicated by the voltage: For high voltages, more work will be done. Specifically, the voltage from a to b is defined to be[7]

$$v_{ab} = \frac{\text{work done by the electrical system in moving } q \text{ from } a \to b}{q} \tag{1.12}$$

The unit of voltage is energy/charge, joules per coulomb in the MKS system, but to honor Count Alessandro Volta (1745–1827), we use the special name volt (V) for this unit.

In our headlight circuit, the voltage between e and f is 12.6 because of the battery. With the switch open, the voltage does not get to the lights (resistances); with the

---
[5] The sign of i_2 would change to -4 A.
[6] No, the physical current is the same, 4 A upward.
[7] Some contexts require taking the limit as $q \to 0$.

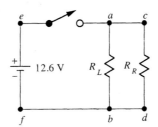

Figure 1.8 Automotive headlight circuit.

switch closed, the voltage is applied to the lights. The lines in Fig. 1.8 represent ideal connections through the wires. After we explore the concept of voltage more thoroughly, we will present the laws that describe how the voltage distributes throughout the circuit with the switch opened and closed.

Voltage as potential. Although our example implies that charges must move for the voltage to exist, such motion is not required. If a bulb were burned out and no current existed, the voltage would still be present. The voltage expresses the *potential* for doing work, that is, it measures how much work would be done if a charge were moved from a to b. It is similar to a gravitational potential, which expresses the work that would be done in a gravitational field if a mass were moved from one point to another. For both a gravitational potential and an electric potential, the path of the mass or charge does not matter. The charge can be moved from a to b through the bulb or it can be moved outside the bulb. In both cases, the work on the hypothetical charge, and hence the voltage, would be the same.

Signs. Both the work and the charge may be positive or negative; hence, voltage must also be a signed quantity. In the case where the battery turns on the headlight, energy is delivered by the electrical circuit to the thermodynamic system (heat and radiation), and in this case, the work done by the electrical system (via q) would be positive. On the other hand, chemical processes in the battery are delivering energy to the electrical circuit; in this case, the work done *by* the electrical system is negative in our definition of the voltage involving the battery. Voltage is thus the work done *by* the electrical system divided by the charge involved in doing that work, taking into account the signs of the work and the charge.

There is yet another way in which the sign of the voltage can be affected because the direction in which the charge is moved also must be considered. If we were to move a hypothetical positive charge from a to b in Fig. 1.8, positive work would be done and the voltage would be positive. If we were to move the hypothetical positive charge from b to a, negative work would be done and the voltage would be negative. To speak meaningfully of voltage as a signed quantity, we must have a clearly defined reference direction to indicate the assumed direction of travel.

EXAMPLE 1.3 The Definition of Voltage

For the battery in Fig. 1.8, $v_{ef} = +12.6$ V. If we move a charge of $+1$ C from e to f by hand, how much work would the electrical system do?

SOLUTION:
By definition, the work done by the electrical system would be

$$W(e \to f) = v_{ef} \times q = +12.6 \times (+1) = +12.6 \text{ J} \qquad (1.13)$$

Thus the charge would pull on our hand, doing positive work on us.

WHAT IF? What if an electron were moved from the negative (f) to the positive (e) terminal of the battery? How much work would be done by the electrical system?[8] Where does this energy come from (or go to)?[9]

subscript notation

Voltage reference directions. We will use two conventions for defining the reference direction of voltages. The more explicit convention uses subscripts, v_{ab}, as in Eq. (1.12), to define the beginning point (a), the ending point (b), and thus the direction traveled (from a to b). By this convention, v_{gh} would be the work done by the electrical system per charge in moving a charge from g to h. Often in a circuit, however, we desire to express the voltage across a single element, as in the present case with the battery or the headlight. In this case, we can mark both ends of the circuit element with polarity symbols: a + at one end and a − at the other end. With this simpler notation, the convention is that the + represents the first subscript and the − the second subscript of the voltage. Figure 1.9 shows the battery/headlight circuit marked in this manner.

physical voltage, physical polarity

Physical voltage. The *physical polarity* of the voltage is the reference direction marking that gives a positive voltage. Thus, the physical voltage is always positive and has polarity markings assigned accordingly. That is, if the voltage with arbitrary +/− reference direction assignment is numerically negative, the physical voltage is polarized in the opposite direction. Most voltmeters have red and black leads, with the black called "common." Such voltmeters will indicate a positive voltage when the common is connected to the minus polarity mark of the physical voltage and the red is connected to the plus polarity mark. On an auto battery, for example, the markings on the battery terminals are those of the physical voltage. Of course, we use the same convention on the battery symbol in circuit theory.

A mechanical analog. Force is a mechanical analog for voltage, in two senses. As external forces impart energy to mechanical systems and thus make things happen, so voltage sources (batteries, for example) impart energy to electrical circuits. Beyond this subjective analogy, however, we will show later in this chapter that equations involving voltage are similar in form to those involving force in mechanical systems.

Kirchhoff's Voltage Law (KVL)

Conservation of Energy

Conservation of energy. Kirchhoff's voltage law (KVL) expresses conservation of electrical energy in electrical circuits. The charges traveling around a circuit transfer energy from one circuit element to another, but do not receive energy themselves on the average. This means that if you were to move a hypothetical test charge around a com-

---

[8] Work $= -12.6 \times (-1.602) \times 10^{-19} = +2.02 \times 10^{-18}$ J.

[9] The energy comes from the battery and originates in the chemical system.

plete loop in a circuit, the total energy exchanged would add to zero. During part of the loop, you would have to push on the charge to move it, but during other parts of the loop, it would pull on you.

Forms of KVL. Because the energy sum is zero, it follows from the definition of voltage that the voltage sum around a closed loop is zero also. This is *Kirchhoff's voltage law* (KVL):

$$\sum_{\text{loop}} \text{voltages} = 0 \tag{1.14}$$

We can apply KVL to the circuit in Fig. 1.8 with these three results:

$$v_{ab} + v_{bf} + v_{fe} + v_{ea} = 0 \tag{1.15}$$

$$v_{ab} + v_{bd} + v_{dc} + v_{ca} = 0 \tag{1.16}$$

$$v_{ef} + v_{fd} + v_{dc} + v_{ce} = 0 \tag{1.17}$$

We note the following:

- Equation (1.15) was written around the loop containing the battery, the switch, and R_L, moving clockwise.
- Equation (1.16) was written around the loop containing the two resistances, moving counterclockwise.
- Equation (1.17) was written around the loop containing R_R, the switch, and the battery, moving counterclockwise.
- In Eq. (1.17) we skipped some intermediate points; for example, we wrote v_{fd} instead of $v_{fb} + v_{bd}$.

The direct connections between a and c, d and b, and b and f imply ideal connections, and hence no voltage, between these points. Therefore,

$$v_{ac} = v_{bd} = v_{bf} = v_{df} = 0 \tag{1.18}$$

and Eqs. (1.15) through (1.17) reduce to

$$v_{ab} + v_{fe} + v_{ea} = 0, \quad v_{ab} + v_{dc} = 0, \quad v_{ef} + v_{dc} + v_{ce} = 0 \tag{1.19}$$

When we apply KVL to voltages whose reference directions are indicated with the $+$ and $-$ polarity convention, we find that we cannot guarantee, as in Eqs. (1.15) through (1.17), that all the signs in the resulting equation will be positive. Thus, we must rewrite KVL in the form

$$\sum_{\text{loop}} \pm v\text{'s} = 0 \tag{1.20}$$

In writing KVL equations with $+$ and $-$ polarity symbols, we write the voltage with a positive sign if the $+$ is encountered before the $-$ and with a negative sign if the $-$ is encountered first as we move around the loop. Applying this rule to Fig. 1.9, we

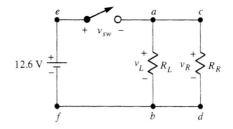

Figure 1.9 Automotive circuit with voltages marked with the +/− convention.

express KVL as

$$-(12.6) + v_{sw} + v_L = 0 \qquad (1.21)$$

Equation (1.21) results from starting at f and proceeding clockwise around the loop including the switch and R_L. The minus sign results because we encounter the $-$ first at the battery. The two plus signs occur because we encounter the $+$ reference marks first for both v_{sw} and v_L. All the signs in Eq. (1.21) result from the voltage reference directions marked in Fig. 1.9, not from the signs of the physical voltages.

EXAMPLE 1.4 Jump-Starting a Car

We have a weak battery with a voltage of 11.5 V and a strong battery with a voltage of 13.6 V. (This includes the benefits of the alternator charging the battery during the jump start.) What would happen in jump-starting the car having the weak battery when the correct connections with the jumper cables are made, and what happens when the incorrect connections are made?

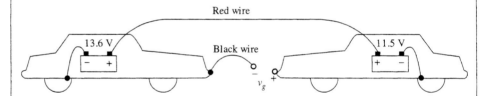

Figure 1.10 Jump-starting a car having a weak battery (correct connections).

SOLUTION:

Figure 1.10 shows the correct connections. According to the recommended procedure, we should first connect the positive terminals, as shown in Fig. 1.10, and then complete the circuit by connecting between points on the auto chassis away from the batteries, to reduce the danger of explosion. Thus, we are left with a voltage across the gap, v_g, before the final connection is made, due to the differing voltages of the two batteries. Using KVL, we determine the gap voltage to be

$$-v_g - 11.5 + 13.6 = 0 \Rightarrow v_g = -2.1 \text{ V} \qquad (1.22)$$

where v_g is the voltage across the gap. The same voltage would exist between the jumper cable and the negative terminal of the battery.

Figure 1.11 shows the incorrect connections, where we connect the negative of one battery

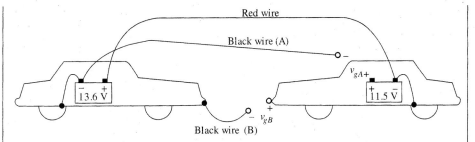

Figure 1.11 Jump-starting a car having a weak battery (incorrect connections).

to the positive of the other battery. From this bad start, we can continue in two ways. If we take the second jumper cable and proceed to connect between the two remaining battery terminals (A), we will have a gap voltage of

$$-v_{gA} + 11.5 + 13.6 = 0 \Rightarrow v_{gA} = +25.1 \tag{1.23}$$

where v_{gA} is the gap voltage with incorrect connection A. This large voltage presents a danger to both batteries and to the person making the connection.

WHAT IF? What if we follow the recommended procedure and connect between the two chassis (B)? What then would be the gap voltage, v_{gB}?[10]

EXAMPLE 1.5 KVL with Mixed Polarity Conventions

Determine the unknown voltages, v and v_{cd} in Fig. 1.12.

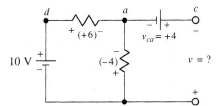

Figure 1.12 The circuit has voltages marked with both the +/− and the subscript notations.

SOLUTION:
Here we have deliberately mixed the reference–direction conventions. We note that all the voltages around the leftmost loop are specified. Let us confirm that the given voltages satisfy KVL. We write a KVL loop equation going clockwise around the loop starting at the bottom of the battery:

$$-(10) + (+6) - (-4) = 0 \tag{1.24}$$

In this equation, the signs outside the parentheses come from the reference directions and the signs inside come from the numerical values of the voltages. Thus, the minus sign is placed

---
[10]The gap voltage, v_{gB}, would be 13.6 V. This is still a dangerous situation.

before the first term because we encounter first the − polarity symbol on the battery, and the next sign is positive because we then encounter the first subscript of the voltage across a resistance. We note that KVL is satisfied.[11]

Let us proceed to determine the unknown v by writing KVL around the rightmost loop. What loop? There is indeed a loop even though no path exists for current, for we could still move our hypothetical test charge around this closed path. Thus, KVL must be satisfied even when there is no path for current. Let us start at point a and go counterclockwise:

$$-(-4) + (v) + v_{ca} = 0$$

$$v = -v_{ca} - 4 = -4 - 4 = -8 \text{ V} \quad (1.25)$$

In the first form of Eq. (1.25), we have written the v_{ca} term with a + sign because we are moving from the first subscript (c) to the second (a). Finally, we can determine v_{cd} by writing KVL from c to d to a to c:

$$v_{cd} + v_{da} + v_{ac} = 0$$

$$v_{cd} = -v_{da} - v_{ac} = -(+6) + (+4) = -2 \text{ V} \quad (1.26)$$

Note that, in the second form of Eq. (1.26), we wrote v_{da} as +6 because d and the + symbol mark the same point in the circuit. That is, the convention for the + and − markings is that we move from the + to the −, which in this case is the same as moving from d to a; hence, v_{da} is +6 V. Note also that $v_{ac} = -v_{ca}$, as shown in what follows.

WHAT IF? What if the 4-V battery were reversed? Find v_{dc} for that condition.[12]

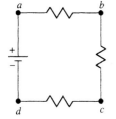

Figure 1.13 Circuit for proving rules of subscript notation.

Two rules for subscripts. As a third application of KVL, we will demonstrate two properties of the subscript notation for voltages. For this purpose, we refer to Fig. 1.13. One application of KVL is to go from a to b and then return to a. The resulting equation is

$$v_{ab} + v_{ba} = 0 \Rightarrow v_{ab} = -v_{ba} \quad (1.27)$$

Thus, we see that reversing the subscripts changes the sign of a voltage. This happens because the physical work done in moving a charge between two points is the negative of the work done in moving the same charge in the opposite direction.

Let us now determine v_{ac} by moving from a to c to b and back to a:

$$v_{ac} + v_{cb} + v_{ba} = 0 \Rightarrow v_{ac} = -v_{ba} - v_{cb} = v_{ab} + v_{bc} \quad (1.28)$$

(We changed the signs in the next-to-last form by reversing the subscripts.) Note the pattern in the subscripts in the last form of Eq. (1.28): The second subscript, b, of the first voltage, v_{ab}, is identical to the first subscript of the second term, v_{bc}. The final form of v_{ac} suggests that the middle point does not matter, that is, drops out. You can

---

[11] What would it mean if KVL were not satisfied around this loop? It would mean that the author, or the printer, had made an error, for a circuit in which KVL is not satisfied is not a circuit at all—it is nonsense, like $1 + 1 = 3$.
[12] The voltage would be 10 V.

verify this for yourself by writing v_{ac} in terms of v_{ad} and v_{dc}. You will get the equation

$$v_{ac} = v_{ad} + v_{dc} \tag{1.29}$$

Again the middle point drops out. This makes sense: It is like saying that the distance from New York to Los Angeles is the distance from New York to St. Louis plus the distance from St. Louis to Los Angeles. Clearly, "St. Louis" is a variable; we could have just as easily said Amarillo or Tucson. These two properties of the subscript notation will prove useful in later chapters.

Check Your Understanding

1. If $v_{ab} = -5$ V, how much energy is required by an *external agent* to move -2 C of charge from b to a?
2. If $v_{ab} = +2$ V and $v_{cb} = -1$ V, find v_{ba} and v_{ca}.
3. A tape player requires three 1.5-V batteries. What voltage is developed if one battery is inserted backward?
4. Is it always true that $v_{ab} + v_{bc} + v_{ca} = 0$?

Answers. (1) $+10$ J; (2) -2 V, -3 V; (3) 1.5 V; (4) yes.

1.5 ENERGY FLOW IN ELECTRICAL CIRCUITS

Voltage, Current, and Power

LEARNING OBJECTIVE 4.
To understand how to use the voltage and current to calculate the power into or out of a circuit element

power

Power from voltage and current. At the beginning of this chapter, we described electrical engineering as the useful control of electrical energy. Usually, energy is being exchanged between parts of a circuit, and the rate of energy flow, the power, must be calculated. We will now show that knowledge of voltage and current everywhere in an electrical circuit allows computation of energy flow (or power) throughout that circuit. Indeed, merely restating the definitions of voltage and current reveals their relationship to power flow. Voltage is the energy exchanged per charge and current is the rate of charge flow. *Power* is defined as the rate of energy exchange; hence, the power is by definition the product of voltage and current:

$$v \left(\frac{\text{work}}{\text{charge}} \right) \times i \left(\frac{\text{charge}}{\text{time}} \right) = \frac{\text{work}}{\text{time}} = \text{power} \tag{1.30}$$

Thus, if we know the voltage across a circuit component and the current through that element, we can determine the power into (or out of) that component by multiplying voltage and current (volts × amperes = watts).

source set, load set

Load sets and source sets. We must consider reference directions when we write a power formula. The sign in the power formula depends on the *combination* of the voltage and current reference directions. We show both possibilities in Fig. 1.14. In a *source set*, Fig. 1.14(a), the current reference direction is directed *out of* the + polarity marking (or the first subscript) of the voltage. This combination is normally used for a source of energy to the circuit such as a battery. In a *load set,* Fig. 1.14(b), the current reference direction is directed *into* the + polarity marking (or the first subscript) of the voltage reference direction. This is the convention normally used with a resistance or

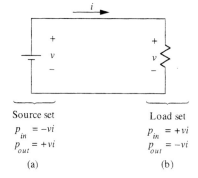

Figure 1.14 Sign convention for power calculations. The source set (a) is normally used with sources such as batteries, and the load set (b) is normally used with passive circuit elements.

Source set
$p_{in} = -vi$
$p_{out} = +vi$
(a)

Load set
$p_{in} = +vi$
$p_{out} = -vi$
(b)

some other element that receives power from the circuit. For both cases in Fig. 1.14, we can speak of the power out of the component and the power into the component as meaningful quantities. For example, although a battery normally gives power to a circuit ($p_{out} = +$), it can be receiving power ($p_{out} = -$ or $p_{in} = +$) when the battery is being charged.

Action and reaction voltage. In Fig. 1.14 all voltages and currents would be numerically positive. The battery voltage causes the current, which in turn causes the voltage across the resistor. Thus causality might be diagrammed as:

$$\text{battery voltage} \Rightarrow \text{current} \Rightarrow \text{resistor voltage} \qquad (1.31)$$

We might call the battery voltage an *action* voltage; it pushes the current. The resistor voltage is a *reaction* voltage; it pushes back against the current. The action voltage of the battery gives energy to the electrical system, provided current flows. The reaction voltage of the resistor represents energy leaving the electrical system.

We must stress that the formulas given in Fig. 1.14 relate solely to reference directions. Any of the numerical values of the voltages, currents, or powers can be positive or negative. The direction of energy flow at a component will be established only after the sign depending on the reference directions is combined with the numerical signs of the voltage and current. The use of load and source sets will be clarified in the example of Fig. 1.15.

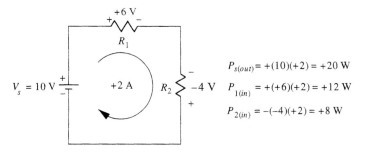

$P_{s(out)} = +(10)(+2) = +20 \text{ W}$
$P_{1(in)} = +(+6)(+2) = +12 \text{ W}$
$P_{2(in)} = -(-4)(+2) = +8 \text{ W}$

Figure 1.15 The battery and R_2 have source sets, but R_1 has a load set.

EXAMPLE 1.6 Calculating the Power In and Out of Circuit Elements

Find the power out of the battery and the power into the two resistances.

SOLUTION:
With the battery, we have a source set because the current reference arrow is out of the + polarity marking. Hence, the battery supplies a power of $+vi$ or $+(+10)(+2) = +20$ W. In the preceding sentence, the + of the "$+vi$" is from the source set of reference directions and the + of the "+20" comes from the combination of the reference directions and the numerical values of the voltage and current. The interpretation of the positive power out of the source is that the battery is delivering energy to the circuit. If the power out of the battery had been negative, we would learn that the battery is being recharged by some other source in the circuit.

Resistance R_1 has a load set and proves to be receiving energy from the circuit. Resistance R_2 is also receiving energy from the circuit because a positive sign is produced by the combination of the minus sign from the reference directions and the minus sign of the voltage. Thus, in calculating power, we must continue to distinguish between signs arising from the reference directions and signs arising from the numerical values of the voltages and currents. Only the meaning of the power variable, "into" or "out of", and the numerical sign of the result allow the final conclusion as to which way energy is flowing at a given instant.

Conservation of Energy

The power out of the source in the previous example, +20 W, is equal to the sum of the powers into the resistances, 12 W + 8 W; hence, electric energy is conserved in the circuit. We call this conservation of *electric* energy because we have accounted for the energy entering and leaving the electrical system represented by the circuit. Certainly total energy (electric plus other forms) would also be conserved; we simply have ignored the nonelectrical aspects of the system. For example, the battery might represent a chemical system, a solar cell, or a small electric generator powered by a windmill. To account for all energy, we would have to study these sources.

Mechanical analogies. We have presented mechanical force, velocity, and displacement as counterparts of voltage, current, and charge, respectively. Here we will review some of the relationships relating these mechanical variables to energy exchanged in a mechanical system. Then we will show the corresponding equations for the electrical variables.

The amount of energy exchanged, dW, by a force f (newtons) operating through a distance, dx (meters), in the same direction would be

$$dW = f\, dx \text{ joules (J)} \tag{1.32}$$

When continuous motion is involved, the rate of energy exchanged, the power, p, would be

$$p = \frac{dW}{dt} = f\frac{dx}{dt} = fu \text{ watts (W)} \tag{1.33}$$

where u is the velocity. Finally, the energy exchanged in a period of time $t_1 < t < t_2$ can be computed by integration:

$$W = \int_{t_1}^{t_2} p\, dt = \int_{t_1}^{t_2} fu\, dt \text{ J} \tag{1.34}$$

Equations (1.32) to (1.34) have electrical counterparts. When a charge dq is moved from a to b, the energy given by the electrical circuit would be

$$dW = v_{ab}\, dq \text{ J} \tag{1.35}$$

When continuous charge flow, a current, is involved, the rate of energy exchanged, the power, p, would be

$$p = \frac{dW}{dt} = v_{ab}\frac{dq}{dt} = v_{ab}i \text{ W} \tag{1.36}$$

where i is the current referenced from a to b. Finally, the energy exchanged in a period of time $t_1 < t < t_2$ can be computed by integration:

$$W = \int_{t_1}^{t_2} p\, dt = \int_{t_1}^{t_2} v_{ab}i\, dt \text{ J} \tag{1.37}$$

EXAMPLE 1.7 Energy to Start Engine

The starter motor of an automobile draws 60 A when turning over the engine. If the voltage is 12.6 V and the engine starts after 10 seconds, what is the power to the starter motor and the energy required to start the engine?

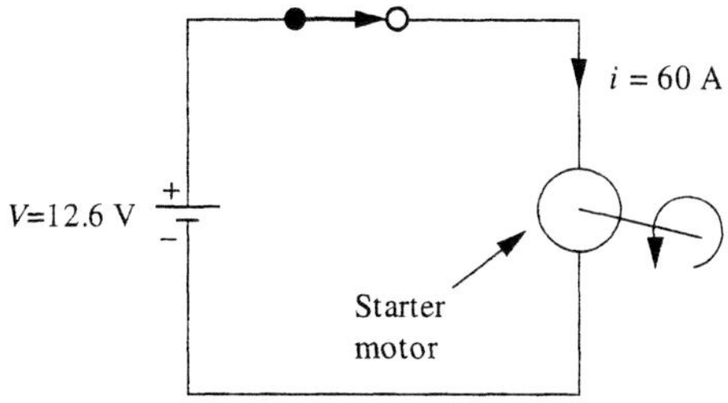

Figure 1.16 Circuit model for the automotive starter circuit.

SOLUTION:
Figure 1.16 shows the circuit. The power out of the battery is

$$P_{\text{out}} = +Vi = +(+12.6)(+60) = 756 \text{ W}$$

$$P_{\text{out of battery}} = P_{\text{into starter}} = p \tag{1.38}$$

The energy required to start the engine would be

$$W = \int_0^{10} p\, dt = 10p = 7560 \text{ J} \tag{1.39}$$

WHAT IF? What if it took 1 minute to start the car and the battery was recharged at a 20-A rate by the alternator? How long would it take to recharge the battery?[13]

---
[13] Three minutes.

Summary. Voltage and current allow us to monitor energy flow in an electrical circuit, are easily measured, and also obey simple laws, KVL and KCL. For these reasons, voltage and current are universally used by electrical engineers in describing the state of an electric network.

In the next section we define resistance and state Ohm's law. This addition allows us to develop basic techniques for analyzing electrical circuits.

Check Your Understanding

1. A battery charger puts 5 A into a 12.6-V auto battery. What is the power into the battery?
2. Find the power *out of* R_1 and R_2 in Fig. 1.15.
3. A 1.5-V flashlight battery puts out 300 mA for 10 minutes. How much energy does this represent?

Answers. (1) 63.0 W; (2) −12 W, −8 W; (3) 270 J.

1.6 CIRCUIT ELEMENTS: RESISTANCES AND SOURCES

Resistances and Switches

In the previous sections, we have defined current and voltage and shown how these describe energy exchanges in an electrical circuit. We have stated Kirchhoff's current law and Kirchhoff's voltage law, which express conservation of charge and electric energy in a circuit, respectively. For illustrative purposes, we have presented some circuit elements—sources, switches, and resistances—without careful definition or explanation. We now define and explain these circuit elements.

Ohm's law. Let us again think about the battery–switch–headlight circuit of Fig. 1.1. We know that chemical action produces charge separation within the battery. This charge separation appears at the battery terminals, and electrostatic forces are experienced by all the charges in the vicinity of the battery, significantly by charges in the wire and the headlights. Because of these forces, electrons move in the wire and headlight, a current exists in the circuit, and the lights glow. Thus the causal chain is

$$\text{battery} \Rightarrow \text{current} \Rightarrow \text{lights heat up} \Rightarrow \text{light emitted} \tag{1.40}$$

LEARNING OBJECTIVE 5.
To understand the relationship between voltage and current in a resistor as described by Ohm's law

It will not surprise you that the greater the voltage, the greater will be the resulting current. For a large class of conductors, the current increases in direct proportion to the voltage. Physical experimentation leads to the following equation, known as *Ohm's law*:

$$i = \frac{v}{R} \quad \text{or} \quad v = Ri \tag{1.41}$$

In Eq. (1.41), R is the *resistance* and has the unit volts per ampere, but to honor Georg Ohm (1787–1854), we use the unit ohm, abbreviated by the uppercase Greek letter omega, Ω. Ohm's law relates voltage and currents for resistive elements. Later, we will discover other voltage–current relationships for capacitors and inductors. The resistance of a piece of wire is directly proportional to its length, to a property of the metal called resistivity, and inversely proportional to its cross-sectional area:

resistance, resistor

$$R = \frac{\ell \rho}{A} \tag{1.42}$$

where ℓ is the length of the wire, ρ is the metal resistivity, and A is the area of the wire. The material from which the wire is drawn is an important factor: Copper is a good conductor; iron not so good. Table 1.1 gives the resistance of some common wires. The resistance of a wire also depends on its temperature, but the resistance often may be considered constant. In electronic circuits, we use resistors made out of a carbon-impregnated binder, high-resistance wire, or metals deposited on nonconducting surfaces. The physical behavior of such resistors leads directly to the circuit theory definition of an ideal resistance, Eq. (1.41).[14]

TABLE 1.1 Properties of Copper Wires

Wire Size (gage)	Diameter (in.)	Resistance (Ω/1000 ft)	Nominal Current Limit (A)	Use
12	0.08081	1.588	23	Household wiring
16	0.05082	4.016	13	Extension cords
24	0.0201	25.7	2	Electronic wiring
30	0.01003	103.2	0.52	Computer circuits

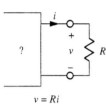

$v = Ri$

Figure 1.17 The circuit symbol for a resistance. Note that the voltage and current reference directions form a load set.

Circuit theory of resistance. The circuit symbol for a resistance, R, is given in Fig. 1.17. Note that the reference directions of the voltage and current form a load set. We could also write Ohm's law for a source set, which would introduce a minus sign, but this is usually avoided.

Ohm's law is presented in graphical form in Fig. 1.18. We have shown voltage to be the independent variable (cause) and current to be the dependent variable (effect). The slope of the line is the reciprocal of resistance. Customarily, Ohm's law is also stated in terms of reciprocal resistance, which is called *conductance*:

conductance

$$i = Gv, \quad \text{where} \quad G = \frac{1}{R} \tag{1.43}$$

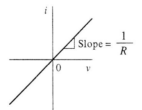

Figure 1.18 Ohm's law in graphical form. We show the voltage as the independent variable (cause) and the current as the dependent variable (effect).

---

[14]We should perhaps distinguish a *resistor* from a *resistance*. A resistor is a device, often cylindrical in shape, exhibiting the property of resistance, placed in an electrical circuit for the purpose of influencing some voltage or current. Wires, light bulbs, heater elements, and resistors all exhibit resistance. However, in common parlance resistor and resistance are synonymous.

The unit of conductance is amperes per volt, but we use siemens (the official unit) or mho (the unofficial but more common term). The symbol for siemens is S, and the common symbol for the mho[15] is the upside-down omega (℧). Because we have a load set, the power into the resistance is $+vi$. We can express the power into a resistance in several ways, as follows:

$$p = +vi = (Ri)i = i^2 R$$
$$= v\left(\frac{v}{R}\right) = \frac{v^2}{R} \text{ watts} \tag{1.44}$$

where v is the voltage across the resistance and i is the current through the resistance. Note that the power into a resistance is always positive. Resistance, therefore, always removes electrical energy from a circuit. Like the headlight circuit, the energy lost to the electrical circuit usually appears as heat (also some light in that case).

EXAMPLE 1.8 — Power in Resistances

What is the maximum allowed voltage across a 1000-Ω, $\frac{1}{2}$-watt resistance?

SOLUTION:
From Eq. (1.44), $v^2/1000 = 0.5 \text{ W} \Rightarrow v = \sqrt{500} = 22.4$ V.

WHAT IF? What if it were a 500-Ω, 2-W resistance and you wanted the maximum allowed current?[16]

short circuit and open circuit

Open circuits and short circuits. Figure 1.19 shows the characteristic of two special resistances. A short circuit ($R = 0$) permits current to flow ($i \neq 0$) without any resulting voltage ($v = 0$), and an open circuit ($R = \infty$) permits voltage ($v \neq 0$) with no current ($i = 0$). In both cases, Eq. (1.44) shows that no power is required for the open or short circuit.

Switches. An ideal switch is a special resistance that can be changed from a short circuit to an open circuit to turn an electrical device ON or OFF. Ideal switches receive no electrical energy from the circuit.[17] Figure 1.20(a) shows a *single-pole, single-throw* switch in its open (OFF) state. Figure 1.20(b) shows a *single-pole, double-throw* switch, which switches one input line between two output lines. Figure 1.20(c) shows a *double-pole, single-throw* switch; the dashed line indicates mechanical coupling between the two components of the switch to cause simultaneous switching. Clearly, switches can have any number of *poles* and *throws*.

---

[15] "Ohm" spelled backward.
[16] The maximum current would be 63.2 mA (milliamperes).
[17] Of course, some energy is required to "switch" the switch between states.

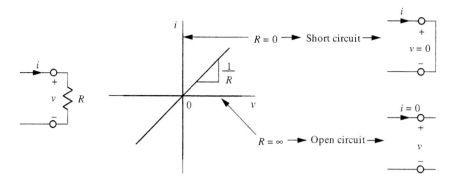

Figure 1.19 Circuit symbol and graphical characteristic of a resistance. The characteristic of a short circuit ($R = 0$) is vertical and that of an open circuit ($R = \infty$) is horizontal.

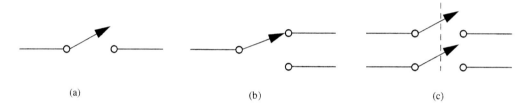

Figure 1.20 Circuit symbols for switches: (a) A single-pole, single-throw switch in the OFF state; (b) a single-pole, double-throw switch; and (c) a double-pole, single-throw switch in the OFF state. The dashed line in (c) indicates mechanical coupling to ensure identical states for both poles.

Voltage and Current Sources

ideal voltage source

Ideal voltage sources. Figure 1.21 shows the circuit symbol, mathematical definition, and graphical characteristic of an ideal general and dc voltage source. The *ideal voltage source* maintains its prescribed voltage, independent of its output current. In general, an ideal voltage source may have positive or negative voltage, and it may be constant or time-varying. In this and the next chapter, we use the battery symbol for a constant (dc) voltage source and always consider the battery voltage to be positive. Note that we have used a source set for the voltage source. Normally, a voltage source would produce a positive current, the physical current, out of the + terminal and thus act as a source of energy for the circuit; but it is possible that some other, more powerful source might force the physical current *into* the + terminal of the voltage source to be positive, thus delivering energy to the source. When this happens for a battery, the current as defined is negative and we say that the battery is being charged.

The dc voltage source of circuit theory models an ideal battery. We recognize that a real battery does not maintain constant voltage under heavy load. Furthermore, real

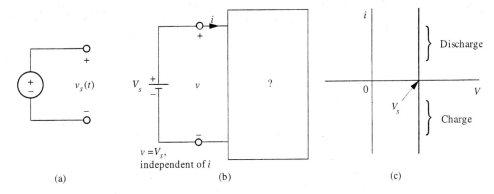

Figure 1.21 (a) General symbol for a voltage source, (b) circuit symbol and mathematical definition for a dc voltage source, and (c) graphical characteristic for a dc voltage source. The source determines the voltage, but the current is determined in part by the load. Note that the voltage and current form a source set, so the power out of the source is positive for positive current.

batteries store a finite amount of energy and need to be recharged or replaced frequently. By contrast, the ideal voltage source can deliver energy without limit. A physical battery often can be represented as an ideal voltage source in a circuit problem.

ideal current source

Ideal current sources. Figure 1.22 shows the circuit symbol, mathematical definition, and graphical characteristic of ideal general and dc current sources. The *ideal current source* produces its prescribed current, independent of its output voltage. Like the ideal voltage source, an ideal current source will deliver any required amount of energy. Unlike the voltage source, there is no physical device at your local hardware store whose electrical properties resemble those of a current source; however, we can build electronic devices that act like current sources, and furthermore the idea is useful in circuit analysis.

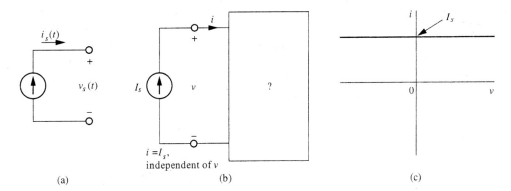

Figure 1.22 (a) General symbol for a current source, (b) circuit symbol and mathematical definition for a dc current source, and (c) graphical characteristic for a dc current source. The source determines the current, but the voltage is determined in part by the load. Note that the voltage and current form a source set, so the power out of the source is positive for positive voltage.

1.6 CIRCUIT ELEMENTS: RESISTANCES AND SOURCES

EXAMPLE 1.9 Voltage and Current Sources

A constant voltage source is connected in series with a current source that produces a pulse of current. The pulse is 0.1 s in duration and triangular in shape, as shown in Fig. 1.23. Find the energy out of the current source.

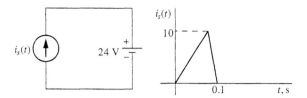

Figure 1.23 The current source provides a triangular pulse of current to the voltage source.

SOLUTION:
Kirchhoff's voltage law requires that the voltage across the current source be +24 V, with the + reference mark at the top. This voltage forms a source set with the current; hence,

$$p_{out} = +(+24)i_s(t) \Rightarrow W_{out} = \int p_{out}\, dt = 24 \int_0^{0.1} i_s(t)\,dt \tag{1.45}$$

But the integral is the area under the triangle in Fig. 1.23:

$$\text{Area} = \tfrac{1}{2}(0.1)(10) = 0.5 \text{ C} \Rightarrow W_{out} = 24 \times 0.5 = 12 \text{ J} \tag{1.46}$$

WHAT IF?
What would be the energy out of the current source if it was oriented with the arrow downward?[18]

Analysis of DC Circuits

Headlight circuit solved carefully. We now have defined sufficient concepts and conventions to analyze the headlight circuit in Fig. 1.9, which we repeat in Fig. 1.24. We have used the + and − polarity notation and assigned reference directions such that the current reference directions are the directions in which current would be numerically positive. The physical current will come out of the + on the battery, divide between the headlights, recombine, and then go into the − on the battery. Note that we have defined voltages v_L and v_R in load sets for the resistances and a source set for the voltage source. Going clockwise around the left loop, we find

$$-12.6 + v_{sw} + v_L = 0 \tag{1.47}$$

---
[18] $W_{out} = -12$ J.

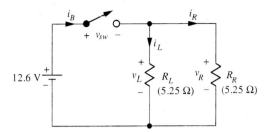

Figure 1.24 Figure 1.9 repeated.

Conservation of Energy

and KVL clockwise around the loop created by the resistors is

$$-v_L + v_R = 0 \tag{1.48}$$

We must solve these equations for two cases: the switch open (OFF) and the switch closed (ON).

With the switch OFF, the solution is trivial. The open switch blocks the voltage from being applied to the headlights; hence, the voltage across the headlights is zero and no current flows. Because $v_L = v_R = 0$, Eq. (1.47) reduces to

$$-12.6 + v_{sw} + 0 = 0 \Rightarrow v_{sw} = 12.6 \text{ V} \tag{1.49}$$

Thus, all the voltage of the battery appears across the switch and no current flows in the resistances representing the headlights.[19] The power out of the battery is zero, and the power into each resistance is zero.

With the switch ON, $v_{sw} = 0$, and Eqs. (1.47) and (1.48) reduce to $v_L = v_R = 12.6$ V. The currents in the headlights follow from Ohm's law:

$$i_L = \frac{v_L}{R_L} = \frac{12.6 \text{ V}}{5.25 \text{ }\Omega} = 2.40 \text{ A} \tag{1.50}$$

Conservation of Charge

and i_R is the same. The current in the switch and battery may be determined from applying Kirchhoff's current law (KCL) to the node where the two resistances connect at the top:

$$-i_B + i_L + i_R = 0 \Rightarrow i_B = i_L + i_R = 2.40 + 2.40 = 4.80 \text{ A} \tag{1.51}$$

Because i_B and the battery voltage form a source set, the power *out of* the battery is

$$p_{out} = +(+12.6)(i_B) = +12.6 \times 4.80 = 60.5 \text{ W} \tag{1.52}$$

The resistance voltages and current form a load set, so the power *into* each resistance is

$$p_{resistances} = +(+12.6)(+2.40) = 30.2 \text{ W} \tag{1.53}$$

Conservation of Energy

We note that the power out of the battery equates to the sum of the powers into the resistances and, because the switch requires no power, conservation of electric energy is satisfied.[20]

---

[19] The reasoning is heuristic and intuitive. Reasoning from the definition of a switch given earlier is surprisingly complicated and lends no insight.

[20] We normally write numerical results to three-place accuracy, but in our calculations we maintain full accuracy. This occasionally leads to apparent errors due to rounding, as here.

Summary. We applied Kirchhoff's laws directly to find the voltage, current, and power for each element in the headlight circuit with the switch OFF and ON. We found all currents positive because we took care to define our current variables according to the directions of the physical current in the circuit. We found all voltages positive because we used load sets for the resistances and switch. The power flow *out of* the battery and the power flow *into* each resistance are physical realities, independent of the reference directions of the voltage and current variables, which are part of a man-made analysis scheme.

Alternative solution based on series and parallel combinations. We may find the voltages, currents, and powers in the circuit by an alternative method that employs equivalent circuits. Figure 1.25 shows the part of the circuit with the resistances. We note that both ends of the resistances are connected together; hence, both resistances have the same voltage, v_p, applied to them. This is called a *parallel connection*, referring not to the geometric positions of the resistances, but rather to the division of the current at the top and the recombining of the currents at the bottom. We have labeled the voltage and current of the parallel connection v_p and i_p, respectively.

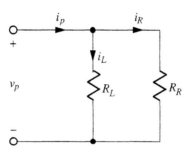

Figure 1.25 The part of the headlight circuit with only the headlights. Our goal is to replace this with a simpler equivalent circuit.

equivalent resistance

We will now show that we may replace the two parallel resistances with a single *equivalent resistance*, that is, a single resistance that has the same external properties (v_p and i_p) as the parallel resistances. Writing KCL in the form that the sum of the currents leaving a node equals the sum of the currents entering, we have

$$i_p = i_L + i_R \tag{1.54}$$

But from Ohm's law, we have

$$i_L = \frac{v_p}{R_L} \quad \text{and} \quad i_R = \frac{v_p}{R_R} \tag{1.55}$$

Substituting Eqs. (1.55) into Eq. (1.54), we obtain

$$i_p = \frac{v_p}{R_L} + \frac{v_p}{R_R} = v_p \left(\frac{1}{R_L} + \frac{1}{R_R} \right) \tag{1.56}$$

Equation (1.56) has the same form of Ohm's law, Eq. (1.41), if we introduce an equivalent resistance, R_{eq}, where

$$\frac{1}{R_{eq}} = \frac{1}{R_L} + \frac{1}{R_R} \Rightarrow R_{eq} = \frac{1}{1/R_L + 1/R_R} \tag{1.57}$$

with the equivalent resistance, R_{eq}, defined in Eq. (1.56), Eq. (1.55) becomes

$$i_p = \frac{v_p}{R_{eq}} \tag{1.58}$$

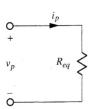

Figure 1.26 The two parallel resistances in Fig. 1.25 are replaced by one equivalent resistance.

Thus we can replace the circuit in Fig. 1.25 by the equivalent circuit shown in Fig. 1.26, where R_{eq} is given by Eq. (1.57). We note

- Equation (1.57) is awkward to write, so we introduce the parallel bars symbol, $R_{eq} = R_L \| R_R$, as a shorthand symbol for this operation. For example,

$$2 \| 5 = \frac{1}{1/2 + 1/5} = 1.43$$

- Equation (1.57) reduces to the product of the resistances over the sum of the resistances, and many students have memorized this formula for parallel resistances. We discourage the use of this formula for two reasons: (1) The formula is valid for two resistances, but not for three or more parallel resistances. Thus, this formula can get you in trouble because it is not general. (2) With a modern calculator, it is easier to calculate with Eq. (1.57) than with the product-over-sum formula. For example, on a reverse-Polish-entry calculator (H-P), we would calculate $2 \| 5$ in the following sequence of keystrokes: 2, 1/x, 5, 1/x, +, 1/x. It takes more keystrokes to execute the product-over-sum formula, because the resistance values need to be stored or entered twice. The product-over-sum formula is useful, however, in mentally checking the magnitude of the combination of two parallel resistances.

- The concept of an equivalent resistance is very general and will be used throughout this book. Thus, the symbol R_{eq} is used like "x" in algebra and is by no means reserved for parallel resistances. So when you read Eq. (1.57) read (or think) "the equivalent resistance *of two parallel resistances* is ..."

Equivalent Circuits

- The resistance in Fig. 1.26 is equivalent to the two resistances in Fig. 1.25 only in certain regards. Specifically, the two circuits are equivalent as far as the external voltage and current, v_p and i_p, are concerned, but information about the internal currents, i_L and i_R, is lost. Later in this section, we show how to recover this lost information.[21]

Equivalent Circuits

- The equivalent resistance defined earlier is one example of an equivalent circuit. An *equivalent circuit* is a circuit that represents in some important respects the properties of another, more complicated circuit. We shall encounter many types of equivalent circuits in our study of electrical engineering: The

---

[21] With the headlights, current i_p divides equally between the two equal resistances.

equivalent resistance of two parallel resistors is simply the first use of an equivalent circuit.

- In our circuit the two resistances represent two identical headlights, each having 5.25-Ω resistance. Thus for us, $R_L \| R_R = 5.25 \| 5.25 = 2.63\ \Omega$.

Series resistances. Because we will need the R_{eq} symbol for another meaning in this section, let us use the symbol $R_p = R_L \| R_R$ to represent the parallel resistances. With this symbol, the circuit reduces to that shown in Fig. 1.27, where R_{sw} represents the resistance of the switch ($R_{sw} = \infty$ for open and $R_{sw} = 0$ for closed). The voltage of the battery is represented by v_B.

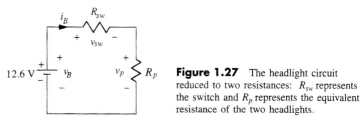

Figure 1.27 The headlight circuit reduced to two resistances: R_{sw} represents the switch and R_p represents the equivalent resistance of the two headlights.

We note that R_{sw} and R_p have the same current, i_B, because there is no place where the current divides. This is called a *series connection* when two circuit elements have the same current; so the battery, switch, and parallel resistances are connected in series. Two resistances in series can be replaced by an equivalent resistance, as we now show. We write KVL around the loop

$$-v_B + v_{sw} + v_p = 0 \Rightarrow v_B = v_{sw} + v_p \qquad (1.59)$$

and introduce Ohm's law for both resistances:

$$v_{sw} = R_{sw} i_B \quad \text{and} \quad v_p = R_p i_B \qquad (1.60)$$

series connection Combining Eqs. (1.59) and (1.60), we find

$$v_B = (R_{sw} + R_p)\, i_B = R_{eq} i_B \qquad (1.61)$$

where $R_{eq} = R_{sw} + R_p$ is the equivalent resistance of R_{sw} and R_p in series connection. Thus, we may determine the battery current as

$$i_B = \frac{v_B}{R_{eq}} \qquad (1.62)$$

For an open switch, $R_{eq} = \infty + R_p$; hence, $i_B = 0$. For a closed switch, $R_{eq} = 0 + R_p$ and $i_B = 12.6/2.63 = 4.80$ A. As before, this current divides equally between the two resistances (headlights) and the power is also as before.

Summary. We have analyzed the automotive headlight circuit by two means. First we applied Kirchhoff's laws directly and treated the switch as a device that either blocked or allowed the voltage to reach the headlights. Then we treated the switch as a resistance that was either zero or infinity, and we used equivalent resistances to represent resistances in parallel and series. Electrical engineers often replace resistances in series or parallel with equivalent resistances. In the next section, we explore and generalize this technique of circuit analysis.

Check Your Understanding

1. Ohm's law is written with a + sign when the reference directions of the voltage across and current through the resistance are related by a load set or source set?
2. A 1000-Ω resistance is rated at 5 W. What is the maximum current that should go through this resistance? The maximum voltage across the resistance?
3. A 5-A current source is connected in series with a 10-V battery, with the current source going into the minus terminal of the battery. What is the power into the battery?

Answers. (1) Load set; (2) 70.7 mA, 70.7 V; (3) −50 W.

1.7 SERIES AND PARALLEL RESISTANCES; VOLTAGE AND CURRENT DIVIDERS

Series Resistances and Voltage Dividers

LEARNING OBJECTIVE 6.
Understand how to combine resistances connected in series and how voltage divides between series resistances

Resistances in series. As stated before, two circuit elements are connected *in series* when the same current flows through them. Figure 1.28 shows a series connection of three resistances and a battery. It is important in our method to anticipate the direction of the physical current, so we have defined the current reference direction such that the current will be positive out of the + terminal of the voltage source; and we put the + and − polarity symbols on the resistance voltages according to load-set conventions. Because i will be numerically positive, the v's across the resistances will also be positive. We can write KVL around the loop, going clockwise.

$$-V_s + v_1 + v_2 + v_3 = 0 \Rightarrow V_s = v_1 + v_2 + v_3 \quad (1.63)$$

Because the v's are positive and their sum is V_s, the battery voltage divides between v_1, v_2, and v_3. To show how this division depends on the values of the resistances, we introduce Ohm's law:

$$V_s = R_1 i + R_2 i + R_3 i$$
$$= (R_1 + R_2 + R_3) i$$
$$= R_{eq} i \quad (1.64)$$

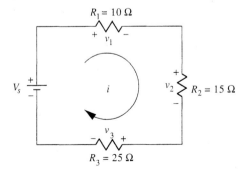

Figure 1.28 The three resistances have the same current and thus are in series. We may replace two or more resistances connected in series with a single equivalent resistance.

where R_{eq} is an equivalent resistance. Thus, resistances connected in series add to an equivalent resistance as

$$R_{eq} = R_1 + R_2 + \cdots + R_n \text{ (all series resistances)} \tag{1.65}$$

Figure 1.29 shows a circuit that is equivalent to that in Fig. 1.28, except that the three resistances have been replaced by R_{eq}.

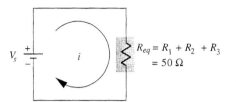

Figure 1.29 The 10, 15, and 25-Ω resistances connected in series are represented by an equivalent resistance of 50 Ω.

Equivalent Circuits

Voltage dividers. We set out to learn how the battery voltage, V_s, divides between the three resistances. We have determined the current:

$$i = \frac{V_s}{R_{eq}} = \frac{V_s}{50} \tag{1.66}$$

We can now obtain the voltage across the individual resistances from Ohm's law applied to the original circuit:

$$v_1 = R_1 i = R_1 \frac{V_s}{R_{eq}} = \frac{R_1}{R_{eq}} \times V_s = \frac{10}{50} V_s \tag{1.67}$$

Formulas for v_2 and v_3 can be written similarly. For example, if V_s were 60 V, v_1 would be 12 V, v_2 would be 18 V, and v_3 would be 30 V. Our results are easily generalized to include an arbitrary number of series resistances; indeed, the voltage across the ith resistance would be

$$v_i = \frac{R_i}{R_{eq}} \times V_s \tag{1.68}$$

where

$$R_{eq} = R_1 + R_2 + \cdots + R_i + \cdots \text{ (all series resistors)}$$

As an intermediate result, we learned that series resistances can be replaced by a single resistance whose value is the sum of the values of the resistances in series.

EXAMPLE 1.10 Voltage Dividers

The circuit of Fig. 1.30 models a flashlight. The shaded boxes represent the batteries, but in this case, we have included the internal resistance of the battery in addition to its internal voltage source. We know that a battery has internal resistance because (1) it becomes hot when used heavily, (2) the battery voltage drops if current is flowing through the battery; and (3) the current it will produce is limited (actually to about 5 A). Calculate the voltage across the 2.5-Ω resistance representing the flashlight bulb when the switch is closed.

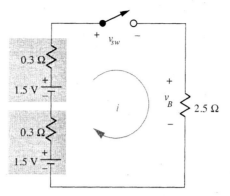

Figure 1.30 Model of flashlight circuit. The shaded boxes represent the two dry-cell batteries, which not only produce voltage, but also have internal resistance. The 2.5-Ω resistance represents the bulb.

SOLUTION:
In the circuit of Fig. 1.30, all elements are in series because they have the same current. Kirchhoff's voltage law for the circuit, starting at the bottom and going clockwise, is

$$-(1.5) + 0.3i - (1.5) + 0.3i + v_{sw} + 2.5i = 0 \tag{1.69}$$

We wrote the equation to show that the voltage from the batteries add to 3.0 V. With the switch closed, Eq. (1.68) gives the voltage across the bulb as

$$v_B = 3.0 \times \frac{2.5}{0.3 + 0.3 + 2.5} = 2.42 \text{ V} \tag{1.70}$$

That the resistances are separated from each other by the voltage sources is irrelevant; the resistances share the same current and hence are connected in series.

WHAT IF? What if the switch was corroded and had a resistance of 1.2 Ω when ON? What then would be the bulb voltage?[22]

Parallel Resistances and Current Dividers

LEARNING OBJECTIVE 7.

Understand how to combine resistances connected in parallel and how current divides between parallel resistances

Parallel resistances. Resistances are said to be connected *in parallel* when they have the same voltage across them. Figure 1.31 shows a parallel combination of three resistances and a current source. We could have defined voltages for each resistance, but because the lines represent ideal connections (wires of zero resistance), the tops of the resistances are connected together and bottoms of the resistances are connected together, as pictured in Fig. 1.6. We were careful to define the currents through the resistances with reference directions such as the currents would be positive, assuming I_s positive. We write KCL for node a:

---

[22] 1.74 V.

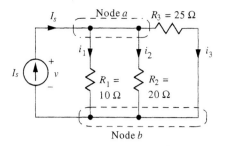

Figure 1.31 Three resistors connected in parallel.

$$-I_s + i_1 + i_2 + i_3 = 0 \Rightarrow I_s = i_1 + i_2 + i_3 \tag{1.71}$$

Because the current reference symbols are directed into the + end of v on the resistances, we have load sets; hence, we introduce Ohm's law, Eq. (1.41), into Eq. (1.71):

$$I_s = \frac{v}{R_1} + \frac{v}{R_2} + \frac{v}{R_3}$$

$$= v\left(\frac{1}{R_1} + \frac{1}{R_2} + \frac{1}{R_3}\right)$$

$$= v\frac{1}{R_{eq}} \tag{1.72}$$

Equivalent Circuits

where R_{eq} is the equivalent resistance of R_1, R_2, and R_3 connected in parallel. The third form of Eq. (1.72) introduces a resistance, R_{eq}, that is equivalent to the three parallel resistances:

$$\frac{1}{R_{eq}} = \frac{1}{R_1} + \frac{1}{R_2} + \frac{1}{R_3} \Rightarrow R_{eq} = \frac{1}{1/R_1 + 1/R_2 + 1/R_3} \tag{1.73}$$

Many students have memorized the special case of Eq. (1.73) that applies to two resistances in parallel—the product over the sum. As stated earlier, we would encourage exclusive use of Eq. (1.73) for two reasons:

1. Equation (1.73) is the correct form regardless of the number of resistances.
2. Equation (1.73) is easier to implement on a calculator than the "product over the sum" for two parallel resistances.

Clearly, Eq. (1.73) represents an extension of Eq. (1.57) for three resistances, and we can use the notation $R_{eq} = R_1 \| R_2 \| R_3$. In this case,

$$R_{eq} = 10 \| 20 \| 25 = \frac{1}{1/10 + 1/20 + 1/25} = 5.26\ \Omega \tag{1.74}$$

Thus, for the purpose of calculating the voltage across the three parallel resistances, we may replace the three resistances with an equivalent resistance of 5.26 Ω. We now determine how the current I_s divides between the three resistances.

Current dividers. We wish to use Eq. (1.72) in determining how the current I_s divides between the three resistances. We can find the voltage from the last form of Eq. (1.72) and solve for the current in, say, R_1 from Ohm's law:

$$v = R_{eq} I_s \Rightarrow i_1 = \frac{v}{R_1} = \left(\frac{R_{eq}}{R_1}\right) \times I_s = \left(\frac{1/R_1}{1/R_1 + 1/R_2 + 1/R_3}\right) \times I_s \quad (1.75)$$

The last form of Eq. (1.75) looks awkward but is the easiest form to implement on a calculator. Substituting the numbers, we learn that the current through R_1 is $0.526\, I_s$, or 52.6% of the total current. You can confirm for yourself that 26.3% passes through R_2 and 21.1% through R_3. These results are easy to generalize: For any number of resistances that are connected in parallel, the current through the ith resistance, R_i, is

$$i_i = \frac{R_{eq}}{R_i} \times I_T = \frac{1/R_i}{1/R_1 + 1/R_2 + \cdots \text{(all parallel resistances)}} \times I_T \quad (1.76)$$

where I_T is the total current entering the parallel combination.

Equation (1.73) shows how to combine resistances in parallel. The addition of reciprocals is awkward to write but easy to accomplish on a calculator. Indeed, adding reciprocals with a calculator is just as simple as adding numbers except that you must hit the 1/x key on the calculator after entering the resistance values. After summing the reciprocals, you then hit the 1/x key to display the equivalent resistance of the parallel combination. If we desire a neat equation, we must use the conductances

$$G_{eq} = G_1 + G_2 + \cdots \text{(all parallel conductances)} \quad (1.77)$$

In terms of conductances, the current divider relationship has the simpler form

$$i_i = \frac{G_i}{G_1 + G_2 + \cdots \text{(all parallel conductances)}} \times I_T \quad (1.78)$$

EXAMPLE 1.11 **Parallel and Series Resistances**

Calculate the voltage across the four resistances in Fig. 1.32.

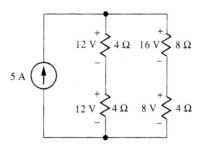

Figure 1.32 The parallel paths each contain two resistances in series.

SOLUTION:

When we have two parallel connections of multiple resistances in a circuit, the voltage across the two parallel combinations divides independently in the two paths. In Fig. 1.32, the equivalent resistance seen by the current source is

$$R_{eq} = (4+4) \parallel (8+4) = 4.8 \, \Omega \tag{1.79}$$

and hence the total voltage across the parallel combination is $5 \times 4.8 = 24$ V with + at the top. This voltage divides independently in the two paths by a routine application of Eq. (1.68), with the results shown in Fig. 1.32.

WHAT IF? What if the 8-Ω resistance were changed to 10 Ω? What then would be the voltage across the 10-Ω resistance?[23] Across each of the two series 4-Ω resistances?[24]

LEARNING OBJECTIVE 8.

To understand how to analyze circuits containing one source and resistances in series and parallel

The full voltage-divider/current-divider technique. In the preceding sections, we learned about series and parallel combinations and how voltage and current divide in them, respectively. Let us now use these concepts to find the unknown voltages in the circuit shown in Fig. 1.33. We first will combine resistances to simplify the circuit. Notice that R_2 and R_3 are connected in parallel. We can replace them with an equivalent resistance of $4 \parallel 6$ or 2.4 Ω. This reduces the network to that shown in Fig. 1.34(a). The 2-Ω resistance is connected in series with the 2.4-Ω resistance and they can be combined into an equivalent resistance of 4.4 Ω, Fig. 1.34(b). Clearly, the voltage across the current source, V_s, is $2 \, \text{A} \times 4.4 \, \Omega = 8.8$ V.

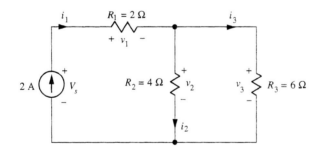

Figure 1.33 Solve for the unknown voltages using the voltage-divider technique.

We will now restore the original circuit. Figure 1.34(c) is the same as Fig. 1.34(a), but because we are planning to divide the 8.8 V between the two resistances, we have defined voltages v_1 and v_4 with the proper polarity markings for that purpose. The original v_1 already has the desired polarity marking, and we introduced a v_4 for the voltage across the 2.4-Ω resistance. "Wait" (you might object), "that is a current source, not a voltage source you are dividing." True, but the voltage created by the current source divides in the series circuit. It does not matter whether we have a 2-A current source pro-

---

[23] 18.2 V.

[24] 12.7 V.

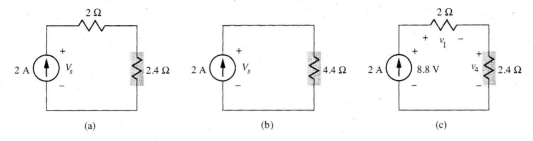

Figure 1.34 Combining resistances in parallel and series simplifies the circuit to the point where the unknown voltage can be determined by a voltage divider.

ducing 8.8 V or an 8.8-V voltage source producing 2 A—the circuit will respond in the same way.[25] So we can divide the 8.8 V whether it is produced by a voltage or a current source. The results are

$$v_1 = 8.8 \times \frac{2}{4.4} = 4.00 \text{ V} \quad \text{and} \quad v_4 = 8.8 \times \frac{2.4}{4.4} = 4.80 \text{ V} \tag{1.80}$$

Because v_4 is the voltage across the parallel combination of the 4- and 6-Ω resistances in the original circuit, v_4 is the same as v_3 and v_2 in the original problem, and i_3 follows from Ohm's law to be 0.80 A. The downward current through the 4-Ω resistance is 4.8 V/4 Ω = +1.2 A.

Reworking with current dividers. We repeat using current dividers instead of voltage dividers. The 2 A passes through R_1, producing 4 V across it, and then divides between the R_2 and R_3. The current through R_3 can be determined directly by current division, as also can be the current through R_2. For example, Eq. (1.76) yields

$$i_3 = 2 \times \left(\frac{1/6}{1/6 + 1/4}\right) = 0.8 \text{ A} \tag{1.81}$$

Note that R_1 has no influence on the division of the current between R_2 and R_3.

From the currents, we can calculate the voltages and complete the problem. For example, Eq. (1.41) yields

$$v_3 = R_3 i_3 = 6 \times 0.8 = 4.8 \text{ V} \tag{1.82}$$

On more complicated circuits, both voltage and current dividers might be used to learn how voltage and current distribute throughout the circuit.

Summary of this method. In this method:

1. We use series and parallel combinations of resistances to simplify a circuit, beginning far away from and coming toward the source.

---

[25] The type of source would matter if we changed one of the resistances in the circuit, for in one case, the source current would remain constant, and in the other case, the source voltage would remain constant.

2. The entire circuit can be reduced to a single resistance across the source, although often the process does not have to be carried that far.

3. The circuit is then restored, voltages and currents being divided progressively to yield the individual currents and voltages throughout the circuit.

Advantages of this method. The genius of this method is that you are guided in the solution by the geometry of the circuit. This important information is neglected if you merely define variables and apply Kirchhoff's and Ohm's laws directly. This method avoids defining unnecessary variables; you merely introduce those needed to perform the voltage and current division. As a final advantage, we might suggest that this method, after you use it a while, will develop your intuition about what is happening in circuits.

Weaknesses of this method. Alongside these advantages, we must place certain limitations. As we presented it, the method works for circuits having only one source, although we will overcome this limitation in Chap. 2. More significantly, this method does not handle the sign for you; you are expected to bring to the problem sufficient insight to know which way the physical current goes. For only one source, this is frequently no problem; but you do have to remember to supply the sign and not look for the mathematics to produce it. Also, this method often requires that you analyze the entire circuit, solving for virtually every voltage and current, to determine any single unknown. Finally, there are circuits where this method fails.

Figure 1.35 shows such a circuit. Simple though this circuit is, no two resistances are in series or parallel. There is no place to start combining resistances. Chapter 2 explains several methods that can solve this circuit.

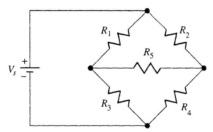

Figure 1.35 This circuit cannot be solved by voltage and/or current dividers because no two resistances are in series or in parallel.

Check Your Understanding

1. Two circuit elements are said to be in parallel if they share the same voltage or current?

2. In a current divider consisting of three resistances, the resistances are in the ratio $1:2:3$. What percent of the total current goes through the largest resistance?

3. What are $2 + 5 \| 7$ and $10 \| 12 + 7 \| (1 + 2)$?

4. Three resistances having the same value, R, are connected together to have an equivalent resistance of $1.5R$. How are they connected?

Answers. (1) Same voltage; (2) 18.2%; (3) 4.92, 7.55; (4) one in series with the other two in parallel.

CHAPTER SUMMARY

We lay the foundation for the entire book in defining the electrical quantities used to describe most electrical systems, voltage and current, stressing that these track energy flow throughout a circuit. We present Kirchhoff's laws as expressions of conservation of charge and electric energy. Resistance is defined and Ohm's law is presented. With the definition of current and voltage sources, we begin to solve simple circuits. More complicated circuits are analyzed by combining series and parallel resistances, and voltage and current dividers are used to determine voltage and current throughout a circuit.

Objective 1: To understand what a circuit is and why circuits are important in electrical engineering. Circuits are the logical place to begin the study of electrical engineering because most electrical devices and applications involve circuits in one form or another. Circuits are relatively easy to understand and require simple mathematics.

Objectives 2 and 3: To understand the definitions of current and voltage and how to use Kirchhoff's laws to express conservation of charge and energy. Voltage and current are used to describe the state of an electric circuit for several reasons: they are easily measured, they describe the flow of energy in the circuit, and they obey simple constraints based on conservation of charge and energy. We stress the necessity of careful attention to signs in writing equations based on these laws because both voltage and current are described by both a magnitude and a sign with respect to a reference direction.

Objective 4: To understand how to use the voltage and current to calculate the power into or out of a circuit element. Power is the product of voltage and current, but again one must pay careful attention to signs, in this case the reference direction for the current relative to the reference direction for the voltage. We recommend using a load set for passive circuit elements such as resistors and using a source set for sources.

Objective 5: To understand the relationship between voltage and current in a resistor as described by Ohm's law. Ohm's law states that the current in a resistor is proportional to the voltage across the resistor. It follows that the power in a resistor is proportional to the square of the current through the resistor and the square of the voltage across the resistor.

Objective 6: To understand how to combine resistances connected in series and how voltage divides between series resistances. Resistors are in series when they have the same current through them. Resistors in series can be combined by adding their resistances. The voltage across resistors connected in series divides between the resistors in proportion to their resistance values.

Objective 7: To understand how to combine resistances connected in parallel and how current divides between parallel resistances. Resistors are in parallel when they have the same voltage across them. Resistors in parallel can be combined by adding the reciprocals of their resistances. The current into two or more resistors connected in parallel divides between the resistors in proportion to the reciprocals of their resistance values.

Objective 8: To understand how to analyze circuits containing one source and resistances in series and parallel. By combining resistors in series and parallel, simple circuits can be reduced to a form where the voltage and current throughout the circuit can be determined with voltage and current dividers.

Chapter 2 presents additional methods for circuit analysis, all based on the definitions and laws presented in this chapter.

GLOSSARY

Charge, p. 5, a property of matter. There are two types of charge, **positive** and **negative**.

Circuit theory, p. 6, taking a given circuit model and solving for voltage, current, power, etc., through the application of known circuit laws.

Conductance, p. 24, the reciprocal of resistance.

Current, p. 7, the movement of charge, usually in a wire or electrical device. If in a period of time Δt, ΔQ coulombs move past a cross-section in a wire in the indicated direction, the current is

$$i = \frac{\Delta Q}{\Delta t} \quad \text{C/s or ampere, A.}$$

Electrical engineering, p. 4, the useful control of electrical energy.

Electrostatic force, p. 5, force between stationary charges, depends on the distance between the charges.

Energy, p. 21, the medium of exchange in a physical system. In mechanics it takes force and movement to do work (exchange energy), and in electricity it takes electrical force and movement of charges to do work (exchange energy). The electrical force is represented by the voltage and the movement of charge by the current in an electrical circuit.

Equivalent circuit, p. 31, a circuit that represents in a specified characteristic the properties of another, more complicated circuit.

Equivalent resistance, p. 30, a single resistance that has the same resistance as a two-terminal network of resistors.

Ideal current source, p. 27, a current source that produces its prescribed current, independent of its output voltage.

Ideal switch, p. 25, a special resistance that can be changed from a short circuit to an open circuit to turn an electrical device ON or OFF.

Ideal voltage source, p. 26, a voltage source that maintains its prescribed voltage, independent of its output current.

Kirchhoff's current law (KCL), p. 10, the constraint imposed at a node by conservation of charge and charge neutrality: the sum of the currents leaving a node is zero at all times. An equivalent statement would be that the sum of the currents entering a node is equal to the sum of the currents leaving that node at any instant of time.

Load set, p. 24, where the current reference direction is directed into the + polarity marking (or the first subscript) of the voltage reference direction. This is the convention normally used with a resistance or some other element that receives power from a circuit.

Magnetic force, p. 5, force between moving charges (that is, between currents) depends on the distance between the charges and their relative velocity.

Node, p. 10, the junction of two or more wires.

Open circuit, p. 25, a "path" that permits no current flow, regardless of the voltage across it.

Parallel connection, p. 30, a connection between two or more circuit elements in which all share the same voltage.

Physical current, p. 8, the magnitude of the current, in the direction in which conventional current is positive. Physical current goes opposite to the direction of electron motion.

Physical polarity of the voltage, p. 14, the reference direction markings that gives a positive voltage. Thus the **physical voltage** is always positive and has polarity markings assigned accordingly.

Power, p. 19, the rate of energy exchange: the electrical power is by definition the product of voltage and current:

$$v\left(\frac{\text{work}}{\text{charge}}\right) \times i\left(\frac{\text{charge}}{\text{time}}\right) = \frac{\text{work}}{\text{time}} = \text{power}$$

Resistance, resistor, p. 23, the circuit property or element that models conversion of electrical energy to nonelectrical energy.

Series connection, p. 32, a connection between two or more circuit elements in which all have the same current.

Short circuit, p. 25, an ideal connection for current flow requiring zero voltage.

Source set, p. 19, where the current reference direction is directed out of the + polarity marking (or the first subscript) of the voltage. This combination is normally used for a source of energy to the circuit, such as a battery.

Supernode, p. 11, a group of nodes in the same circuit, including any circuit elements connected between the nodes. Kirchhoff's Current Law must be obeyed at a supernode.

Voltage, p. 12, the work done by the electrical system in moving a charge from one point to another in a circuit, divided by the charge.

PROBLEMS

Section 1.3: Current and Kirchhoff's Current Law

1.1. The allowed safe current for a No. 12 copper wire, 0.08081 inch in diameter, is 23 A.
  (a) How many coulombs of charge pass a cross-section of the wire in 1 minute?
  (b) At 23-A direct current, how fast are the electrons traveling in mm/s?

1.2. In a No. 10 copper wire 0.1019 inch in diameter, electrons are moving 1 ft every minute. What is the magnitude of the current in the wire?

1.3. The current in a wire is zero for $t < 0$ but begins increasing at a rate of 5 amperes/second at $t = 0$, as shown in Fig P1.3.
   (a) At what time is the current 7.2 A?
   (b) At what time has 12 C of charge passed a cross-section of the wire?

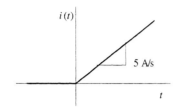

Figure P1.3

1.4. For the circuit in Fig. P1.4:
   (a) Write KCL for node a using the first form of KCL on page 10.
   (b) Write KCL for node b using the third form of KCL on page 10.
   (c) Add or subtract the equations resulting from KCL at nodes a and b to show that i_1 and i_4 are equal.

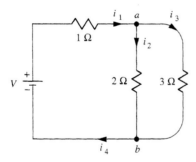

Figure P1.4

1.5. Use KCL to determine i and i_{ab} in the circuit shown in Fig. P1.5.

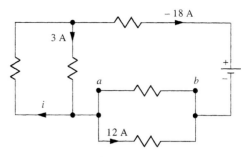

Figure P1.5

1.6. For the circuit in Fig. P1.6, compute the unknown currents using KCL.

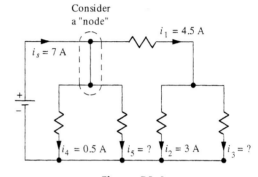

Figure P1.6

1.7. In a simulated lightning bolt in a laboratory, the current increases linearly from 0 to 1500 A in 1 μs and then decreases linearly back to 0 in 4 μs, shown in Fig. P1.7, making a 5 μs total for the duration of the lightning. What is the total charge required?

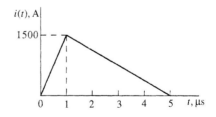

Figure P1.7

Section 1.4: Voltage and Kirchhoff's Voltage Law

1.8. How much work is done by a 12.6-V battery if 1 microcoulomb of positive charge is pulled off the negative terminal and placed on the positive terminal?

1.9. If $v_{xy} = +120$ V and a -4-C charge is moved by an external agent from y to x, how much work is done on the external agent by the electrical system?

1.10. In a cathode-ray tube (CRT), a beam of electrons is accelerated through a certain voltage (V_{CRT}), then allowed to drift through an evacuated space until it strikes the CRT surface, making a small spot of light. If $|V_{CRT}| = 9500$ V, what velocity do the electrons acquire from the voltage source? *Hint:* Equate the final kinetic energy of an individual electron with the electrical energy delivered by the source to one electron.

1.11. (a) Write KVL to determine v_{ab} in the circuit Fig. P1.11.
   (b) If a battery (voltage source) were inserted between c and d to make $v_{ab} = 12.3$ V, draw the circuit showing the battery polarity and voltage magnitude.

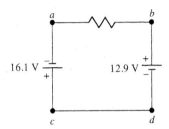

Figure P1.11

1.12. For a circuit, $v_{ab} = 9$ V, $v_{cb} = -5.2$ V, $v_{dc} = -11$ V, and $v_{de} = -4$ V. Using the rules for adding subscripts, determine v_{ba}, v_{bd}, and v_{ae}. *Hint:* Draw a "circuit" showing the lettered points to aid in visualizing the patterns.

1.13. For the circuit shown in Fig. P1.13, find V_1, v_{ad}, v_{bc}, and $v_{ac} + v_{ce}$.

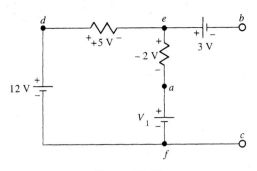

Figure P1.13

1.14. For the circuit shown in Fig. P1.14, use KCL and KVL to determine i_1, i_2, v_{ad}, and v_x.

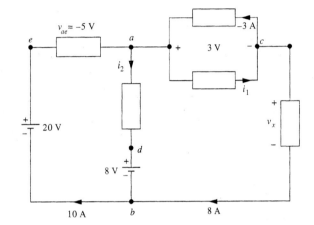

Figure P1.14

Section 1.5: Energy Flow in Electrical Circuits

1.15. The MKS unit for energy is the joule, which is 1 watt-second, but the unit in common use by electrical utilities is the kilowatt-hour (kWh).
   (a) How many joules are there in 10 kWh?
   (b) At 9.35 cents/kWh, how many joules can one buy for a penny?

1.16. For the circuit shown in Fig. P1.16:
   (a) Use KCL and KVL to find v_2, v_3 and i.
   (b) Calculate the power *into* every resistance and the power *out of* every source. Show that energy is conserved, that is, that $p_{out} = p_{in}$.

1.17. The circuit shown in Fig. P1.17 represents a battery charger charging a battery.
   (a) Find the power into the battery being charged.

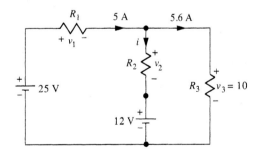

Figure P1.16

   (b) What is the time it would take to impart 1 kilocoulomb (10^3 C) to the battery?

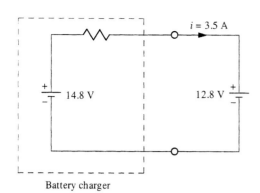

Battery charger

Figure P1.17

(c) What energy is given to the battery in this period of time?

1.18. In Fig. P1.18, the power into the 5-V battery is $+10$ W. Determine the power into R and the power out of the 9-V battery.

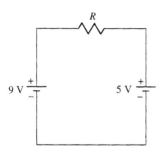

Figure P1.18

1.19. In the circuit in Fig. P1.19 determine the following:
(a) v_{ab}.
(b) i.
(c) The sum of the powers into R_1, R_2, and the 2-V battery.

1.20. In the circuit shown in Fig. P1.20, give the values for the following:
(a) v_{ab}.
(b) v_x.
(c) The power out of the 4-V battery.

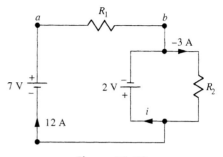

Figure P1.19

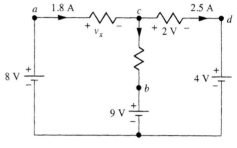

Figure P1.20

1.21. For the circuit shown in Fig. P1.21, compute the unknown voltages and currents. (You supply the voltage and current variables for the unknowns, including their reference directions.) Compute the power *into* the two resistances and the power *out of* both sources. Show conservation of power.

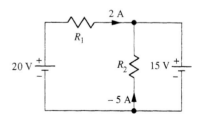

Figure P1.21

Section 1.6: Circuit Elements: Resistors, Switches, and Sources

1.22. A resistor capable of handling safely 5 W is to be placed across the terminals of a 10-V voltage source. What range of resistances will not exceed the 5-W limit?

1.23. Find the value of R_1, R_2, and R_3 in Fig. P1.16.

1.24. A car radio designed to operate from a 6.3-V system uses 4.5 A of current, as shown in Fig. P1.24.

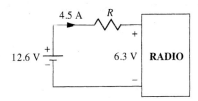

Figure P1.24

(a) What resistance should be placed in series with this radio if it is to be used in a 12.6-V system?
(b) What should be the power rating of this resistance?

1.25. A voltage source, current source, and resistance are shown in series in Figure P1.25.
(a) For $I_s > 0$, find the value of V_s that makes the power into the voltage source equal to the power into the resistance. The required V_s is a function of I_s and R.
(b) For $I_s < 0$, show that no positive value of V_s will satisfy the conditions given in (a). Explain why this is true.

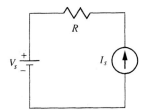

Figure P1.25

1.26. Figure P1.26 shows a voltage source and a switch.
(a) Is the switch single- or double-pole?
(b) Is the switch single- or double-throw?
(c) What is the output voltage for the switch in position a and in position b?

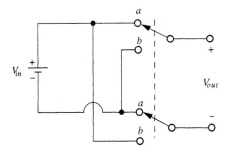

Figure P1.26

1.27. Figure P1.27 shows a voltage source, light bulb, and two double-throw, single-pole switches.
(a) For the light bulb to be ON, what are the possible settings for the switches?
(b) Assuming that the light is ON, what happens to the light as a consequence of changing either switch?
(c) Assuming that the light is OFF, what happens to the light as a consequence of changing either switch?

Note: This is (for some reason) called a "three-way" switch and is often used in a long hall or stairway so that a light can be turned ON or OFF from either end of the hall.

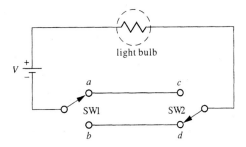

Figure P1.27

1.28. Figures 1.28(a) and 1.28(b) show a resistance and a voltage source or current source, respectively. Plot of $V(R)$ and $I(R)$ vs. R for the range $1 < R < 10\,\Omega$ in both cases. *Note:* Commonly, we consider $V(R)$ to mean that V depends on R, so V cannot be constant as R is varied. But actually $V(R) =$ constant is a perfectly good mathematical function that happens to have a zero derivative everywhere.

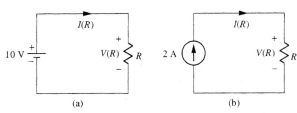

Figure P1.28

1.29. Figure 1.29 shows circuits containing voltage and current sources and resistances. Some circuits are valid (non contradictory) for all values of the variables, some are valid for no values, and some are valid for specific values of the variables. Classify each and tell what, if any, values of the V's, I's, and R's lead to circuits compatible with the definitions of the sources and resistance.

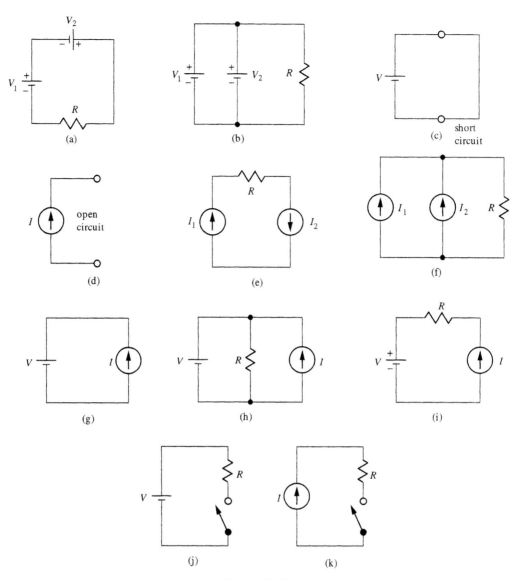

Figure P1.29

Section 1.7: Series and Parallel Resistances; Voltage and Current Dividers

1.30. (a) Three resistors connected in series have resistance values in the ratio 1:1.5:3.3 and combine to an equivalent resistance of 150 Ω. What is the smallest resistor?
(b) The three resistors, still in series, are placed across a 300-V dc voltage source. What is the voltage that will appear across the largest resistor?

1.31. (a) Two series resistances are to work as a voltage divider, with the smaller receiving 30% of the total voltage. What are the resistors given that their equivalent resistance is 200 Ω?
(b) If both resistors are 2-watt resistors, what is the maximum total voltage the voltage divider can handle without exceeding this rating for either resistor?

1.32. For the circuit shown in Fig. P1.32, the arrow indicates that the resistor R is variable; hence, the voltage across R, $v(R)$, will depend on the value of R, as the notation indicates. Determine the function and plot $v(R)$ for $0 < R < 40\,\Omega$, and find the value of R to make the voltage 5.5 V.

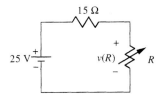

Figure P1.32

1.33. (a) Find the equivalent resistance of the parallel combination shown in Fig. P1.33.
(b) If 10 A enters the parallel combination, referenced in at a and out at b, what is the current referenced downward in the 6-Ω resistance?
(c) What is the current referenced upward in the 2-Ω resistance?
(d) What is the voltage v_{ab} for this current?

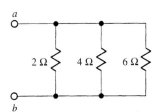

Figure P1.33

1.34. Consider three parallel resistances, R_1, R_2, and R_3, with a current of I_T A passing through the parallel combination. Show that the power in the equivalent resistance, $R_{eq} = R_1 \| R_2 \| R_3$, is the correct value, that is, the sum of the powers in the resistances in the original circuit.

1.35. For the circuit shown in Fig. P1.35 determine the following:
(a) Find R such that $i_R = 0.5$ A.
(b) What value of R gives a power of 833 W in R?
(c) What is the maximum possible value of the power in R?

1.36. Design a current divider that has an equivalent resistance of 100 Ω and divides the current in the ratio of 5:2. This problem is summarized in Fig. P1.36.

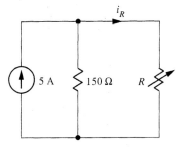

Figure P1.35

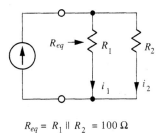

$R_{eq} = R_1 \| R_2 = 100\,\Omega$

$2i_1 = 5i_2$

Figure P1.36

1.37. (a) Evaluate $R_{eq} = 8 + (3+4) \| 6\,\Omega$.
(b) Draw the circuit corresponding to this expression.

1.38. (a) What resistance in parallel with 105 Ω reduces the equivalent parallel resistance to 91.5 Ω?
(b) What is $1 \| 2 \| 4 \| 8 \| 16 \| \cdots$?
Hint: Write the expression for the total conductance. The result should be a familiar geometric series that is easily summed.

1.39. As the variable resistor in Fig. P1.39, R, varies from a short to an open circuit, the equivalent resistance should vary between 30 and 75 Ω.
(a) Design R_1 and R_2 to accomplish this result.
(b) Find R to give $R_{eq} = (30 + 75)/2\,\Omega$.

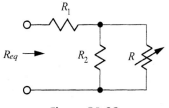

Figure P1.39

1.40. The resistance R in Fig. P1.40 is variable from a short circuit to an open circuit.
(a) Find the maximum and minimum equivalent resistance at the input.
(b) What value of R gives the average between R_{max} and R_{min}?

1.41. For the circuits shown in Fig. P1.41, find the indicated unknowns using voltage and current dividers.

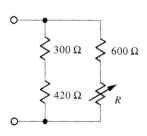

Figure P1.40

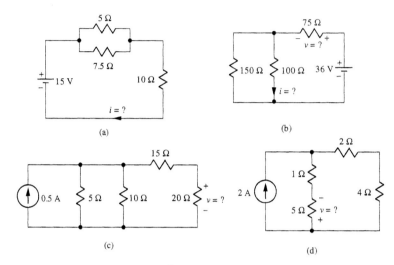

Figure P1.41

1.42. For the circuits in Fig. P1.42, find the indicated unknowns using voltage and current dividers.

1.43. For the circuit shown in Fig. P1.43, find the value of R that makes the power into the 100-Ω resistor to be 10 W.

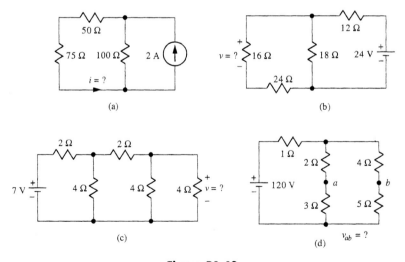

Figure P1.42

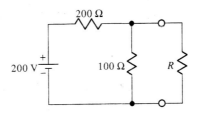

Figure P1.43

General Problems

1.44. In the circuit shown in Fig. P1.44, all voltages are dc. The boxes represent unknown (and in some cases, weird) circuit elements. The current in No. 1 is also dc, as shown. The voltage across No. 1 is $v_{ab} = +40$ V. The current in No. 3 is the pulse of current shown in the figure and is described by the equation $i_3(t) = 8(t - 2t^2)$, $0 < t < 0.5$, and $i_3 = 0$ at all other times.

(a) Find the work done by the electrical system in moving an electron from c to b.

(b) What is the minimum power into No. 2 as time varies?

(c) How much charge is moved through No. 3 by current pulse $i_3(t)$?

(d) Find the electrical energy contributed to the circuit by No. 3 during the time period $0 < t < 0.5$ s.

1.45. In a semiconductor such as silicon, current is carried both by conduction electrons and by mobile "holes" that behave like positive charges having a charge equal in magnitude to that of the electron. Find the magnitude and direction of the current in a silicon diode of cross-sectional area 10^{-9} m^2 with electrons (n_e of 10^{23} electrons/m^3) traveling with 0.3 mm/s to the right and holes (n_h of 1.5×10^{22} holes/m^3) traveling to the left at 0.1 mm/s.

1.46. A lightning bolt might carry a current of 10^4 A and last for about 1 ms, as shown in Fig. P1.46.

(a) How much charge is exchanged between ground and cloud in such a lightning bolt?

(b) How many raindrops had to fall, each having a deficiency of 10 electronic charges, to create the initial charge separation neutralized by the lightning?

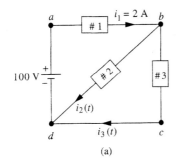

(a)

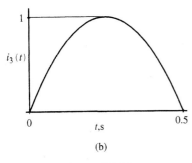

(b)

Figure P1.44

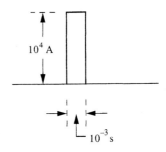

Figure P1.46

1.47. A battery can be rated in "A-h," ampere-hours. For example, a 60 A-h battery ought to put out 60 A for 1 h, or 1 A for 60 h, or some other combination multiplying to 60.

(a) How many coulombs of charge may one hope to get out of a fully charged 90 A-h battery?

(b) How many electrons would pass through the wire discharging such a fully charged battery?

1.48. A fully charged auto battery has a voltage of 12.6 V. As the battery discharges, the voltage drops. The voltage will drop to about 10.0 V as total discharge is approached, at which point the voltage drops to essentially zero if you try to draw current from the battery. Consider the case where we are discharging a 60-A-h battery over a 6-h period at a uniform 10-A rate. Assume the battery voltage drops from 12.6 to 10.0 V at a linear rate.

(a) Describe the voltage as a function of time during discharge.

(b) Compute the total energy output of the battery for this discharge rate.

(c) Show that the energy output is independent of the discharge rate. *Hint:* Let T be the time for discharge in hours and $60/T$ be the current.

1.49. Figure P1.49 shows the voltage and current for a semiconductor switch closure. Note the switch is not ideal in that voltage and current do not change instantaneously. This means that the switch will dissipate some energy during its operation.

(a) How much energy is used up in each switching operation? (Assume a source set of reference directions.)

(b) If the same energy were used in opening the switch, and the switch went through 5000 cycles each second, what is the average power lost in the switch? Each cycle involves opening and closing the switch.

1.50. Problems based upon circuit symmetry. The equivalent resistances in Fig. P1.50 may be solved readily using circuit symmetry. For example, in circuit (a), the circuit has even symmetry about the center line of the bridge; hence, any current that enters at the top divides equally. Consequently, the voltage across the middle resistance is zero, no current passes through it, and the left and right resistances are actually in series (have the same current). Using this approach, find the indicated equivalent resistance for all three circuits.

1.51. These problems are based upon infinite ladder circuits. The equivalent resistances of the circuits in Fig. P1.51 may be determined by considering the properties of infinity. In both cases, if you remove the first two resistances (the first rung of the ladder), the remaining circuit has an equivalent resistance that is closely related to that of the original problem. Find the equivalent resistance in parts (a) and (b).

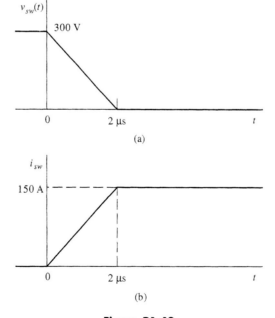

Figure P1.49

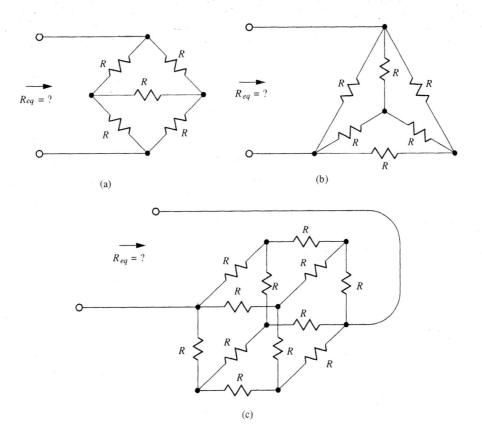

Figure P1.50

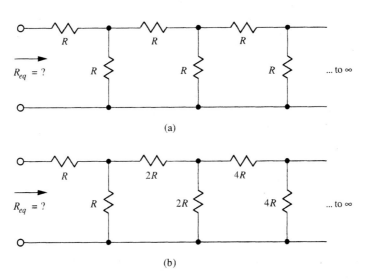

Figure P1.51

Answers to Odd-Numbered Problems

1.1. (a) 1380 C; (b) 0.384 mm/s.

1.3. (a) 1.44 s; (b) 2.19 s.

1.5. $i = -15$ A, $i_{ab} = 6$ A.

1.7. 3.75 mC.

1.9. +480 J.

1.11. (a) -29 V; (b) 41.3 V, + at left.

1.13. 9 V, -3 V, $+4$ V, $+2$ V.

1.15. (a) 3.6×10^7 J; (b) 3.85×10^5 J/cent.

1.17. (a) 44.8 W; (b) 286 s; (c) 12,800 J.

1.19. (a) $+9$ V; (b) -15 A; (c) 84 W.

1.21. (a) $p_{sources} = 40 + 45$ W; (b) $p_{resistors} = 10 + 75$ W.

1.23. $R_1 = 3$ Ω, $R_2 = 3.33$ Ω, $R_3 = 1.79$ Ω.

1.25. (a) $V_s = I_s R$; (b) same formula, but violates polarity convention for voltage sources.

1.27. (a) For ON,

SW1	SW2
a	c
b	d

(b) it goes OFF; (c) it comes ON.

1.29. (a) Valid for all V_1, V_2, and R; (b) valid for $V_1 = V_2$; (c) valid for $V = 0$, or not valid; (d) valid for $I = 0$, or not valid; (e) valid for $I_1 = I_2$; (f) valid for all I_1, I_2, and R; (g) valid for all V and I; (h) valid for all V, I, and R; (i) valid for all V, I, and R; (j) valid for all V, R, and switch positions; (k) valid for closed switch only.

1.31. (a) 60 Ω and 140 Ω; (b) 23.9 V.

1.33. (a) 1.09 Ω; (b) 1.82 A; (c) -5.45 A; (d) 10.9 V.

1.35. (a) 1350 Ω; (b) 74.9 Ω or 300 Ω; (c) 937.5 W.

1.37. 11.2 Ω.

1.39. (a) 70 Ω, 45 Ω; (b) 45 Ω.

1.41. (a) 1.15 A; (b) 0.160 A, $v = 20$ V; (c) 0.870 V; (d) -5.00 V.

1.43. 60.2 Ω.

1.45. -5.05×10^{-9} A to the right.

1.47. (a) 324,000 C; (b) 2.02×10^{24} electrons.

1.49. (a) 0.0150 J; (b) 150 W.

1.51. (a) $1.62R$; (b) $1.78\,R$.

CHAPTER 2

The Analysis of DC Circuits

Superposition

Thévenin's and Norton's Equivalent Circuits

Node-Voltage Analysis

Loop-Current Analysis

Controlled Sources

Chapter Summary

Glossary

Problems

objectives

1. To understand the principle of superposition and use superposition to analyze circuits with multiple sources
2. To understand the origin of Thévenin's equivalent circuit and be able to derive the Thévenin equivalent circuit of a network with a load
3. To understand the significance of impedance level in interactions between circuits
4. To understand how to analyze a circuit using nodal analysis
5. To understand how to analyze a circuit using loop currents
6. To understand how to choose the most efficient manner for analyzing a circuit
7. To understand the need for controlled sources in circuit models and their effect in circuit analysis techniques

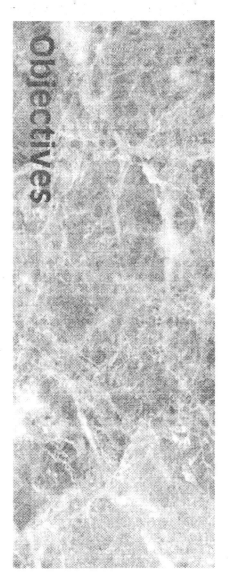

Simple circuits can be analyzed by simple methods, but not all circuits are simple. The methods of this chapter, although illustrated on relatively simple circuits, are general enough to handle complicated circuits.

The method of Thévenin equivalent circuits, furthermore, yields an insight that figures explicitly or implicitly into virtually every electrical design.

Introduction. In Chap. 1, we defined voltage and current in terms of energy, charge, and time. We showed that knowing the voltage and current throughout a circuit allows the engineer to know how energy is entering, exiting, and moving throughout the circuit. We presented Kirchhoff's voltage and current laws as statements of conservation of energy and charge, expressed in terms of voltage and current. We gave special attention to voltage and current reference directions in the writing of circuit equations.

We then defined voltage sources, current sources, and resistances. We discussed Ohm's law, which relates the voltage across a resistance to the current through the resistance. We solved simple circuits involving sources, switches, and resistances. Finally, we showed that resistances in series (having the same current) and resistances in parallel (sharing the same voltage) can be combined into equivalent resistances that simplify circuit analysis. Voltage and current dividers were used to find how the voltage and current distribute in simple circuits.

The method of circuit analysis summarized on page 39 works well for simple circuits. In Chap. 2, we add methods of circuit analysis that are useful in more complicated circuits.

2.1 SUPERPOSITION

Superposition Illustrated

Demonstration of the principle. The principle of superposition extends the method based upon resistance combinations taught in Chap. 1. Before giving a formal definition, we will illustrate the principle of superposition. We will solve for v_2 in the circuit in Fig. 2.1. Because the circuit has two sources, we cannot use voltage and current dividers. Thus, we will apply Kirchhoff's voltage and current laws, and Ohm's law. These are

$$\text{Ohm's law:} \quad v_1 = i_1 R_1 \quad \text{and} \quad v_2 = i_2 R_2 \tag{2.1}$$

$$\text{KVL:} \quad -v_1 + v_2 + V_s = 0 \tag{2.2}$$

$$\text{KCL:} \quad -I_s + i_1 + i_2 = 0 \tag{2.3}$$

First, we eliminate i_1 and i_2 with Ohm's law, so Eq. (2.3) becomes

$$\frac{v_1}{R_1} + \frac{v_2}{R_2} = I_s \tag{2.4}$$

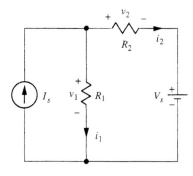

Figure 2.1 Solve for v_2 using KVL and KCL.

When we eliminate v_1 between Eqs. (2.2) and (2.4), we obtain the following result:

$$v_2 = \frac{I_s - (V_s/R_1)}{(1/R_1)+(1/R_2)} = I_s(R_1 \parallel R_2) - V_s \times \frac{R_2}{R_1 + R_2} \qquad (2.5)$$

Some observations based on the result. The first form of Eq. (2.5) emerges from the algebra; the second form can be interpreted from the results of Chap. 1. Examination of Eq. (2.5) suggests the following:

1. There are two components of v_2, one for each source. This means that one part of v_2 is caused by the current source and the other part is caused by the voltage source.

2. The part of the voltage due to the current source is what would be caused by the current source acting alone, provided the voltage source is replaced by a short circuit. The current source sees two resistances in parallel if the voltage source is replaced by a short circuit. By *short circuit,* we mean an ideal connection having no voltage across it.

short circuit

3. The part of the voltage due to the voltage source is what would be caused by the voltage source acting alone, provided the current source is replaced by an open circuit. By *open circuit,* we mean that no path exists for current flow. Note that if there is no current flow through the current source, the two resistances are connected in series and the voltage-divider form, the second term in Eq. (2.5), is easily identified.

open circuit

4. The total voltage is the sum of the separate effects of the two sources, provided the polarity of the effects is considered. In this case, the current source produces a voltage with a polarity the same as the reference direction of v_2 and it appears with a positive sign. The voltage source produces a voltage with a polarity opposite to that of v_2, and this component appears in the summation with a negative sign.

These observations suggest the principle of superposition, which we now state.

Principle of Superposition

LEARNING OBJECTIVE 1.

To understand the principle of superposition and use it to analyze circuits with multiple sources

Principle stated. The principle of superposition may be stated as follows: *The response of a circuit due to multiple sources can be calculated by summing the effects of each source considered separately, all others being turned OFF.* By OFF, we mean that current sources are replaced by open circuits and voltage sources are replaced by short circuits.

turned-OFF sources

Turned-OFF sources. The concept of turning-OFF a source is important and bears elaboration. We start with the graphical presentation of the $i - v$ characteristics of a resistance in Fig. 2.2. Because the slope of the resistance characteristic is $1/R$, a zero-resistance line would have an infinite slope. Thus, any amount of current can pass without producing a voltage. On the other hand, the line for an infinite resistance would have zero slope, implying zero current no matter how large the voltage. In Fig. 2.2, we have called zero resistance a short circuit and infinite resistance an open circuit. The circuit symbol for a short circuit is merely a line, and the circuit symbol for an open circuit is a break in the line, indicating that there is no path for current.

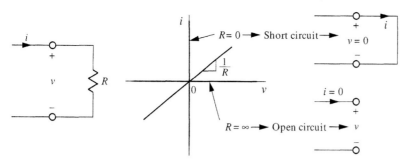

Figure 2.2 The characteristic of a short circuit is vertical and the characteristic of an open circuit is horizontal.

From our earlier definition of a voltage source, we know that a voltage source establishes a certain voltage at its terminals, independent of the current. In Fig. 2.3, this is represented by a vertical line at V_S. It follows that a voltage source of zero volts would be represented by a vertical line of infinite slope through the origin, which is the same as the characteristic of a short circuit. Thus, a turned-OFF voltage source is a short circuit.

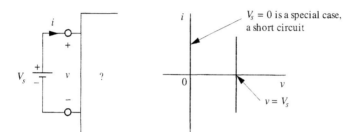

Figure 2.3 The voltage source has a vertical characteristic. Zero voltage ($V_s = 0$) is equivalent to a short circuit.

Figure 2.4 shows the definition of a current source. This is a horizontal line of constant current. A current source of zero value produces a horizontal line through the origin, the same characteristic as an open circuit. Thus, a turned-OFF current source is identical to an open circuit.

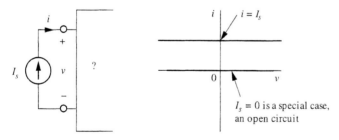

Figure 2.4 The current source has a horizontal characteristic. Zero source current ($I_s = 0$) is equivalent to an open circuit.

Benchmark example. We will use the principle of superposition to solve for the voltage across the 3-Ω resistance in the circuit shown in Fig. 2.5. We call this our "benchmark example" because we will solve this problem by every method presented in this chapter.

Turn ON 4-A source. There are three sources, so we will have to analyze three circuits, each having a single source. In Fig. 2.6, we calculate the voltage due to the 4-A current source, v_4. We have turned OFF the 5-A current source, replacing it with an open circuit; and we have turned OFF the 6-V source, replacing it with a short circuit. The re-

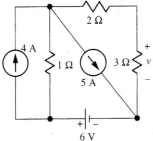

Figure 2.5 We will calculate the contributions of each source to v and add all results.

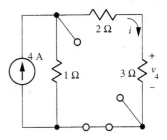

Figure 2.6 The circuit with the 4-A source ON and the others OFF.

sulting circuit shows the 2- and 3-Ω resistances connected in series with each other and together in parallel with the 1-Ω resistance. We use a current divider, Eq. (1.76),

$$i = 4 \text{ A} \times \frac{1/(2+3)}{1/(2+3) + 1/1} = 2/3 \text{ A} \tag{2.6}$$

and then calculate the voltage across the 3-Ω resistance from Ohm's law:

$$v_4 = 2/3 \text{ A} \times 3 \text{ Ω} = 2.00 \text{ V} \tag{2.7}$$

The polarity resulting from the 4-A source is the same as the original polarity markings for v. Hence this contribution will appear with a + sign in the final summation.

Turn ON 5-A source. Figure 2.7 shows the circuit with the 4-A and 6-V sources turned OFF. From the perspective of the 5-A source, the 2- and 3-Ω resistances are in series and together in parallel with the 1-Ω resistance. The voltage resulting from the 5-A source, v_5, can be calculated with the current-divider approach just used, but for practice, we will use voltage dividers. The voltage across the 3-Ω resistance is

$$v_5 = \underbrace{5 \times 1 \| (2+3)}_{\text{voltage across parallel combination}} \times \underbrace{\frac{3}{2+3}}_{\substack{\text{portion across} \\ \text{3-Ω resistor}}}$$

$$= 5 \times 0.833 \times 0.6 = 2.50 \text{ V} \tag{2.8}$$

Here the polarity is opposite to the reference direction of v, so we use a minus sign in the final summation.

Turn ON 6-V source. Finally, we calculate in Fig. 2.8 the contribution of the voltage source. With both current sources turned OFF, the three resistances are connected in series and the resulting voltage due to the 6-V source, v_6, is readily calculated by voltage division.

$$v_6 = 6 \times \frac{3}{1+2+3} = 3.00 \text{ V} \tag{2.9}$$

Note that the sign of v_6 will be positive in the final summation because the polarity of the voltage due to the voltage source is the same as that of v.

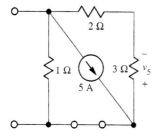

Figure 2.7 The circuit with the 5-A source ON and the others OFF. Here the voltage produced is opposite to the reference direction in Fig. 2.5.

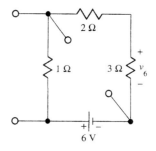

Figure 2.8 The circuit with the 6-V source ON and the others OFF.

The principle of superposition states that we can now obtain the total voltage across the 3-Ω resistance by adding up the contributions of the three sources, supplying the signs from the polarity produced from each source:

$$v_3 = +v_4 - v_5 + v_6 = +2.00 - 2.50 + 3.00 = +2.50 \text{ V} \quad (2.10)$$

EXAMPLE 2.1 Reducing the Current to Zero

Find the value of I_s to reduce the voltage across the 4-Ω resistor to zero in Fig. 2.9.

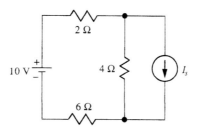

Figure 2.9 The voltage and current sources cause voltages of opposite polarities across the 4-Ω resistor.

SOLUTION:
The voltage due to the 10-V source is

$$v_{10} = 10 \times \frac{4}{2 + 4 + 6} = 3.33 \text{ V} \quad (2.11)$$

with the + at top. The voltage due to the current source is

$$v_{I_s} = I_s \times 4 \| (6 + 2) = I_s \times 2.67 \text{ Ω} \quad (2.12)$$

with the + at the bottom. To add to zero, the two must be equal:

$$3.33 = I_s \times 2.67 \Rightarrow I_s = 1.25 \text{ A} \quad (2.13)$$

WHAT IF? What if the 2- and 6-Ω resistors were exchanged?[1]

---

[1] It would make no difference to the voltage across the 4-Ω resistor.

Limitations of superposition. Superposition works because Kirchhoff's laws and Ohm's law are linear equations. In all KVL and KCL equations, voltage and current appear in the first power; there are no squares or square roots or functions as e^v.

Superposition does not work for power calculations, however, because power calculations involve multiplication of voltage by current, or squaring a variable like i^2R. Thus, you will not get the correct total power by adding the power due to each source considered separately. Superposition can be used indirectly for power calculations, however, because the total current through a resistance can be computed using superposition and this total current can then be used to calculate correctly the power in that resistance. You can confirm for yourself the failure of superposition to compute power correctly by calculating the power in the 3-Ω resistance due to each source acting alone and testing to determine whether these powers add up to the true power calculated from the total voltage given in Eq. (2.10). Neither will superposition yield correct answers for circuits containing *nonlinear* electronic devices such as diodes or transistors.

Summary. Superposition is valid for computing the effects of individual sources on the voltages or currents throughout a circuit. As such, it allows us to extend the method of voltage and current dividers to circuits having multiple sources. Because this method develops intuition and thus aids in design, superposition is often used by circuit designers.

On the other hand, this method has weaknesses. One weakness is that you often have to analyze the circuit, perhaps several times, to calculate a single unknown current or voltage. With the simple examples we have given, this is fairly easy but in a large circuit this method becomes an ordeal. It would be much better to solve for the unknown without having to calculate every voltage and current along the way. Another weakness is the multiple solutions—much better to solve for the effects of all sources at once. Yet another weakness is that this method is somewhat unsystematic—many decisions have to be made along the way. The methods to be presented in the remainder of this chapter eliminate these weaknesses.

Check Your Understanding

1. Is an infinite resistance equivalent to a turned-OFF voltage or current source?
2. A turned-OFF voltage source is equivalent to a resistance of what value?
3. Does superposition of powers give the correct value for computing the power out of a dc voltage source?

Answers. (1) Current source; (2) zero ohms; (3) yes. Generally you *cannot* superpose powers but it works in this case because the voltage is the same in each calculation.

2.2 THÉVENIN'S AND NORTON'S EQUIVALENT CIRCUITS

Example to Justify the Concept

load

We will analyze the circuit shown in Fig. 2.10 by a method that you will surely think strange. We are to determine the current in the variable load resistor R_L as a function of that resistance. A *load* is an element in a circuit of particular importance; in one sense, the circuit exists to supply the load.[2]

LEARNING OBJECTIVE 2.
To understand the origin of Thévenin's equivalent circuit and be able to derive the Thévenin equivalent circuit of a network with a load

Our method uses an equivalent circuit first proposed in the 1880s by a French telegrapher named M. L. Thévenin. Thévenin's equivalent circuit leads to one of the most useful ideas of electrical engineering, namely, the idea of the "output impedance"[3] of a circuit. This concept influences the thinking of electrical engineers in much of the work they do. We now justify the method as we analyze the circuit in Fig. 2.10. Later we will distill the method into a simple procedure.

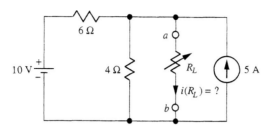

Figure 2.10 Solve for the current in R_L as a function of R_L, $i(R_L)$.

- **Remove the load resistor.** Yes, the first step is to dismantle the circuit you are trying to analyze. In the lab, you could literally cut out the resistor; on paper, you merely erase it or redraw the circuit without it. Figure 2.11 gives the resulting circuit.

open-circuit voltage

- **Measure** (in the lab) or calculate (on paper) the open-circuit voltage between a and b, v_{ab}. We call v_{ab} the *open-circuit voltage* because this voltage appears between a and b with the load removed, that is, with the circuit "open" between a and b. This is a good opportunity for you to calculate v_{ab} by superposition. You should get $v_{ab} = 4 + 12 = 16$ V.

- **Connect** to b, but not to a, a voltage source equal to the open-circuit voltage (a 16-V battery in this case), as shown in Fig. 2.12. The voltage across the gap, $v_{aa'}$, is now zero. If you do not believe that, write KVL around a to b to a' and back to a and you ought to get zero for $v_{aa'}$.

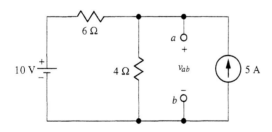

Figure 2.11 The open-circuit voltage is $v_{ab} = 16$ V.

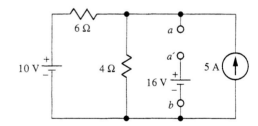

Figure 2.12 With the 16-V source inserted, $v_{aa'} = 0$ V.

---

[2] If you are making toast, the power system exists for the sake of the toaster.
[3] *Impedance* is a generalized form of resistance. The idea is defined precisely in Chap. 4. For now, we introduce the word as a synonym for resistance.

- **Reconnect** the load resistor between a and a'. Because there is no voltage across the gap, replacing the resistor will disturb nothing and the voltage across the resistor will remain at zero. *Hence, no current will exist in the load resistor.* This is an important conclusion and deserves careful attention.

    By replacing the load, we have restored our original circuit, except that now we have inserted a voltage source in series with the load. That voltage source has a polarity *opposing* the flow of current through the load and has the exact magnitude to prevent any current from flowing. We might say that it "bucks out" the current in the load. In hydraulics, it would be like inserting a centrifugal pump to stop fluid flow.

- **Superposition.** Normally, we would calculate the total current in the modified circuit as the combined results of all three sources, but this time we will distinguish between the original sources and the inserted source, as indicated in the equation

$$i_{total} = i_{original} - i_{inserted} = 0 \tag{2.14}$$

In Eq. (2.14), $i_{original}$ stands for the current through the load from the original sources in the circuit, the 10-V battery and the 5-A current source; and $i_{inserted}$ stands for the current due to the open-circuit voltage source that we inserted, a 16-V battery in this case. The minus sign comes from the polarity with which we inserted the battery, namely, so as to oppose the original current.

Conclusion. Equation (2.14) shows that the two components must be equal. The component due to the original sources is what we set out to calculate in the first place. Because this is equal to the current due to the inserted source, we can calculate the current due to one source instead of calculating the current due to the original sources (two in this case).

When we use superposition, we turn all the sources OFF except the source we are considering at the moment. Hence, we calculate the effect of the inserted source by turning OFF the original sources. This produces the circuit shown in Fig. 2.13. Now R_L is seen to be connected in series with an output impedance of $R_{eq} = 4 \parallel 6 = 2.4 \, \Omega$ and the resulting current is thus

$$i(R) = \frac{V_{open\ circuit}}{R_L + R_{eq}} = \frac{16}{R_L + 2.4} \tag{2.15}$$

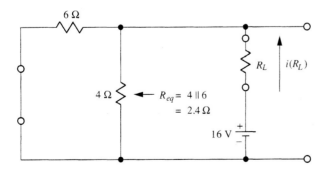

Figure 2.13 With the original sources OFF, the circuit is reduced to an equivalent resistance.

Equation (2.15) gives the current in R_L as a function of R_L, which was what we set out to find.

This was an easy problem to begin with, so you may wonder why we solved it by this roundabout method. The point is this: *We would have derived the same simple equation as Eq. (2.15) no matter how complicated the original circuit.* There could have been hundreds of sources and thousands of resistors in the original circuit and we still would have reduced the circuit to two parameters, an open-circuit voltage and an output impedance.

Thévenin's Equivalent Circuit

Equivalent Circuits

Basic concept. Equation (2.15) suggests the simple equivalent circuit shown in Fig. 2.14. The Thévenin equivalent circuit consists of a voltage source, V_T, in series with an output impedance. The magnitude and polarity of V_T are identical to the open-circuit voltage at a–b, the terminals of the load. The output impedance, R_{eq}, is computed at the load terminals with all sources in the circuit turned OFF.

Let us solve another problem using Thévenin's equivalent circuit, this time skipping the justifying steps. The circuit in Fig. 2.15 is drawn with everything in a box except the load. We wish to calculate the voltage across that load resistor. We replace the circuit in the box with the simpler circuit shown in Fig. 2.14.

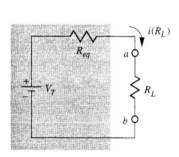
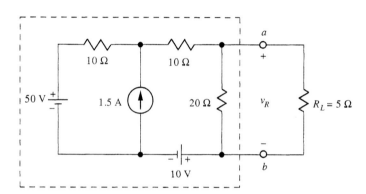

Figure 2.14 Equivalent circuit suggested by Eq. (2.15).

Figure 2.15 Replace the circuit in the box by a Thévenin equivalent circuit.

Thévenin voltage calculation. First, we calculate the Thévenin voltage source, V_T, the open-circuit voltage between terminals a and b with R_L removed. We use the method of superposition to find the current in the 20-Ω resistor and, from that, the open-circuit voltage. The current referenced downward with the 50- and 10-V voltage sources ON and the 1.5-A current source OFF would be

$$i_V = \frac{50 - 10}{10 + 10 + 20} = 1.00 \text{ A} \tag{2.16}$$

The downward current due to the 1.5-A current source acting alone would be

$$i_I = 1.5 \times \frac{1/(10 + 20)}{1/(10 + 20) + (1/10)} = 0.375 \text{ A} \tag{2.17}$$

Hence, the current referenced downward with all sources ON would be 1.375 A and the open-circuit voltage would be 1.375 A × 20 Ω = 27.5 V with the + at the top. The polarity is important because we require that the two circuits in the box in Figs. 2.15 and 2.16 be fully equivalent to the load. This requires that the polarity in the Thévenin equivalent circuit be identical to that in the original circuit.

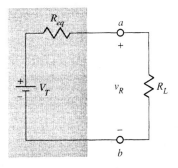

Figure 2.16 The Thévenin equivalent circuit consists of the open-circuit voltage in series with the output impedance.

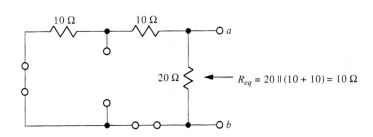

Figure 2.17 Turning OFF the three sources leaves a series-parallel combination.

Computing R_{eq}, the output impedance. To make this computation, we turn OFF all three sources within the box, with the results shown in Fig. 2.17. Thus, with $R_{eq} = 10\ \Omega$, we now have the Thévenin equivalent circuit shown in Fig. 2.18.

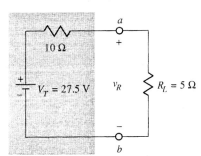

Figure 2.18 The Thévenin equivalent circuit for the circuit of Figure 2.15.

The solution to our original problem is now easy:

$$v_R = 27.5 \times \frac{5}{5 + 10} = 9.17\ \text{V} \tag{2.18}$$

Benefits of the Thévenin equivalent circuit. Of course, with modest effort, we could have computed this result directly from the original circuit in Fig. 2.15. With this method, however, we gain the freedom to ask many additional questions such as: What value of R_L makes the voltage 30 V? What value of R_L withdraws the most power from the circuit? These questions lead to mathematical complexities with the original circuit, but can be answered simply with the Thévenin equivalent circuit. The answer to the first question is that no value of R_L will give 30 V. The investigation of the second question leads to an interesting and important result, to which we will soon turn.

 **Equivalent Circuits**

Equivalent circuits. First, let us consider further what we mean by a Thévenin *equivalent* circuit. The equivalent circuit replaces the circuit within the box only for effects *external* to the box. We can no longer ask questions about the circuit in the box after we have replaced it by an equivalent circuit. For example, if we are interested in the current in the 20-Ω resistor or the total power consumed by the resistors in the box, the equivalent circuit is useless.

Maximum power transfer. Let us now investigate the question of maximizing the power in R_L in Fig. 2.15. We let load resistance R_L be the independent variable, the power in the load be the dependent variable P, and find $P(R_L)$ using basic circuit techniques. The power in R_L as a function of R_L, $P(R_L)$, is

$$P(R_L) = i^2 R_L = \frac{V_T^2 R_L}{(R_L + R_{eq})^2} = \frac{(27.5)^2 R_L}{(R_L + 10)^2} \tag{2.19}$$

impedance match

To maximize $P(R_L)$ as given in Eq. (2.19), we take the derivative with respect to R and set it equal to zero. You can confirm that the maximum (or minimum) occurs at R_L equals R_{eq} (10 Ω). This is called an *impedance match*, when the load is equal to the output impedance of the circuit. To confirm that we have a maximum and not a minimum, a second derivative can be taken, but an easier way in this case is to make a simple sketch of $P(R_L)$ to show that we have found a maximum. Figure 2.19 shows a plot of Eq. (2.19).

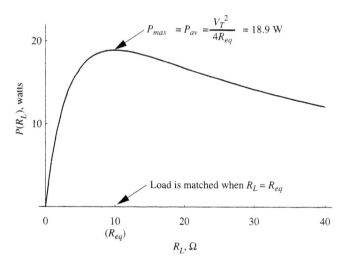

Figure 2.19 The power in R_L is maximum when $R_L = R_{eq}$.

Importance of maximum power transfer. Obtaining the maximum power out of a circuit is important because often we deal in electronics with small amounts of power and wish to make full use of the power that is available. On a TV set, for example, we pull out the "rabbit ears" antenna to receive power from radio waves originating at a transmitter many miles away. The antenna does not collect much power, so the TV receiver is designed to make maximum use of the power provided by the antenna. Although our results were derived for a simple battery and resistor, they can be applied to a TV antenna. Figure 2.19 shows that we should design the receiver input circuit, represented here by a load resistor, to match the output impedance of the antenna for withdrawing the maximum power from the antenna.

optimum, optimize

Optimization. When a design is the best possible under the constraints, the design is said to be *optimum*. Thus, the TV receiver input is *optimized* when its input impedance is matched to the output impedance of the antenna because this gives maximum power to the receiver.

EXAMPLE 2.2 **Loudspeaker analysis**

Consider a stereo system that puts out 25 watts into each of two 8-Ω speakers. What would be the voltage provided by this amplifier to each speaker?

SOLUTION:
We assume that the speakers are adequately represented by 8-Ω resistances and that the speakers and amplifier output impedances are matched. Thus, the circuit we propose as a model for each channel is shown in Fig. 2.20.

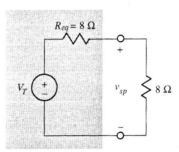

Figure 2.20 Circuit model for one channel of the stereo output circuit.

We have used a general symbol for the voltage source because the audio voltage would certainly not be a constant voltage.[4] Because we have assumed that the load and output impedance of the amplifier are matched, this requires that R_{eq} be also 8 Ω. From the maximum in Fig. 2.19, we calculate the Thévenin voltage to be

$$\frac{V_T^2}{4 \times 8} = 25 \text{ watts/channel} \Rightarrow V_T = 28.3 \text{ V} \qquad (2.20)$$

The voltage across the speaker will be half this value because the Thévenin voltage divides equally between the output impedance and the 8-Ω speaker resistance. Hence, the answer is 14.1 V.

WHAT IF? What if 16-Ω speakers were used with this same stereo? What then would be the power in each speaker?[5]

---

[4] In Chap. 4 of *Foundations of Electronics*, we study the appropriate model for such an audio signal. For now, we are merely interested in the voltage representing the output.

[5] The power would drop to 22.2 watts. This slight decrease shows the benefits of optimizing the amplifier/speaker power transfer. In general, when a system is optimized, performance is insensitive to changes such as using 16-Ω instead of 8-Ω speakers.

Norton's equivalent circuit. An American engineer named E. L. Norton (1889–1983) came up with a similar equivalent circuit. Norton's equivalent circuit consists of a current source connected in parallel with an output impedance, R_{eq}, as shown in Fig. 2.21. The derivation of the values for the impedance and the current source, I_N, is similar to that for Thévenin's circuit and will not be repeated here. Indeed, the output impedance is the same as before, namely, the resistance of the circuit presented to the load after all internal sources are turned OFF. The Norton current source has a magnitude identical to the current that would flow in a short circuit of the output terminals.

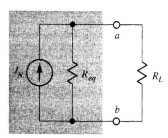

Figure 2.21 The Norton equivalent circuit appears in the box.

To illustrate, we will find the Norton equivalent circuit for the circuit of Fig. 2.15. The value of the output impedance, R_{eq}, is the same as before, 10 Ω. The value of the Norton current source, I_N, can be determined by replacing the load with a short circuit. This gives the circuit in Fig. 2.22. The short circuit effectively removes the 20-Ω resistor from the circuit because it forces its voltage, and hence its current, to be zero. The current flowing through the short circuit is easily calculated. The result is 2.75 A from a to b in the short circuit; hence, the Norton equivalent circuit is that shown in Fig. 2.23. From this simple circuit, v_R is seen to be $2.75 \times 10 \parallel 5 = 9.17$ V, as before. Notice that the polarity of the current source must produce the physical current from a to b.

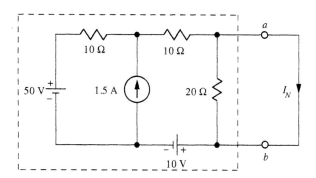

Figure 2.22 Short the output to find I_N. The 20-Ω resistor is effectively removed for the calculation of I_N.

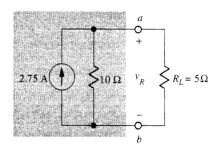

Figure 2.23 Norton equivalent circuit.

Relationship between Thévenin's and Norton's equivalent circuits. If two circuits are equivalent to the same circuit, they must be equivalent to each other. Thus, if the Norton circuit in Fig. 2.21 is open-circuited, the voltage must be V_T, the same as the voltage source in the Thévenin circuit of Fig. 2.16. You can see that this proves true in the example, because 2.75 A × 10 Ω = 27.5 V. In general, it must be true that

$$V_T = I_N R_{eq} \tag{2.21}$$

Equation (2.21) is useful in theoretical work, but it also can be applied in the laboratory to find the output impedance of a source. In practice, we may be unable to get inside and turn OFF internal sources, for example, in a battery or an electronic circuit, but we can measure the open-circuit output voltage and the current that flows when the output terminals are shorted. From the measured voltage and current, we can compute the output impedance

$$R_{eq} = \frac{V_T}{I_N} \tag{2.22}$$

Graphical Interpretation. Figure 2.24(b) shows the output characteristic of any linear circuit, Fig. 2.24(a). The intercept on the voltage axis ($i_{out} = 0$) is the Thévenin, or open-circuit voltage, and the intercept on the current axis ($v_{out} = 0$) is the Norton, or short-circuit current. The slope is the negative of the reciprocal of the output impedance, as required by Eq. (2.22).

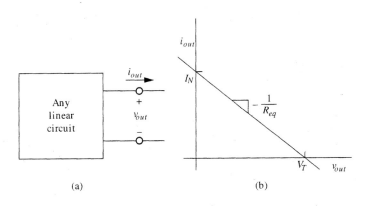

Figure 2.24 (a) An arbitrary linear circuit, with the output voltage and current indicated. (b) The output characteristic of the circuit in graphical form. The intercepts are the Thévenin voltage and the Norton current, and the slope is the negative of the reciprocal of the output impedance.

available power

Available power. The maximum power that can be extracted from a circuit is called the *available power*, P_{av}, and is given by

$$P_{av} = P(R_{eq}) = \frac{V_T^2}{4R_{eq}} \tag{2.23}$$

as shown in Fig. 2.19. Note that sources with low output impedance can supply large amounts of power to an external load.

EXAMPLE 2.3 Available Power from an Auto Battery

The open-circuit voltage of a standard automotive battery is 12.6 V, and the short-circuit current is approximately 300 A. What is the available power from the battery?

SOLUTION:

From Eq. (2.22) we find the output impedance of the battery

$$R_{eq} = \frac{V_T}{I_N} = \frac{12.6}{300} = 0.042 \; \Omega \tag{2.24}$$

Therefore, the available power from Eq. (2.23) is

$$P_{av} = \frac{V_T^2}{4R_{eq}} = \frac{(12.6)^2}{4 \times 0.042} = 945 \; W \tag{2.25}$$

WHAT IF? What if you compute the available power from one of the flashlight batteries in Fig. 1.30?[6]

loading of a circuit, voltmeter

Loading of a circuit. We also are concerned about the *loading* of circuits, which occurs when the output voltage is changed significantly by a load. Figure 2.25(a) shows a voltage divider creating a voltage v and a *voltmeter* to measure the voltage. The voltmeter has an input resistance of 10 MΩ, and hence connecting it will change the voltage to be measured. We may determine the loaded voltage, v', from the Thévenin equivalent circuit in Fig. 2.25(b). The Thévenin voltage for the voltage divider, with the voltmeter for a load is

$$V_T = 10 \times \frac{500}{500 + 800} = 3.85 \; V \tag{2.26}$$

This is the voltage that an ideal voltmeter would measure. The output impedance is $R_{eq} = 800 \| 500 = 308 \; k\Omega$. With the nonideal voltmeter connected, the loaded voltage is

$$v' = 3.85 \times \underbrace{\frac{10 \; M\Omega}{10 \; M\Omega + 308 \; k\Omega}}_{0.970} = 3.73 \; V \tag{2.27}$$

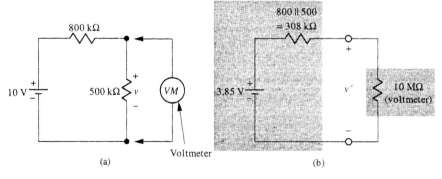

Figure 2.25 (a) The amount of loading from the meter depends on the output impedance of the circuit at the point of measurement and the resistance of the meter; (b) Thévenin equivalent circuit showing the effect of meter loading.

---

[6] The available power is 1.875 W.

The loaded voltage, which is what the voltmeter will indicate, proves to be 3% lower than the unloaded voltage of 3.85 V, which is the voltage we are seeking to measure with the voltmeter. In practice, we can correct for loading error if we know the output impedance of the circuit and the input impedance of the meter.

Impedance Level

Impedance Level

impedance level

LEARNING OBJECTIVE 3.

To understand the significance of impedance level in interactions between circuits

Definition of impedance level. Impedance level is a broad concept that is better illustrated than defined. The resistance of a load is its impedance level, and the output impedance of a circuit to a load is the impedance level of the circuit seen by that load. Thus, the *impedance level* describes the approximate ratio of voltages to currents in a circuit or portion of circuit. The interaction of a circuit with a load depends on the relative values of their respective impedance levels.

Circuit model. Consider the Thévenin equivalent circuit shown in Figure 2.26, bearing in mind that this represents the most general circuit/load interaction possible. We now investigate the effect of the impedance level of circuit and load upon the transfer of power, voltage, and current.

Power. We have already shown that when the impedance level of the load is equal to the output impedance of the circuit, the power in the load is optimized, that is, is maximum. Hence, it would be fitting to say that this condition yields a transfer of power to the load. Figure 2.19 shows that the amount of power depends reciprocally on the output impedance level of the source and will be relatively insensitive to the load impedance, especially if the load impedance is greater than the source impedance.

Voltage. The voltage across the load in Fig. 2.26 is determined by voltage division to be

$$v_L = V_T \times \frac{R_L}{R_L + R_{eq}} = V_T \times \frac{1}{1 + R_{eq}/R_L} \quad (2.28)$$

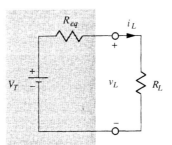

Figure 2.26 This circuit models all circuit/load interactions.

When $R_L \gg R_{eq}$, we have $v_L \approx V_T$. Hence, when the impedance level of the load is large relative to that of the source, the voltage transferred to the load will be approximately equal to the open-circuit voltage and will be relatively insensitive to the load-impedance level. Hence, this condition transfers voltage to the load. This is the desired condition when we measure voltage.

Current. When $R_L \ll R_{eq}$, the voltage to the load will be small, but the current will be

$$i_L = \frac{V_T}{R_L + R_{eq}} = \frac{V_T}{R_{eq}} \times \frac{1}{1 + R_L/R_{eq}} = I_N \times \frac{1}{1 + R_L/R_{eq}} \qquad (2.29)$$

ammeter

Hence, when the load impedance is small compared with that of the source, the current will be approximately the Norton current, independent of R_L. Hence, current is transferred to the load in this condition. Thus, a meter to measure current, an *ammeter*, should have an impedance level that is low compared with the impedance level of the circuit to be measured.

Importance of impedance level. Consideration of impedance level has application for many circumstances. Consider the power system supplying energy for lighting, electric motors, etc. These appliances are designed to operate at a prescribed voltage; hence, the impedance level of the power system should be small compared with all the loads placed on it. Indeed, because all loads are placed in parallel, the output impedance of the power system should be small compared with the parallel combination of all loads on the system at any given time. If this were not the case, the voltage of the power system would fluctuate seriously according to the load on the system.

Everyone has had the experience of being shocked by sliding across a car seat in cold, dry weather. The spark can jump as much as 5 mm, which means that at least 15,000 V is developed. But the impedance level of the "circuit" is extremely high relative to the impedance of your body and therefore the voltage is not transferred to your body, just the current, which is extremely small. Hence, no harm is done. Of course, the opposite would be true if you were struck by lightning.

We showed before that in electronics, the impedance levels are controlled to optimize transfer of power between parts of the electronic circuit. Hence, in power, electronics, metering, and, indeed, many other contexts, the impedance levels of the various parts of the circuits are critical to performance.

Source Transformations

source transformations

Transformations between Thévenin's and Norton's equivalent circuits can be used in connection with other methods of circuit analysis. For example, in Figure 2.15, we may transform the 50-V source in series with the 10-Ω resistance into a 5.0-A current source in parallel with 10 Ω. Thus, the circuit becomes that shown in Fig. 2.27, and we may combine the two current sources to a single 6.5-A current source. Next, we may convert the 6.5-A source in parallel with 10 Ω to a 65-V voltage source in series with 10 Ω and then combine the 65-V source with the opposing 10-V source to yield Fig. 2.28. Clearly, $v_{ab} = V_T = 27.5$ V. When we switch back and forth between Thévenin and Norton equivalent circuits, these are *source transformations*.

Benchmark example, solved with equivalent circuits. Figure 2.29(a) gives the benchmark example of Fig. 2.5, with the 3-Ω resistor marked as the load. When we remove the load, KCL at the upper node shows a physical current of 1 A up in the 1-Ω resistor, and the open-circuit voltage therefore will be

$$V_T = +6\,\text{V} - 1\,\text{A} \times 1\,\Omega = +5\,\text{V} \qquad (2.30)$$

and the output impedance is clearly 2 Ω + 1 Ω; thus, the problem reduces to that in Fig. 2.29(b). Clearly, the unknown voltage is +2.50 V. If you refer back to the previ-

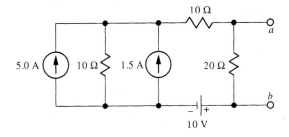

Figure 2.27 Same circuit as Fig. 2.15, but with 50-V source transformed into a current source.

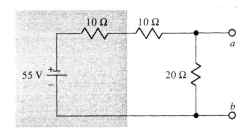

Figure 2.28 After several source transformations, the Thévenin voltage can be determined by inspection.

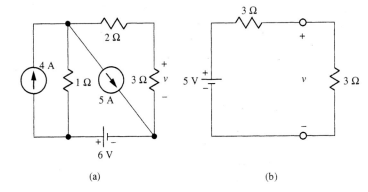

(a)

(b)

Figure 2.29 (a) Benchmark example, Fig. 2.5, with the 3-Ω resistor marked as the load; (b) the reduced circuit.

ous solution of this problem using superposition, you will see the benefits of using the Thévenin equivalent circuit.

Comparison of equivalent circuits with voltage and current divider techniques. In closing this section, we wish to compare the method of equivalent circuits with our first method, voltage and current dividers. With voltage and current dividers, the strategy was to combine resistors until we have represented the entire circuit to the *source* as a single equivalent resistance. We then restore the original circuit, dividing voltage and current as we go. Eventually, we can calculate the current or voltage across a particular resistor of interest. If there are multiple sources, we must repeat this process for each source and use superposition.

The methods of equivalent circuits and source transformations work in the opposite direction. Here we represent the entire circuit, including multiple sources, to the *load* of interest. We bring everything, as it were, to a particular point in the circuit that has special importance. For this reason, the method of equivalent circuits plays an important role in both design and analysis of electrical circuits.

Check Your Understanding

1. A 1.5-V dry cell has a maximum (short-circuit) current of 300 mA. What is the internal resistance of the battery?
2. A circuit has an output voltage of 20 V and a short-circuit current of 0.5 A. What is the maximum power that can come out of this circuit?

3. Normally, do we adjust a load for maximum power in a power circuit or an electronics circuit?

4. A circuit has a variable load, R_L. Power measurements show that the power into R_L is maximum at 10 W with $R_L = 2\,\Omega$. What would be the open-circuit voltage, V_T?

Answers. (1) 5 Ω; (2) 2.5 W; (3) in electronics—power circuits require constant voltage; (4) 8.94 V.

2.3 NODE-VOLTAGE ANALYSIS

Basic Idea

A systematic method. The methods presented thus far in Chaps. 1 and 2 work well for simple circuits and are especially useful to designers. But the time comes when one must analyze a large circuit containing many circuit elements, and for this purpose, we will develop two systematic methods of circuit analysis. These methods develop sets of linear equations to be solved simultaneously for the unknown voltages or currents, a task for which computers are well suited. To keep the mathematics within bounds, however, we will limit our examples to relatively simple circuits.

Node voltages. Node-voltage analysis, or nodal analysis, is based on Kirchhoff's current law. We write KCL equations at all nodes except one, but we write these equations in such a way that current variables are never formally defined. We avoid defining current variables by expressing the currents in terms of the "node voltages" in a special way. Figure 2.30 shows a simple circuit with two voltage sources and a resistor. The point at the bottom we have denoted r and the points at the ends of the resistor we have called a and b. Using KVL and Ohm's law, we may express i_{ab} thus:

$$-v_{ar} + v_{ab} + v_{br} = 0 \Rightarrow v_{ab} = v_{ar} - v_{br} \qquad (2.31)$$

$$i_{ab} = \frac{v_{ab}}{R_{ab}} = \frac{v_{ar} - v_{br}}{R_{ab}} \qquad (2.32)$$

reference node, node voltage

We will change the appearance of Eq. (2.32) by simplifying our notation: We drop the r in the voltage subscript. The *reference node*, labeled r, is like an elevation datum in surveying: We measure all elevations (voltages) relative to it.[7] When we talk about the *node voltage* at a, v_a, we are to understand that we are talking about v_{ar}, the voltage be-

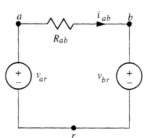

Figure 2.30 The reference node is marked r.

---

[7]When we speak of the elevation of St. Louis as being 413 ft or at the Salton Sea as being −287 ft, we are talking about the elevation relative to mean sea level, to which we assign an elevation of zero.

tween node a and the reference node, which is thus assigned a voltage of zero. Later, we will discuss how to choose the reference node; here we are interested in the pattern of subscripts in Eq. (2.32). With the change in notation, the form becomes

$$i_{ab} = \frac{v_a - v_b}{R_{ab}} \qquad (2.33)$$

Pattern of subscripts. Equation (2.33) can be stated in words as follows: The current from a to b is the voltage at a, minus the voltage at b, divided by the resistance between a and b.[8] The pattern established in Eq. (2.33) is so simple and intuitive that with it we can express currents without defining current variables. That is, we can keep the "current from a to b" part in our heads and write on the paper the voltage at a, minus the voltage at b, divided by the resistance between a and b, the right-hand side of Eq. (2.33).

EXAMPLE 2.4 Constrained Nodes

Determine the current through the 7-Ω resistor, referenced toward the left, in Fig. 2.31.

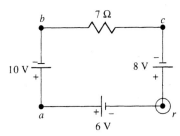

Figure 2.31 Determine the current in the resistor, referenced toward the left, $c \to b$.

constrained node

SOLUTION:
This current would be the voltage at c, minus the voltage at b, divided by 7 Ω. We may use KVL to determine v_c and v_b. We begin at b, go to r, writing v_{br}, then to a and back to b: $v_{br} - (6) + (10) = 0$; hence, $v_{br} = v_b = -4$ V. To find v_c, we go from c to r, around the battery and then return through the 8-V battery: $v_{cr} + (8) = 0$; or $v_{cr} = v_c = -8$ V. So the current toward the left through the 7-Ω resistor would be:

$$\frac{v_c - v_b}{R_{cb}} = \frac{-8 - (-4)}{7} = -0.571 \text{ A} \qquad (2.34)$$

In this circuit, the voltages at all nodes are constrained to the reference node by the voltage sources. In general, a *constrained node* is a node whose voltage is constrained to the voltage of another node by a voltage source. In this example, the nodes are constrained to the reference node and hence their node voltages are known absolutely.

---

[8] This applies only when there is a resistor between a and b.

> **WHAT IF?** What if node a were the reference node, zero volts? Find v_c, v_b, and i_{cb}.[9]

Node-Voltage Technique

LEARNING OBJECTIVE 4.

To understand how to analyze a circuit using nodal analysis

We will find the voltage across the 3-Ω resistor in Fig. 2.32(a). We will analyze this circuit through the node-voltage method, which consists of the following steps:

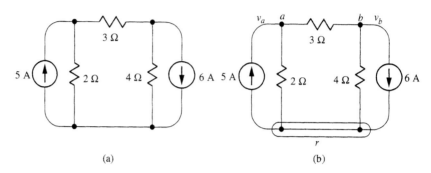

(a)   (b)

Figure 2.32 (a) Find the voltage across the 3-Ω resistor. (b) Circuit with labeled nodes. The unknowns are v_a and v_b.

1. **Define a reference node.** The circuit has three nodes. We may choose any of them as the reference node. Often, the node with the most wires is chosen as the reference node. In this case, the node at the bottom will be chosen. We mark the reference node with r, as shown in Fig. 2.32(b).

2. **Count the independent nodes.** Here we have three nodes, the reference node plus two independent nodes.[10] Thus, two equations will be written.

3. **Label the independent nodes.** We have already labeled the reference node; now we label the other nodes, as shown in Fig. 2.32(b). The node voltages, v_a and v_b, are our secondary unknowns from which we will calculate the primary unknown, the voltage across the 3-Ω resistor. In nodal analysis, we calculate the node voltages; then we calculate from these the specified unknown in the problem, a current or voltage or power of interest.

Conservation of Charge

4. **Write KCL in a special form.** The special form is

$$\sum \text{currents leaving the node in resistors}$$
$$= \sum \text{currents entering the node from current sources} \quad (2.35)$$

---

[9] $v_c = -14$ V, $v_b = -10$ V, $i_{cb} = -0.571$ A, the same as before.
[10] We define independent nodes on page 82.

But we avoid defining current variables by expressing currents with the node voltages:

$$\frac{v_a - (0)}{2} + \frac{v_a - v_b}{3} = +(5) \tag{2.36}$$

In Eq. (2.36), the left side represents the currents leaving node a in the two resistors connected directly to node a, and the right side represents the current entering from the 5-A source. The current flowing through the 2-Ω resistor to the reference node is merely the node voltage, v_a, divided by the resistance between node a and the reference node. We wrote the (0) for the voltage of the reference node as a reminder. The current from a to b through the 3-Ω resistance is derived from the pattern in Eq. (2.33). Similarly, we can write KCL for node b with the node voltages:

$$\frac{v_b - v_a}{3} + \frac{v_b - (0)}{4} = -(+6) \tag{2.37}$$

where the first term on the left side represents the current flowing from node b to node a. This is the negative of the second term in Eq. (2.36). This change in sign occurs because we are now expressing the current referenced in the opposite direction from before. Note also that the current source term on the right side has a negative sign because we are summing the currents *entering* the node from the sources. Equations (2.36) and (2.37) are rewritten:

$$\begin{aligned}(\tfrac{1}{2} + \tfrac{1}{3})v_a - (\tfrac{1}{3})v_b &= 5 \\ -(\tfrac{1}{3})v_a + (\tfrac{1}{3} + \tfrac{1}{4})v_b &= -6\end{aligned} \tag{2.38}$$

We may solve these equations by any method, such as Cramer's rule for determinants, with the result

$$v_a = 2.44 \text{ V} \quad \text{and} \quad v_b = -8.89 \text{ V} \tag{2.39}$$

5. **Compute the primary unknown.** We can now calculate the original unknown from the resulting node voltages. The voltage across the 3-Ω resistor was specified, but we cannot calculate it without a polarity marking. If we wish the + polarity symbol at node a, we have the marking shown in Fig. 2.33. We can determine v_3 by writing KVL around the loop *rabr*:

$$\begin{aligned}v_{ra} + v_3 + v_{br} &= 0 \\ v_3 = -v_{ra} - v_{br} &= v_{ar} - v_{br} = v_a - v_b\end{aligned} \tag{2.40}$$

The second line of Eq. (2.40) was converted via Eq. (1.27) to a form where r was the second subscript and then r was dropped. Thus, v_3 with the polarity marking of Fig. 2.33 is $2.44 - (-8.89)$ or 11.3 V. Had we marked the + at node b, we would have followed a similar procedure. You can confirm for yourself that this would have reversed the right sides of Eq. (2.40) and resulted in v_3' being -11.3 V, where the primed v_3 has the + at node b.

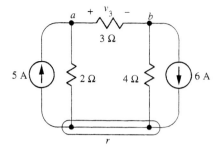

Figure 2.33 Using v_a and v_b to determine v_3.

Some Refinements

How to handle voltage sources. Nodal analysis deals routinely with current sources. The circuit of Fig. 2.34, however, is currently beyond the reach of the method. In Fig. 2.34, we have identified a reference node and labeled the other two nodes a and b; we will find v_a and v_b. To find the voltage at node a, we will write KVL from r to a and back to r through the voltage source:

$$v_{ra} + 10 = 0 \Rightarrow v_{ar} = v_a = +10 \text{ V} \qquad (2.41)$$

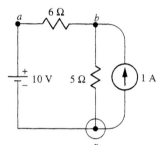

Figure 2.34 Node voltage problem with a voltage source. The unknowns are v_a and v_b.

The voltage source constrains v_a to be +10 V. Thus, when we write KCL for node b, we treat v_a as known rather than unknown:

$$\frac{v_b - (+10)}{6} + \frac{v_b - (0)}{5} = +(+1) \qquad (2.42)$$

The solution for v_b is 7.27 V.

We may also use a source transformation to convert the voltage source to a current source. The 10-V source in series with the 6-Ω resistance in Fig. 2.35(a) can be transformed to the Norton form in Fig. 2.35(b) using Eq. (2.21). In this form, the current sources can be added and the parallel resistances can be combined; so the voltage across the circuit is

$$v_{br} = v_b = \left(\frac{10}{6} + 1\right)(6 \parallel 5) = 7.27 \text{ V} \qquad (2.43)$$

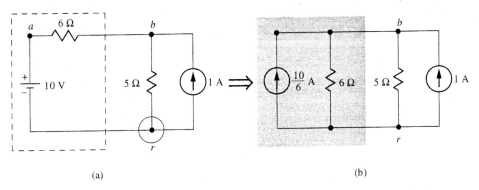

Figure 2.35 A source transformation allows solution without using constrained nodes. Note that the number of independent nodes does not change.

EXAMPLE 2.5 Constrained Nodes

Find the power in the 4-Ω resistor in Fig. 2.36.

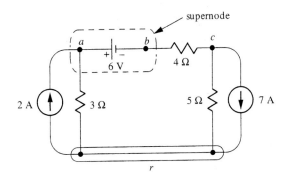

Figure 2.36 Nodes a and b are constrained by the 6-V source.

SOLUTION:

Here we have one node constrained to another independent node rather than to the reference node. The voltages at nodes a and b are unknown, but they are not independent. If we knew the voltage at either of them, we could determine the voltage at the other; thus, it would be inappropriate to treat them as independent unknowns. We can determine the relationship between them by writing KVL from r to a to b and back to r:

$$v_{ra} + 6 + v_{br} = 0 \Rightarrow -v_a + 6 + v_b = 0 \tag{2.44}$$
$$v_b = v_a - 6$$

In the second line of Eq. (2.44), we have written the equation with v_a as our unknown and v_b expressed from v_a. We can now write KCL for node c:

$$\frac{v_c - (0)}{5} + \frac{v_c - (v_a - 6)}{4} = -(+7) \tag{2.45}$$

In the second term on the left side, we write $(v_a - 6)$ for v_b in expressing the current referenced out of node c in the 4-Ω resistor.

The two constrained nodes are treated as a supernode, as shown in Fig. 2.36, and KCL is written as

$$\frac{v_a - (0)}{3} + \frac{(v_a - 6) - v_c}{4} = +(+2) \tag{2.46}$$

We now have two equations in two unknowns, v_a and v_c, which yield -2.75 V and -20.4 V, respectively. From the second term in Eq. (2.46), we calculate the current in the 4-Ω resistor to be

$$\frac{(-2.75 - 6) - (-20.4)}{4} = 2.92 \text{ A} \tag{2.47}$$

The power in the 4-Ω resistor is

$$P_R = i_R^2 R = (2.92)^2(4) = 34.0 \text{ W} \tag{2.48}$$

We could alternatively transform the 6-V source in series with the 4-Ω resistor to a 1.5-A current source in parallel with a 4-Ω resistance. The source transformation permits routine application of nodal analysis.

Summary. A voltage source will establish a constraint between two nodes. We can express the node voltage at one end of a voltage source in terms of the node voltage at the other end plus or minus the source value, depending on the polarity of the source. To determine the sign, we may have to write KVL around a loop containing the reference node and the two constrained nodes. Constrained nodes are treated as a supernode when KCL equations are written.

independent node

Independent nodes. The analysis of Fig. 2.36 shows that a first count of the nodes of a circuit could overestimate the number of unknown node voltages to be determined. If there are voltage sources, some nodes will be constrained and the number of unknowns, and equations to be solved, will be reduced. An *independent node* is a node whose voltage cannot be derived from the voltage of another node. When we analyze a circuit using the method of node voltages, we will have as many unknowns (and equations to solve) as we have independent nodes.[11]

Here is a rule for counting independent nodes: Turn OFF all sources and count the nodes that remain separated by resistors. The number of independent nodes is one less than the number of remaining nodes. Turning OFF all sources means that voltage sources are replaced by short circuits and current sources are replaced by open circuits. The second step in node-voltage analysis is to turn OFF all sources and determine the number of independent nodes. The third step is to label only the independent nodes, because these are the only secondary unknowns.

---

[11] The number of independent nodes may be reduced by combining resistances in series. If the primary unknown disappears, a voltage divider can find the voltage at the node that vanished.

EXAMPLE 2.6 Counting the independent nodes on the benchmark example

Count the independent nodes for the benchmark example circuit in Fig. 2.37(a).

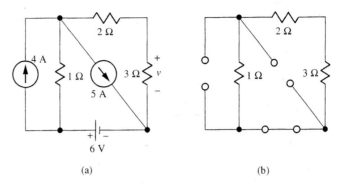

Figure 2.37 (a) Benchmark example circuit. (b) The circuit with sources OFF. There are three nodes, of which two are independent, but one of these may be eliminated by combining the series resistances.

SOLUTION:

When we turn OFF the sources, we have the circuit in Fig 2.37(b). There are three nodes, two of which are independent. However, if we add the two series resistors to make a 5-Ω resistor and determine the unknown voltage with a voltage divider, we can get by with solving one nodal equation. Thus, we define the top node as node a and write KCL in nodal form:

$$\frac{v_a - (+6)}{1} + \frac{v_a - (0)}{2 + 3} = +4 - 5 \Rightarrow v_a = \frac{6 - 1}{1.2} = 4.17 \text{ V} \quad (2.49)$$

and hence from the voltage divider,

$$v = 4.17 \times \frac{3}{2 + 3} = 2.50 \text{ V} \quad (2.50)$$

WHAT IF?

What if we had not combined resistors?[12]

Critique

A general method. Nodal analysis is our first systematic method for analyzing circuits. It can be implemented somewhat routinely and always works. It is probably the favorite method of electrical engineers for analyzing electronic circuits because most

---

[12] Then we would have had to solve two equations in two unknowns, but the answer would have come out the same.

circuit components are connected to the electronic ground, which is used for a reference node.

The physical meaning of the node voltages is clarified by considering a voltmeter. The black voltmeter lead, often marked "common," should be attached to the reference node. If the red voltmeter lead is then touched to the nodes in the circuit, the voltmeter will indicate the node voltages.[13] Thus, node voltages are easily measured. Loop currents, the next method we present, also qualifies as a popular and powerful method, but measuring currents involves disconnecting the circuit and hence the loop currents are not easily measured. As you will see, the loop currents may not actually exist in the circuit either.

What about KVL? We might pause to ask: How can we analyze a circuit without considering KVL? This is certainly what we appear to do when we use the method of node voltages. *Answer*: Kirchhoff's voltage law refers to individual voltages around a loop. Node voltages, on the other hand, are all referred to the same point. Hence, KVL does not apply directly.

ground node

The reference node versus "ground." When the node voltage method is presented in books and used in practice, the reference node is often called the *ground node* and given the symbol ⏚. Strictly speaking, the ground in an electrical circuit identifies the point that is physically connected via a thick wire to the moist earth, usually for safety purposes. We discuss grounding in Chap. 6; here we only comment on the relationship between the reference node of nodal analysis and the physical ground of an electrical system.

The grounded portion of an electrical circuit usually has many wires connecting to it, and hence the electrical ground is often designated the reference node in a nodal analysis. But this is mere coincidence: The reference node and the ground are totally different concepts. We have avoided calling the reference node the "ground" to establish the concept of the reference node independent of the concept of electrical grounding. You should be aware, however, that many people use the terms interchangeably when discussing nodal analysis.

potential difference, potential rise, potential drop

Node voltages and electrical potential. The analogy we made earlier between node voltages and elevations above mean sea level has a factual basis. Elevation is a measure of gravitational potential, and the node voltages are a measure of the electrical potential of the various nodes in an electrical circuit, relative to the reference node. What we have defined as the voltage between two points can also be called the *potential difference* between the points. Likewise, we can define *potential rises* and *potential drops* in a circuit; for example, the potential rises across a battery (going from − to +) and drops across a resistor (going from + to −). However, we will not speak of potential rises and drops in our development of circuit theory. Our definition of voltage is that of a potential drop.

Check Your Understanding

1. What is the name of the point in a circuit defined to have zero volts?

---

[13] This assumes a modern electronic voltmeter that indicates the sign of the voltage. An old-fashioned analog meter might have to be reversed for negative node voltages.

2. If $v_{rb} = +5$ V, where r is the reference node, what is the node voltage at node b?

3. If node a is connected to the reference node by a 40-V voltage source with the $-$ at node a, what is v_a?

4. Two nodes are said to be constrained together when connected by a voltage source or a current source. Which?

Answers. (1) The reference node, not "ground"; (2) -5 V; (3) -40 V; (4) voltage source.

2.4 LOOP-CURRENT ANALYSIS

LEARNING OBJECTIVE 5.
To understand how to analyze a circuit using loop currents

Simple Method of Loop-Current Analysis

Introduction. The method of loop currents is similar to nodal analysis, except that the variables are currents, not voltages, and the equations are based on KVL. The goal is to avoid defining unnecessary variables and to systematically write simultaneous equations for the unknown loop currents.

We will present the method by analyzing the circuit shown in Fig. 2.38(a); we are to solve for the voltage across the 2-Ω resistance using loop currents. We give step-by-step instructions:

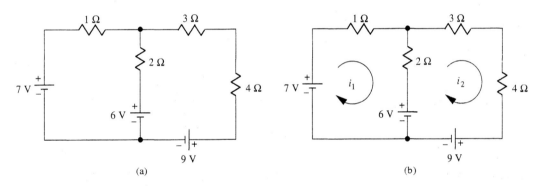

Figure 2.38 (a) The voltage across the 2-Ω resistor is to be determined. (b) The same circuit with loop currents i_1 and i_2 defined.

independent loop, loop current

1. **Define and label loop currents.** Label the independent loops 1, 2, 3, ... and define loop currents i_1, i_2, i_3, ... going clockwise around the loops. An *independent loop* is a loop that does not pass through a current source. A *loop current* flows in a closed path, following a loop. In this circuit, both loops are independent loops and we get the picture of Fig. 2.38(b).

Conservation of Energy

2. **Write KVL** going with the currents around each loop in a special form:

$$\sum \text{voltages across resistors in loop} = \sum (+ \text{ or } -) \text{ voltage sources in loop} \quad (2.51)$$

On the right side of Eq. (2.51), we use $+$ if the voltage aids the loop current for that loop and $-$ if the voltage source opposes the loop current for that loop.

3. **Solve for the loop currents** and from them compute the currents, voltages, or powers required. Note that because we write one KVL equation for each loop and we have one loop current for each loop, we always get the same number of equations and unknowns. We will now follow this plan for the circuit in Fig. 2.38(a). The loop currents have already been defined in Fig. 2.38(b). The KVL equation for loop 1 is

$$i_1(1) + (i_1 - i_2)(2) = +(+7) - (+6) \tag{2.52}$$

The first term in Eq. (2.52) is the voltage across the 1-Ω resistance. We are going with the current and we are using a load set for the voltage across the resistance; thus we automatically get a + sign for that voltage. The second term is the voltage across the 2-Ω resistance. The downward current in that resistance is $i_1 - i_2$, the difference between the two loop currents. Because i_1 is referenced down and i_2 is referenced up, the current in the resistance is their difference. We are going with i_1, so we write the voltage across that resistance as $+i \times R$, where i is the current referenced in the direction we are going, $i = i_1 - i_2$. Thus, the + sign in the second term is automatic, just like the + sign on the first term. The $+(+7)$ on the right side is due to the 7-V source. It has a + sign because that source tends to force the loop current in that loop (i_1) in its positive direction. The $-(+6)$ is due to the 6-V source. It has the minus sign outside the parentheses because it opposes i_1.

Of course, there are not two physical currents in the 2-Ω resistance, one going up and the other going down. The loop "currents" are mathematical variables that may or may not be identical to a current somewhere in the circuit. In this case, i_1 is the current in the 1-Ω resistance, i_2 is the current in the 3- and 4-Ω resistances, and $i_1 - i_2$ is the current referenced downward in the 2-Ω resistance.

The KVL equation for the second loop is

$$i_2(3) + i_2(4) + (i_2 - i_1)(2) = +(+6) - (+9) \tag{2.53}$$

We are now going with i_2 around loop 2. The first two terms in Eq. (2.53) are due to i_2 going through the 3- and 4-Ω resistances. We automatically get + signs because we are going in the same direction as the loop current. When we get to the 2-Ω resistance, we write the current as $i_2 - i_1$ because now we are going up, in the same direction as i_2. This term in Eq. (2.53) is the negative of the corresponding term in Eq. (2.52). The sign is changed because in both cases we are writing voltages across resistances going clockwise: in loop 1 this requires going downward through the 2-Ω resistance and in loop 2, this requires going upward.

We now have two equations in two unknowns:

$$(1 + 2)i_1 - (2)i_2 = +1 \tag{2.54}$$

$$(-2)i_1 + (3 + 4 + 2)i_2 = -3 \tag{2.55}$$

Therefore, i_1 is 0.130 A and i_2 is -0.304 A.

4. Compute the primary unknown. This result is not the end of the problem, however, for we set out to calculate the voltage across the 2-Ω resistance. We were silent about the reference direction of this voltage because we wanted to consider both possibilities. If we put the + polarity symbol at the top of the 2-Ω resistance, the voltage would be $+(i_1 - i_2)\,(2\,\Omega)$ or $[0.130 - (-0.304)]\,(2) = +0.870$ V. If we put the + polarity symbol at the bottom of the 2-Ω resistance, the voltage would be $(i_2 - i_1)(2\,\Omega)$ or $[(-0.304) - (+0.130)](2) = -0.870$ V.

EXAMPLE 2.7 A Bridge Circuit

Write the loop-current equations for the circuit in Fig. 1.35, redrawn in Fig. 2.39.

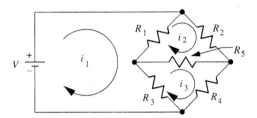

Figure 2.39 Bridge circuit from Chap. 1. No two resistors are in series or parallel, but loop current analysis works well.

SOLUTION:
This is often called a bridge circuit, and its analysis is beyond the method of series and parallel resistance combinations. The three loop currents are defined in Fig. 2.39. The KVL equation for loop 1 is

$$R_1(i_1 - i_2) + R_3(i_1 - i_3) = V \qquad (2.56)$$

or

$$i_1(R_1 + R_3) - (R_1)i_2 - (R_3)i_3 = V \qquad (2.57)$$

Inspection of Eq. (2.57) shows a simple pattern in the signs and coefficients of the three currents. The sign of the i_1 term is positive because we are going around loop 1 in the same direction as i_1. The other terms have minus signs because all the loop currents are referenced clockwise and thus go opposite directions in the resistances that are common to two loops. Similarly, the coefficient multiplying i_1 is the sum of the resistances in loop 1, and the coefficients multiplying the other currents are the negatives of the resistances in common between loop 1 and the loops of those currents. Note that Eqs. (2.54) and (2.55) follow these patterns. Thus, we may write the equation around the second loop as

$$-(R_1)i_1 + (R_1 + R_2 + R_5)i_2 - (R_5)i_3 = 0 \qquad (2.58)$$

The zero is put on the right-hand side of the equation because there are no voltage sources in loop 2.

WHAT IF? What if you had to produce the third equation for loop 3 on an exam. What would you write?[14]

---

[14] $-(R_3)i_1 - (R_5)i_2 + (R_3 + R_5 + R_4)i_3 = 0.$

Some Extensions and Fine Points

How to handle current sources.
Perhaps you have noticed that, as it now stands, we cannot incorporate current sources into our loop-current method. Current sources require a modest extension of the standard procedure. Consider the circuit of Fig. 2.40. The circuit has a current source; but, doing the obvious thing, we have defined and labeled our loop currents in the standard way. The current source would constrain the two loop currents passing through it, for by definition of a current source we must require that

$$i_2 - i_1 = 2 \text{ A} \tag{2.59}$$

Having two unknowns, we require another equation. The second equation comes from KVL around the outer loop:

$$5i_1 + 8i_2 - 10 = 0 \tag{2.60}$$

In writing Eq. (2.60), we have departed from the standard procedure in two ways: We went around the outer loop, which does not follow any single loop current; and we wrote KVL with all terms on the left side of the equation, which is the way we originally learned to write KVL equations. Equations (2.59) and (2.60) yield $i_1 = -0.462$ A and $i_2 = 1.538$ A. We note that i_1 is the current in the voltage source and the 5-Ω resistance, and i_2 is the current in the 8-Ω resistance. Note that $i_2 - i_1 = 2$ A, as required by the current source.

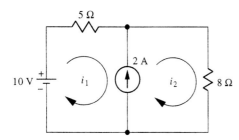

Figure 2.40 The current source constrains i_1 and i_2.

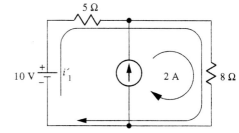

Figure 2.41 Loop current i_1' is routed to avoid the current source.

constrained loop current

The solution just presented gives the correct answer but is not the recommended way to handle current sources. We now present another solution that uses the concept of a *constrained loop current*, which is a loop current that is known because it passes through a current source. Figure 2.41 shows the same circuit with one unknown loop current defined, i_1', and one known loop current of 2 A defined to flow around the right loop. "Wait," you should say, "that current will divide at the top and go both ways." Yes, that is true, but bear with us until we finish and then you will understand this approach. Because we now have only one unknown, we need only one equation, which comes from KVL around the loop of i_1':

$$5i_1' + 8(i_1' + 2) - 10 = 0 \tag{2.61}$$

In Eq. (2.61) the term for the 8-Ω resistance includes the effect of both the unknown and the constrained loop currents. They add because both are referenced in the same direction through that resistance. The solution of Eq. (2.61) is routine: $i_1' = -0.462$ A. Using this and the constrained loop current, we calculate the current in the 8-Ω resistance to be $i_1' + 2 = 1.538$ A. Thus, by using the constrained loop concept to handle the current source, we get the same results for the physical currents as we obtained before with the first, more straightforward method.

EXAMPLE 2.8 Redirecting the constrained current

Constrain the 2-A loop current from the source to flow through the 5-Ω resistance and voltage source, as shown in Fig. 2.42, and solve for the physical currents again.

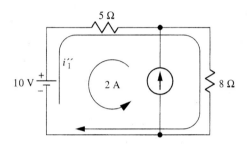

Figure 2.42 Again the loop current misses the current source. The constrained loop current is rerouted.

SOLUTION:
The KVL equation around the loop of i_1'' is

$$5(i_1'' - 2) + 8i_1'' - 10 = 0 \tag{2.62}$$

The solution of Eq. (2.62) is routine: $i_1'' = 1.538$ A, and the current referenced left to right in the 5-Ω resistance is $i_1'' - 2 = -0.462$ A. Again, we have the same answers for the physical currents.

WHAT IF? What if you split the 2-A current and have two constrained loops of 1 A each?[15]

We have shown *how* the constrained loop current works, but *why* does it work? Who are we to make the current from the current source go wherever we wish? *Answer*: We are the ones defining variables in the problem. We are defining variables by numbering loops and by drawing currents going clockwise around these loops; and when there are current sources, we are defining variables by defining the paths in which those currents flow. Look at Figs. 2.41 and 2.42. The unknown loop current *appears* to be the same in both, but by changing the path of the 2-A current, we changed the *meaning*

---

[15] It works, but there is no advantage to doing it that way.

of the unknown loop current. Thus, we can handle current sources by defining their currents to flow in certain paths, modifying our interpretation of the unknown loop currents accordingly.

Counting independent loops. The first step in our standard procedure was to number the independent loops. We identify independent loops by turning OFF all sources, as we did when we wished to identify independent nodes in the method of node voltages. For example, when we turn OFF both sources in Fig. 2.40, the current source becomes an open circuit and the voltage source becomes a short circuit. We are left with one loop containing two resistances. Thus, we have one independent loop, requiring one unknown and one KVL equation.

loop currents, mesh currents

Loop currents and mesh currents. What we have been using up to now are mesh currents, a special class of loop currents. In circuit terminology, a *loop* is any closed path. A *mesh* is a special loop, namely, the smallest loop one can have. A mesh is thus a loop that contains no other loops. In the fuller sense of the loop-current method, we can define the loop currents with great freedom, allowing them to go wherever we wish within certain guidelines. For our relatively simple circuits, the guidelines are that we must define the correct number of currents and that we must go through each resistance with at least one loop current.

We will rework the circuit in Fig. 2.38(a) as an example of this more general loop-current method. The circuit is redrawn in Fig. 2.43 with true loop[16] currents drawn as shown. We defined one current clockwise and the other counterclockwise to show the generality of the method. We write KVL around the loops of the unknown currents with the following results:

$$-7 + 1\,(i'_1 - i'_2) + 2i'_1 + 6 = 0 \tag{2.63}$$

$$+7 - 9 + (3 + 4)i'_2 + 1\,(i'_2 - i'_1) = 0 \tag{2.64}$$

Therefore, $i'_1 = 0.435$ A and $i'_2 = 0.304$ A. Because i'_1 is the only current through the 2-Ω resistance, we must solve for it only to compute the voltage across that resistance, the original unknown. The result is $0.435\,\text{A} \times 2\,\Omega = 0.870\,\text{V}$, + at top, which agrees with our earlier result.

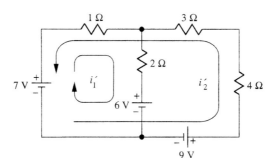

Figure 2.43 These are true loop currents, not mesh currents.

---

[16] Actually i'_1 is also a mesh current, but is called a loop current here because it is used in the more general method.

The more general loop-current method is useful when we wish to determine only one current or voltage, because we can define all the unknown loop currents to avoid that path except one loop current. The resulting equations can then be solved for that one loop current, which will be the desired current. The trouble with the generalized loop-current method is that all the nice symmetries and automatic sign patterns of the mesh-current method vanish, and once again we are required to pay careful attention to signs.

branch currents

How we can ignore KCL? With the loop-current method, we solve for the currents of a circuit by writing only KVL equations. How can we ignore KCL? *Answer*: KCL applies to *branch currents*, currents that flow from one point to another through a direct path. We can ignore KCL because we defined the currents to flow in complete loops rather than from one point to another in the circuit. In Fig. 2.44, we show two loop currents passing through a node. If we wrote KCL for such a node, each loop current would contribute two equal and opposite terms: Each loop current must add to zero at every node. Thus, KCL is satisfied automatically when we use loop currents to describe the circuit; we have only to satisfy KVL to find the solution.

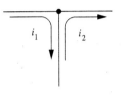

Figure 2.44 Node with two loop currents.

EXAMPLE 2.9 Benchmark Example

Solve the benchmark example by loop-current analysis.

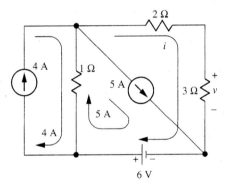

Figure 2.45 The benchmark example with loop currents defined. We have constrained currents such that only one current passes through the 3-Ω resistor.

SOLUTION:
Figure 2.45 shows the circuit marked for loop-current analysis. Note that we have one independent and two constrained loops. KVL around the loop is

$$2i + 3i - 6 + 1 \times (i + 5 - 4) = 0 \Rightarrow i = \frac{6 + 4 - 5}{2 + 3 + 1} = \frac{5}{6} \text{ A} \quad (2.65)$$

and the unknown voltage is $3i = 2.50$ V, as before.

Summary of Methods of Circuit Analysis

LEARNING OBJECTIVE 6.
To understand how to choose the most efficient manner for analyzing a circuit

Which method to use? We now have four methods for analyzing circuits: the method of current and voltage dividers augmented by superposition, Thévenin and Norton equivalent circuits, the method of node voltages, and the method of loop currents. How can we select the best one to use on a given circuit? Here are some of the factors to consider:

- **Design or analysis?** Are we trying to design or analyze the circuit? If the circuit is completely specified and we are trying to determine some aspect of its response, say, the power out of a source or the voltage across some resistance, the nodal or loop methods are favored. These are general methods of analysis, which solve for the entire circuit response all at once. Special attention is not given toward the effect of a single resistance or source; rather, everything is incorporated into the equations at the beginning.

  On the other hand, if you are designing the circuit in some regard, the methods of voltage and current dividers or equivalent circuits are favored. These methods focus on the effects of individual resistances and sources at specific points in the circuit. Design must, of course, deal in such details and these methods are well suited for allowing the designer to control the way voltage and current distribute throughout a circuit.

- **Number of equations?** How many equations must be solved? It is possible for a circuit to have more independent nodes than loops, or vice versa. The next example illustrates this point.

- **Fine points.** Finally, there are numerous minor considerations that would suggest a method if those discussed fail to dictate a choice. If there are many current sources, nodal analysis is favored, but if many voltage sources, loop currents might be easier. If the unknown is a voltage, nodes might be best, but if the unknown is a current, loops might be more efficient. If there is only one source and the circuit is not too complicated, the method of voltage and current dividers is favored. These decisions come easily as a result of much experience in circuit analysis. As a beginner, you will have to practice the various methods on a number of circuits before you develop the ability to choose the most efficient method.

EXAMPLE 2.10 Loop Currents or Node Voltages?

Find i in the circuit shown in Fig. 2.46.

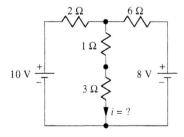

Figure 2.46 The circuit has two independent loops but only one independent node.

SOLUTION:
This circuit has two independent loops but only one independent node if we combine the series resistors. Thus, we favor nodal analysis for simpler mathematics.

Choosing the node at the bottom for the reference node, we designate the voltage at the top node as v and the nodal equation is

$$\frac{v-(+10)}{2} + \frac{v-(0)}{1+3} + \frac{v-(+8)}{6} = 0 \tag{2.66}$$

The solution is $v = 6.91$ V, and the current downward is this voltage divided by $4\,\Omega$, or 1.73 A.

Check Your Understanding

1. How many independent loops are there in the circuit in Fig. 2.15 with the load removed?
2. What is the name of a loop current passing through a current source?
3. If a circuit has five independent nodes and two independent loops, what method of analysis does this suggest?
4. In the standard method for mesh (loop) current analysis, what is the coefficient of the mesh current in the equation for that mesh?

Answers. (1) One independent loop; (2) a constrained loop current; (3) loop currents (two equations in two unknowns); (4) the sum of the resistances around its mesh.

2.5 CONTROLLED SOURCES

LEARNING OBJECTIVE 7.

To understand the need for controlled sources in circuit models and their effect on circuit analysis techniques

controlled source

Introduction. When the analysis of electronic circuits became important in circuit theory, controlled sources were added to the family of circuit elements. Table 2.1 shows the four types of controlled sources. Note that a rhombic shape indicates a controlled source.

In this section, we will address the questions: What are controlled sources? What is the significance of controlled sources? and How does the presence of controlled sources affect circuit analysis?

What are controlled sources?

By *source* we mean a voltage or current source in the usual sense. By *controlled* we mean that the strength of such a source is controlled by some circuit variable elsewhere in the circuit.

A two-port circuit. To illustrate the nature of controlled sources, we will analyze the two-port network in Fig. 2.47. This circuit is a two-sided voltage divider: if voltage is applied at v_1, then it will be divided by R_3 and R_2 and appear at v_2, and if voltage is applied at v_2, it will be divided by R_3 and R_1 and appear at v_1. Note: we have referenced the currents, i_1 and i_2, into the two-port to preserve symmetry.

Replacing the circuit. If the circuit in Fig. 2.47 were embedded in a larger circuit, Kirchhoff's laws could still be written, perhaps in the form of nodal equations, and the

TABLE 2.1 Names, Circuit Symbols, and Definitions for the Four Possible Types of Controlled Sources

Name	Circuit symbol	Definition and units
current-controlled voltage source		$v_2 = r_m i_1$ r_m = transresistance units = ohms
current-controlled current source		$i_2 = \beta i_1$ β = current gain, dimensionless
voltage-controlled voltage source		$v_2 = \mu v_1$ μ = voltage gain, dimensionless
voltage-controlled current source		$i_2 = g_m v_1$ g_m = transconductance units = Siemans (mhos)

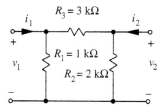

Figure 2.47 A two-way voltage divider two-port circuit. A two-port is a circuit or system with input and output.

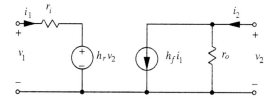

Figure 2.48 A two-port described by hybrid parameters. We will determine the parameters to make this circuit equivalent to that in Fig. 2.47.

resulting equations solved. Our approach, however, will be to replace the circuit in Fig. 2.47 by that in Fig 2.48.

Here we have separated input and output into two separate circuits: the input has the form of a Thévenin equivalent circuit and the output the form of a Norton equivalent circuit. This mixed arrangement leads to the name "hybrid two-port" for this equivalent circuit, and the four parameters, r_i, h_r, r_o, and h_f are called hybrid parameters.

Controlled sources. The apparent separation of input and output in Fig. 2.48 is misleading, for input and output are coupled, both ways, by the controlled sources. The voltage-controlled voltage source in the input circuit, $h_r v_2$, expresses the effect of the output voltage, v_2, on the input circuit, and the current-controlled current source in the output circuit expresses the effect of the input current on the output circuit.

The hybrid equations. The circuit equations for the hybrid two-port are easily written: KVL in the input circuit

$$v_1 = r_i i_1 + h_r v_2 \tag{2.67}$$

and KCL in the output circuit

$$i_2 = \frac{v_2}{r_o} + h_f i_1 \tag{2.68}$$

Input current and output voltage are treated as independent variables to express the dependent variables v_1, the input voltage, and i_2, the output current. This choice is somewhat arbitrary but follows naturally from the form of the hybrid two-port.

Determining the hybrid parameters. We may determine the hybrid parameters in Eqs. (2.67) and (2.68) by performing thought experiments on the original circuit. In Eq. (2.67), the parameter r_i is the ratio

$$r_i = \left.\frac{v_1}{i_1}\right|_{v_2 = 0 \text{ (shorted output)}} \tag{2.69}$$

Thus r_i is the input impedance to the network with the output shorted. If in Fig. 2.47 we short the output ($v_2 = 0$) and observe the input impedance of the resulting circuit, we observe

$$r_i = R_1 \| R_3 = 1 \text{ k}\Omega \| 3 \text{ k}\Omega = 750 \text{ }\Omega \tag{2.70}$$

Equation (2.70) follows because shorting the output places R_3 in parallel with R_1, and R_2 is eliminated. Similarly the reverse voltage gain, h_r, would be

$$h_r = \left.\frac{v_1}{v_2}\right|_{i_1 = 0 \text{ (open input)}} \tag{2.71}$$

Thus we open the input and relate v_1 and v_2 from the voltage divider of R_1 and R_3.

$$h_r = \frac{R_1}{R_1 + R_3} = \frac{1}{1 + 3} = 0.250 \tag{2.72}$$

The same type of reasoning shows that r_o is the output impedance with the input open circuited

$$r_o = \left.\frac{v_2}{i_2}\right|_{i_1 = 0 \text{ (open input)}} = R_2 \| (R_1 + R_2) = 1333 \; \Omega \tag{2.73}$$

and the forward current gain, h_f, is derived with the output shorted

$$h_f = \left.\frac{i_2}{i_1}\right|_{v_2 = 0 \text{ (shorted output)}} = -\frac{\frac{1}{R_3}}{\frac{1}{R_3} + \frac{1}{R_1}} = -0.250 \tag{2.74}$$

Thus the equations describing the hybrid two-port equivalent circuit to the circuit in Fig. 2.47 are

$$\begin{aligned} v_1 &= 750 i_1 + 0.250 v_2 \\ i_2 &= \frac{v_2}{1333} - 0.250 i_1 \end{aligned} \tag{2.75}$$

Summary. Our analysis of the circuit in Fig. 2.47 was undertaken to show a role for controlled sources in the analysis of circuits. Although the circuit we modeled is simple, the hybrid equivalent and our analysis are general and could be applied to any two-port, no matter how complicated.

A transistor model. Controlled sources are normally introduced to model amplifying devices such as transistors. In this context the controlled source represents the physical mechanisms internal to the transistor that couple input and output and give the transistor its remarkable properties. In the transistor a controlled voltage or current source may depend on the controlling variable in a nonlinear manner. However, the nonlinear relationship is frequently linearized to examine the effects of small variations about some dc value.

Notation. When we linearize, we will use the customary notation of small letters to represent general and time-variable voltages and currents and large letters to represent constants such as the dc value or the peak value of a sinusoid. On subscripts, large letters represent the total voltage or current and small letters represent the small-signal component. Thus the equation,

$$i_B(t) = I_B + I_b \cos \omega t \tag{2.76}$$

means that the total base current of the transistor is the sum of a constant and a small signal component, which is sinusoidal with an amplitude of I_b. These relationships are illustrated in Fig. 2.49.

Modeling a transistor with controlled sources. To illustrate the use of controlled sources in modeling transistors, we will develop a circuit model for the bipolar junction transistor (BJT). In Figure 2.50 we show the standard symbol for an npn BJT with base (B), emitter (E), and collector (C) identified, and voltage and current variables defined.

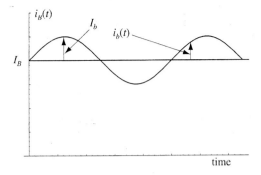

Figure 2.49 The time-average or dc value is often larger than the signal, $i_b(t)$. This allows nonlinear devices such as transistors to be analyzed by linearizing the equations.

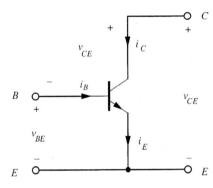

Figure 2.50 An *npn* BJT in the common-emitter configuration.

We have shown the common-emitter configuration, with the emitter terminal shared to make input and output terminals.

Modeling the base characteristic. The base current, i_B, ideally depends upon the base-emitter voltage, v_{BE}, by the relationship,

$$i_B = I_0 \left\{ \text{Exp}\left[\frac{v_{BE}}{V_T}\right] - 1 \right\} \tag{2.77}$$

where I_0 and V_T are constants. The base current depends on the base-emitter voltage only, but in a nonlinear manner. We can represent this current by a voltage-controlled current source, but the more common representation would be that of a nonlinear conductance, $G_{BE}(v_{BE})$, where

$$G_{BE}(v_{BE}) = \frac{i_B}{v_{BE}} \tag{2.78}$$

Linearized model. We now model the effects of small changes in the base current. If the changes are small, the nonlinear nature of the conductance can be ignored and the circuit model becomes a linear conductance (or resistor). Mathematically this conductance arises from a first-order expansion of the nonlinear function in Eq. (2.77). Thus, if $v_{BE} = V_{BE} + v_{be}$, where v_{BE} is the total base-emitter voltage, V_{BE} is a (large) constant voltage and v_{be} is a (small) variation in the base-emitter voltage, then the first two terms in an expansion are

$$i_B = I_0\left\{Exp\left[\frac{V_{BE} + v_{be}}{V_T}\right] - 1\right\} \approx \underbrace{I_0\left\{Exp\left[\frac{V_{BE}}{V_T}\right] - 1\right\}}_{I_B} + \underbrace{\frac{I_0}{V_T}Exp\left[\frac{V_{BE}}{V_T}\right] \times v_{be}}_{i_b}$$

(2.79)

The base current is approximated by the sum of a constant term and a term that is first-order in the small variation in base-emitter voltage, v_{be}. The multiplier of this small voltage is the linearized conductance, g_{be}. If we were interested only in small changes in currents and voltages, only this conductance would be required in the model. Thus the input (base-emitter) circuit can be represented for the small-signal base variables, i_b and v_{be}, by either equivalent circuit in Fig. 2.51:

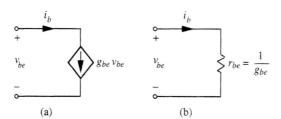

Figure 2.51 Equivalent circuits for the base circuit.

The voltage-controlled current source, $g_{be}v_{be}$, can be replaced by a simple resistor because the small-signal voltage and current are in the same branch.

Modeling the collector-emitter characteristic. The collector current, i_C, can be represented by one of the Eber and Moll equations

$$i_C = \beta I_0\left\{Exp\left[\frac{v_{BE}}{V_T}\right] - 1\right\} + I_0'\left\{Exp\left[\frac{v_{BC}}{V_T}\right] - 1\right\} \quad (2.80)$$

where β and I_0' are constants. If we restrict our model to the amplifying region of the transistor, the second term is negligible and we may express the collector current as

$$i_C = \beta I_0\left\{Exp\left[\frac{v_{BE}}{V_T}\right] - 1\right\} = \beta \times i_B \quad (2.81)$$

Thus for the ideal transistor, the collector-emitter circuit may be modeled by a current-controlled current source, which may be combined with the results expressed in Eq. (2.77) to give the model shown in Fig. 2.52.

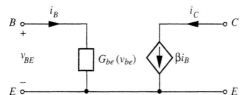

Figure 2.52 Equivalent circuit for BJT.

Small-signal models for the transistor. Using the technique of small-signal analysis, we may derive either of the small-signal equivalent circuits shown in Fig. 2.53.

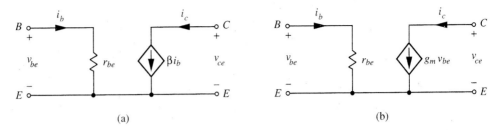

Figure 2.53 Two BJT small-signal equivalent circuits ($g_m = \beta/r_{be}$).

Hybrid two-port model. The small signal characteristics of the *npn* transistor in its amplifying region is better represented by the hybrid parameter equivalent circuit shown in Fig. 2.54.

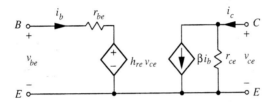

Figure 2.54 Full hybrid parameter model for small-signal BJT.

Note we have introduced a voltage-controlled voltage source to model the influence of the (output) collector-emitter voltage on the (input) base-emitter voltage, and we have placed a resistor, r_{ce}, in parallel with the collector-current source to show the influence of the collector-emitter voltage on the collector current.

The four parameters in Fig. 2.54 (r_{be}, h_{re}, β, and r_{ce}) are the hybrid parameters describing the transistor properties, although our notation differs from that commonly used. The parameters in the small-signal equivalent circuit depend on the operating point of the device, which is set by the time-average voltages and currents (V_{BE}, I_C, etc.) applied to the device. All of the parameters are readily measured for a given transistor and operating point, and manufacturers commonly specify ranges for the various parameters for a type of transistor.

EXAMPLE 2.11 Transistor amplifier

Consider a transistor in the common emitter configuration with $r_{be} = 1000\ \Omega$, $h_{re} =$ negligible, $h_{fe} = \beta = 120$, and $r_{ce} = 10,000\ \Omega$. This transistor is placed in a circuit with input and output properties as shown in Fig. 2.55.

Find the current gain of the circuit, $A_i = i_{load}/i_{in}$.

SOLUTION:
The negligible reverse voltage gain permits the circuit to be analyzed from input to output. The base current is

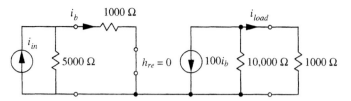

Figure 2.55 Transistor amplifier. The input circuit is in the Norton form, and the output load is a 1000 Ω resistor.

$$i_b = i_{in} \times \cfrac{\cfrac{1}{1}}{\cfrac{1}{1} + \cfrac{1}{5}} = \cfrac{i_{in}}{1.2} \qquad (2.82)$$

and the current source splits between the output impedance and the load:

$$i_{load} = -100 i_b \times \cfrac{\cfrac{1}{1}}{\cfrac{1}{1} + \cfrac{1}{10}} = \cfrac{-100 i_b}{1.1} = -\cfrac{100 i_{in}}{1.1 \times 1.2} = -75.8 i_{in} \qquad (2.83)$$

The current gain is thus -75.8.

WHAT IF? What if you want the voltage gain, defined as the voltage across the load resistor divided by the voltage at the transistor input?[17]

What is the significance of controlled sources?

Commonplace wisdom in engineering education and practice is that information and techniques that are presented visually are more useful than abstract, mathematical forms. Equivalent circuits are universally used in describing electrical engineering systems and devices because circuits portray interactions in a universal, pictorial language. This is true generally, and it is doubly necessary when circuit variables interact through the "mysterious" coupling modeled by controlled sources. This is the primary significance of controlled sources: that they represent unusual couplings of circuit variables in the universal, visual language of circuits.

A second significance is illustrated by our equivalent circuit of the *npn* bipolar transistor, namely, the characterization of a class of similar devices. For example, the parameter β in Eq. (2.81) gives important information about a single transistor, and similarly for the range of β for a type of transistor. In this connection, controlled sources lead to a vocabulary for discussing some property of a class of systems or devices, in this case the current gain of an *npn* BJT.

---

[17] The voltage gain is -90.9.

How does the presence of controlled sources affect circuit analysis?

The presence of nonreciprocal elements, which are modeled by controlled sources, affects the analysis of the circuit. Simple circuits may be analyzed through the direct application of Kirchhoff's laws to branch circuit variables. Controlled sources enter this process by defining relationships between branch circuit variables. Thus controlled sources add no complexity to this basic technique.

The presence of controlled sources negates the advantages of the method that uses series and parallel combinations of resistors for voltage and current dividers. The problem is that the couplings between circuit variables that are expressed by controlled sources make all the familiar formulas unreliable.

When superposition is employed, the controlled sources are left ON in all cases as independent sources are turned ON and OFF, thus reflecting the kinship of controlled sources to the circuit elements. In principle little complexity is added; in practice, the repeated solutions required by superposition entails much additional work when controlled sources are involved.

The classical methods of nodal and loop (mesh) analysis incorporate controlled sources without great difficulty. For purposes of determining the number of independent variables required, the controlled sources are treated as ordinary voltage or current sources. The equations are then written according to the usual procedures. Before the equations are solved, however, the controlling variables must be expressed in terms of the unknowns of the problem. For example, let us say we are performing a nodal analysis on a circuit containing a current-controlled current source. For purposes of counting independent nodes, the controlled-current source is treated as an open circuit. After equations are written for the unknown node voltages, the current source will introduce into at least one equation its controlling current, which is not one of the nodal variables. The additional step required by the controlled source is that of expressing the controlling current in terms of the nodal variables.

The parameters introduced into the circuit equations by the controlled sources end up on the left side of the equations with the resistors rather than on the right side with the independent sources. Furthermore, the symmetries that normally exist among the coefficients are disturbed by the presence of controlled sources.

The methods of Thévenin and Norton equivalent circuits continue to be very powerful with controlled sources in the circuits, but some complications arise. The controlled sources must be left ON for calculation of the Thévenin (open-circuit) voltage or Norton (short-circuit) current and also for the calculation of the output impedance of the circuit. This usually eliminates the method of combining elements in series or parallel to determine the output impedance of the circuit, and one must either determine the output impedance from the ratio of the Thévenin voltage to the Norton current, Eq. (2.22), or else excite the circuit with an external source and calculate the response.

CHAPTER SUMMARY

Chapter 2 builds on Chap. 1 by giving a variety of methods for analyzing circuits. The various methods are compared for their advantages and disadvantages. The concept of impedance level was introduced to explain the interaction between parts of a circuit.

Objective 1: To understand the principle of superposition and use superposition to analyze circuits with multiple sources. Superposition is based on the linearity of Ohm's and Kirchhoff's laws. Using superposition, we can examine the effects of multiple voltage and current sources at a prescribed point in a circuit. A key concept is turning OFF a source.

Objective 2: To understand the origin of Thévenin's equivalent circuit and be able to derive the Thévenin equivalent circuit of a network with a load. Thévenin's equivalent circuit is used to describe the interaction of a circuit with a load. The equivalent circuit consists of a voltage source, the open-circuit voltage, in series with a resistor, the output impedance. A complicated problem can often be reduced to two simpler problems by using Thévenin's equivalent circuit.

Objective 3: To understand the significance of impedance level in interactions between circuits. The interaction between a circuit and a load depends on the output impedance level of the circuit relative to the input impedance level of the load. When the source impedance level is small relative to the load, voltage will be transferred to the load; when the source impedance level is large relative to the load, current will be transferred to the load; and when impedance levels are roughly the same, power will be transferred.

Objective 4: To understand how to analyze a circuit using nodal analysis. In nodal analysis, all voltages are defined relative to a reference node, which is assigned a voltage of zero. Equations are based on Kirchhoff's current law in a special form, resulting in simultaneous equations for the voltages at all nodes. This method of circuit analysis is systematic and efficient, especially when the circuit contains many loops but few nodes, as is common in electronic circuits.

Objective 5: To understand how to analyze a circuit using loop currents. Loop currents are mathematical variables defined to flow in closed loops. Equations for loop currents are based upon Kirchhoff's voltage law, although voltage variables are not defined. This method is systematic and efficient when the circuit contains voltage sources and few loops.

Objective 6: To understand how to choose the most efficient manner for analyzing a circuit. Our methods for circuit analysis now include voltage and current dividers plus superposition, Thévenin equivalent circuits, nodal analysis, and loop-current analysis. These methods are compared for ease of mathematics, for suitability in analysis or design, and for relative popularity and importance among practitioners.

Objective 7: To understand the need for controlled sources in circuit models and their effect in circuit analysis techniques. Controlled sources model electrical coupling effects that are unusual. Circuits containing controlled sources are best analyzed by formal techniques, in which the controlling variables are expressed in terms of the formal unknowns.

The methods of this chapter are used throughout the remainder of the book to formulate circuit equations. In Chap. 3, we use them to formulate differential equations describing circuits containing resistors, inductors, and capacitors.

GLOSSARY

Available power, p. 71, the maximum power that can be extracted from a circuit.

Constrained loop current, p. 88, a loop current that is known because it passes through a current source.

Constrained node, p. 77, a node whose voltage is constrained to the voltage of another node by a voltage source.

Controlled source, p. 93, a voltage or current source controlled by a voltage or current elsewhere in the circuit.

Ground node or ground, p. 84, the point in an electrical circuit that is physically connected via a thick wire to the moist earth, usually for safety purposes.

Impedance level, p. 73, the approximate ratio of voltages to currents in a circuit or portion of circuit. The interaction of a circuit with a load depends on the relative values of their respective impedance levels.

Impedance match, p. 68, a state in which the load impedance is equal to the output impedance of the circuit, required for maximum power transfer.

Independent node, p. 82, a node whose voltage cannot be derived from the voltage of another node.

Load, p. 64, is an element in a circuit of particular importance; in one sense the circuit exists to supply the load.

Loading of a circuit, p. 72, the change in output voltage when a load or meter is attached. Could also apply to a change in current when an ammeter is inserted.

Loop, p. 90, any closed path in an electrical circuit.

Loop current, p. 85, a current that flows in a closed path.

Mesh, p. 90, a loop that contains no other loops, the smallest possible loops.

Open circuit, p. 59, a "path" where no current flows, regardless of the voltage.

Open-circuit voltage, p. 64, the voltage between two terminals with the load removed.

Optimum, p. 69, when a design is the best possible in some specified characteristic under the constraints.

Potential difference, p. 84, the voltage between two points in a circuit. Our definition of voltage is equivalent to a potential drop.

Reference node, p. 76, the points in an electrical circuit from which all node voltages are measured. By definition, the node voltage of the reference node is zero.

Short circuit, p. 59, an ideal connection requiring zero voltage.

Source transformations, p. 74, a technique for circuit analysis that requires repeated conversions between Thévenin and Norton equivalent circuits.

Turned-OFF current source, p. 59, an open circuit, equivalent to infinite resistance.

Turned-OFF voltage source, p. 59, a short circuit, equivalent to zero resistance.

PROBLEMS

Section 2.1: Superposition

2.1. Find the current in the 20-Ω resistor in Fig. P2.1 using superposition.

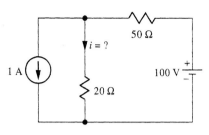

Figure P2.1

2.2. In Fig. P2.2, find i using the principle of superposition.

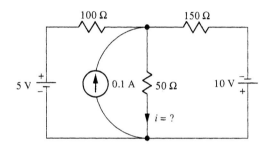

Figure P2.2

2.3. For the circuit in Fig. P2.3, determine the current in the 10-Ω resistor with the reference direction shown using the principle of superposition.

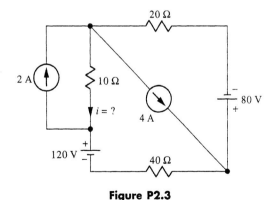

Figure P2.3

2.4. For the circuit shown in Fig. P2.4, determine the power into the 15-Ω resistor using superposition.

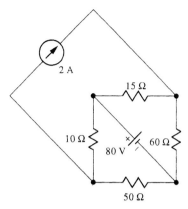

Figure P2.4

2.5. For the circuit shown in Fig. P2.5, determine the following:

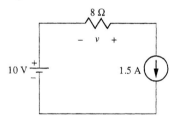

Figure P2.5

(a) Find v using superposition.
(b) Determine the power out of the current source.

2.6. For the circuit shown in Fig. P2.6, determine the following:

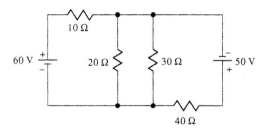

Figure P2.6

(a) Find the current in the 20–Ω resistance due to the 60-V source. (Do not consider the current due to the 50-V source.)
(b) With both sources operating, find the ratio of the current in the 30-Ω resistance to the current in the 20-Ω resistance.

Section 2.2: Thévenin's and Norton's Equivalent Circuits

2.7. Place a connection (a short circuit, no resistor) between a and a' in Fig. 2.12. Solve for the current flowing down from a to a'. The answer should be zero current, as argued in Section 2.2.

2.8. Develop a Thévenin equivalent circuit for the part of the circuit shown in the box in Fig. P2.8. Use this equivalent circuit to solve for i, as shown.

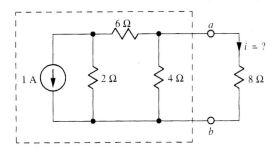

Figure P2.8

2.9. For the circuit in Fig. P2.9, determine the following:

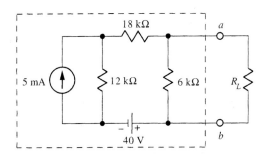

Figure P2.9

(a) Replace the circuit in the box by a Thévenin equivalent circuit.
(b) Find v_{ab} for $R_L = 3$ kΩ.
(c) What value of R_L receives maximum power from the circuit?
(d) What value of R_L makes the current in the 6-kΩ resistor equal 0.1 mA?

2.10. A black box was connected to a variable resistor and the power in the resistor was measured as the resistance was varied. The results are shown in Fig. P2.10. From this graph, determine the Thévenin equivalent circuit for the circuit in the black box.

2.11. Rework Problem P2.8, this time using a Norton equivalent circuit for the portion of the circuit in the box.

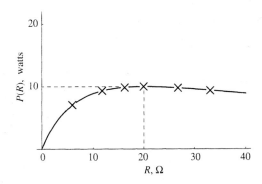

Figure P2.10

2.12. A mysterious black box is found in the electrical engineering lab. A curious student measured the output voltage to be 6.3 V. Then he shorted the output through an ammeter (consider the ammeter to have zero resistance), which indicated a current of 126 A. Give the Norton equivalent circuit for the box. How much power can be gotten out of the box if a variable resistor is connected to its terminals and adjusted for maximum power?

2.13. For the circuit shown in Fig. P2.13, determine the following:
(a) Find the value of R to receive maximum power.
(b) For the value of R in part (a), find the power out of the 12-V source.

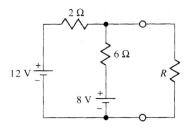

Figure P2.13

2.14. The starter motor on a car draws 75-A starting current, which lowers the battery voltage from 12.6 to 9.1 V. What would be the battery voltage if it were being charged at 30 A?

2.15. For the circuit shown in Fig. P2.15, determine the following:
(a) Find the current in the 3-Ω resistor.
(b) What resistance, replacing the 3-Ω resistor, would draw one-half the current in part (a)?

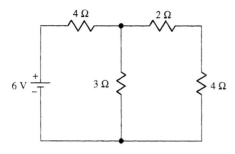

Figure P2.15

2.16. For the circuit shown in Fig. P2.16, determine the following:

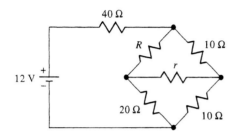

Figure P2.16

(a) What value of resistance, R will reduce the current through r to zero?
(b) For the value of R in part (a), what is the current through the voltage source?

2.17. A circuit has a variable load, R. Power measurements show that the power into R is maximum at 100 W with $R = 6\ \Omega$. What current would flow if R were replaced by a 3-Ω resistor?

2.18. What value of I_s reduces the circuit in Fig. P2.18 to a simple resistor as seen by the output terminals?

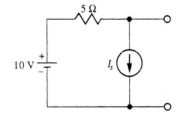

Figure P2.18

2.19. For the circuit shown in Fig. P2.19, find I_s such that the current in the 120-Ω resistance is zero.

2.20. Using a Thévenin equivalent circuit, find R in Fig. P2.20 such that $i = 0.5$ A.

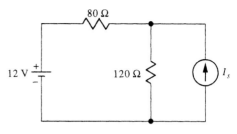

Figure P2.19

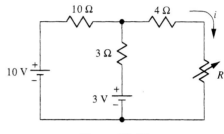

Figure P2.20

2.21. For the circuit in Fig. P2.21, find V to make the current in the 5-Ω resistor equal zero.

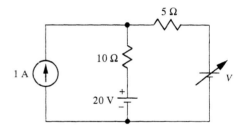

Figure P2.21

2.22. Show that the available power from a circuit is equal to the product of one-half the Thévenin voltage with one-half the Norton current. Express the output impedance in terms of the available power and the open-circuit voltage.

2.23. A student is testing a circuit containing batteries and resistors. The output voltage is 6.26 V when measured with a good (assume ideal) voltmeter, but 6.05 V when a 600-Ω resistor is connected across the output terminals. What current would result if the output were short-circuited?

2.24. The circuit in Fig. P2.24 shows the circuit for a voltage source and "potentiometer," which is a resistor with a sliding contact. The variable x varies from 0 to 1, and the open-circuit voltage increases in proportion with x.

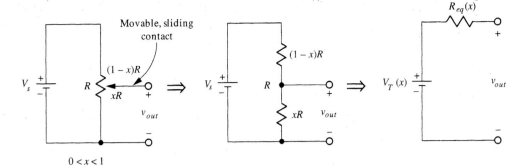

Figure P2.24

(a) Determine a Thévenin equivalent circuit for the voltage source, with the voltage and output impedance functions of x.
(b) What is the maximum output resistance if the total resistance of the potentiometer is R?
(c) If the load resistor were equal to the resistance of the potentiometer, R, what would x have to be to give an output voltage of $0.5V_s$?

2.25. A circuit, as shown in Fig. P2.25(a), has a variable load and ideal meters to monitor load voltage and current. The table in Fig. P2.25(b) shows partial results of a series of tests. Fill in the blank spaces in the table with the missing data.

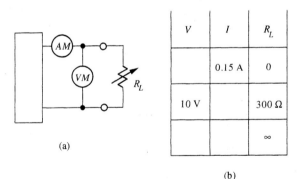

V	I	R_L
	0.15 A	0
10 V		300 Ω
		∞

(a)

(b)

Figure P2.25

2.26. A voltmeter has an impedance level (input resistance) that is 100 times the impedance level of the circuit it is measuring (the output resistance, with the meter as load). Find the % error in the meter reading, where

$$\% \text{ error} = \frac{\text{true voltage} - \text{measured voltage}}{\text{measured voltage}}$$

2.27. A circuit is represented by a Thévenin equivalent circuit, as shown in Fig. P2.27. With no ammeter, the current would be $i = V/R_{eq}$. Find the impedance level of the meter in terms of R_{eq} to cause a 1% error in the measurement. The definition of % error is given in the previous problem.

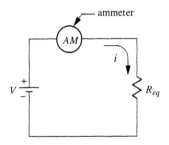

Figure P2.27

2.28. A voltmeter has a 1-MΩ (megohm = 10^6 Ω) impedance level (input resistance). Find the voltage it would measure for the Thévenin voltage between a and b in Fig. P2.9 with no load other than the voltmeter.

Section 2.3: Node-Voltage Analysis

2.29. (a) For the circuit shown in Fig. P2.29, solve for v_a and v_b using node-voltage techniques.
(b) Find the voltage across the three resistors (you supply reference directions), and show that KVL is satisfied.
(c) Now double v_a and repeat part (b). Note that KVL is still satisfied, even if *incorrect* node voltages are used. Of course, KCL would be violated.

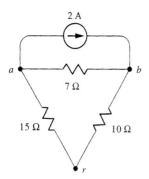

Figure P2.29

2.30. For the circuit shown in Fig. P2.30, write the KCL equation for node b using the node-voltage patterns and solve for $v_{br} = v_b$. Check your solution using the voltage-divider method.

2.31. For the circuit in Fig. P2.31, determine the following:
  (a) Write KVL to show $v_{ar} = v_a = +5$ V and $v_{cr} = v_c = -10$ V.
  (b) Find the current downward in the 50-Ω resistor by first solving for v_b using nodal analysis.
  Hint: Nodes a and c are constrained to the reference node.

2.32. Using node-voltage analysis, solve for the indicated unknowns in Fig. P2.32. *Hints*: In parts (a) and (b),

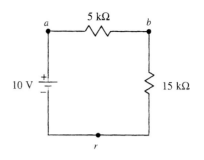

Figure P2.30

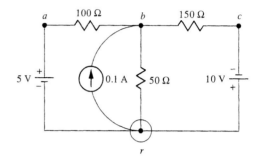

Figure P2.31

note that kV, mA, and volts make a consistent set of units; in part (c), find v_a, then v_{10} from the voltage divider.

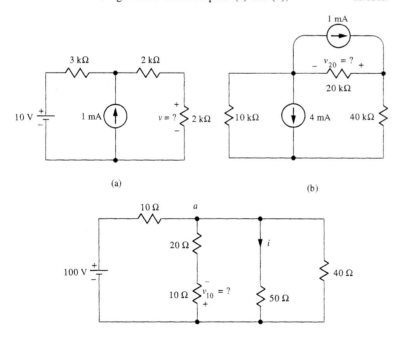

Figure P2.32

2.33. For Fig. P2.33, write the nodal equations using the notation given. Do not solve the equations.

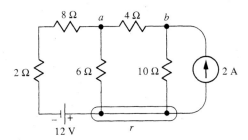

Figure P2.33

2.34. For Fig. P2.34, determine the current in the 100-Ω resistor with the reference direction given. Use nodal analysis.

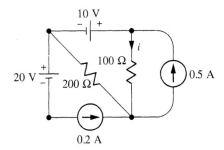

Figure P2.34

2.35. For the circuit shown in Fig. P2.35, determine the following:

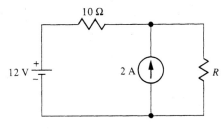

Figure P2.35

(a) Using nodal analysis, find the current in the 10-Ω resistor for $R = 6\,\Omega$.
(b) Derive a Thévenin equivalent circuit with R as the load.
(c) Find the value of R that gives 10% of the current that would flow if the load were replaced by a short circuit.

2.36. Write the nodal equations for the circuit of Fig. P2.36 in the form
$$v_a(\quad) + v_b(\quad) = (\quad)$$
$$v_a(\quad) + v_b(\quad) = (\quad)$$

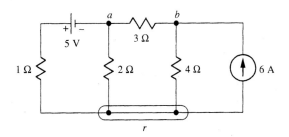

Figure P2.36

Section 2.4: Loop-Current Analysis

2.37. Solve for all branch currents in the circuit of Fig. P2.37 using loop-current analysis.

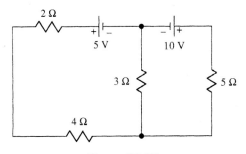

Figure P2.37

2.38. Find the power out of the 10-V source in Fig. P2.38 using loop currents to analyze the circuit.

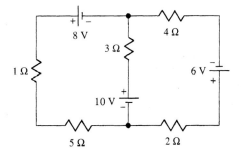

Figure P2.38

2.39. Solve for the current referenced downward in the 20-Ω resistor in the circuit of Fig. P2.39 using the constrained loop concept to handle the current source.

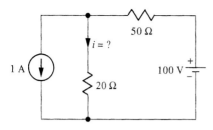

Figure P2.39

2.40. Solve for i in Fig. P2.40 using the loop-current method.

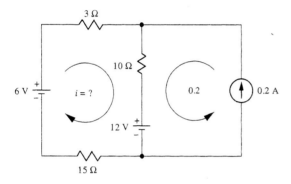

Figure P2.40

2.41. For the circuit of Fig. P2.41, solve for i_4 using mesh-current variables to analyze the circuit. Note that you may direct the constrained loops to miss the 4-Ω resistor. Be sure to count independent loops before defining variables.

2.42. Solve for the power out of the 80-V source in the circuit of Fig. P2.42, using loop-current variables. Remember to count independent loops first.

2.43. True loop (not mesh) currents are defined in Fig. P2.43 satisfying the rules given in Sec. 2.4.

General Problems

2.44. (a) Find the current in the 40-Ω resistor in Fig. P2.44 caused by the voltage sources. (Ignore the current caused by the current sources.)
(b) Find the current in the 40-Ω resistor in Fig. P2.44 caused by the current sources.

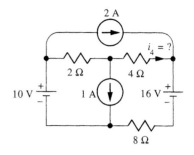

Figure P2.41

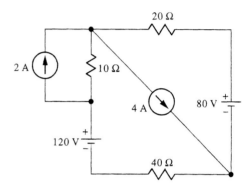

Figure P2.42

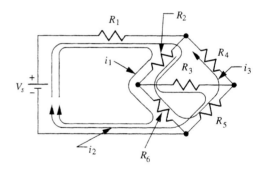

Figure P2.43

Write the KVL equations for the circuit using these loop variables. You are not required to solve the resulting equations.

(Ignore the current caused by the voltage sources.)
(c) Find the current in the 40-Ω resistor in Fig. P2.44 caused by all sources acting simultaneously.

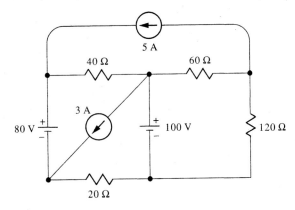

Figure P2.44

2.45. For Fig. P2.45, find R such that the power into the 200-Ω resistor is 12.5 W.

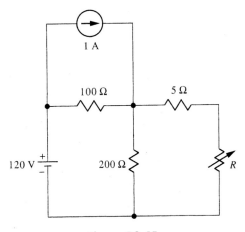

Figure P2.45

2.46. A 16-Ω loudspeaker draws maximum power from the output of its amplifier, which is capable of producing 25 W in the speaker. What would be the power produced in an 8-Ω speaker? Represent the loudspeakers as resistors of 16 and 8 Ω, respectively.

2.47. An electric stove (dc or ac, it does not matter, because the same power formulas apply) requires 230 V for the line voltage. The stove uses two heater elements that can be switched in one at a time or placed in series or parallel, making four heating temperatures. If the highest setting requires 3000 W and the lowest 500 W, what are the powers for the two intermediate settings?

2.48. A voltmeter must draw some current from the circuit it is measuring in order to operate (see Fig. P2.48).

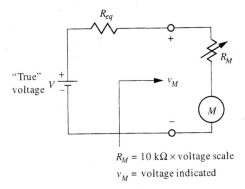

$R_M = 10 \text{ k}\Omega \times \text{voltage scale}$
$v_M = \text{voltage indicated}$

Figure P2.48 Metering a voltage. The impedance of the meter, R_M, depends on what voltage scale is used.

The amount of current it draws depends on the meter scale and is specified in terms of the "ohms per volt" of the meter. For example, a 10-kΩ/V meter would have a resistance of 10 kΩ on the 1-V scale, 100 kΩ on the 10-V scale, and so on. For this problem, assume that we are measuring with a 10-kΩ/V meter.
(a) If we measure 5 V on the 10-V scale, what current does the meter draw from the circuit?
(b) If the output impedance of the circuit is 600 Ω and we measure 5 V on the 10-V scale, what is the true open-circuit voltage of the source, that is, what would be measured by an ideal meter that drew no current from the circuit?
(c) With our 10-kΩ/V meter, on an unknown circuit we measure 24 V on the 30-V scale but 30 V on the 100-V scale. Explain the reason for this apparent discrepancy and determine from these measurements the Thévenin equivalent circuit for the source we are measuring.

2.49. The ladder network shown in Fig. P2.49, if terminated with the proper value of R_t, has the property that the input current is divided by 2 at each node, as shown. What should be the value of R_t for this property? What would be the input resistance to the ladder if the R, $2R$ pattern were continued infinitely?

2.50. After Norton died and went to heaven, he chanced to encounter Thévenin one day. They got into a discussion about whose equivalent circuit was better. To maintain peace, they proposed the "Thevenort" circuit shown in Fig. P2.50.
(a) Give values of V_h, I_h, and R_h that correspond to an open-circuit voltage (V_{oc}) of 5 V and a short-circuit current (I_{sc}) of 2 A. (The answer is not unique.)

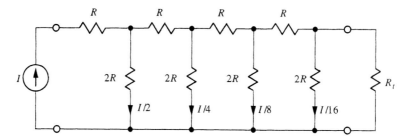

Figure P2.49

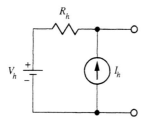

Figure P2.50

(b) Give general relationships that relate V_h, I_h, and R_h to the open-circuit voltage (V_{oc}) and the short-circuit current (I_{sc}).

2.51. Find the power in the 100-Ω resistor in Fig. P2.51 by a method of your own choosing.

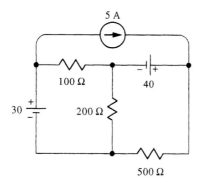

Figure P2.51

2.52. We have two batteries, A and B. Battery A has a voltage of 13.8 V and a short-circuit current of 350 A. Battery B has a voltage of 13.2 V and a short-circuit current of 220 A. The batteries may be connected in series or in parallel to give power to a single resistor. What connection, series or parallel, has the most power-producing capability, and what is the maximum power that can be obtained from that connection?

2.53. For the circuit in Fig. P2.53, find v_{ab} by any method.

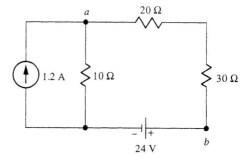

Figure P2.53

2.54. For the circuit in Fig. P2.54, find v_{ab} by any method.

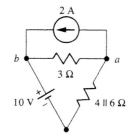

Figure P2.54

2.55. For the circuit in Fig. P2.55, find v_{ab} by any method.

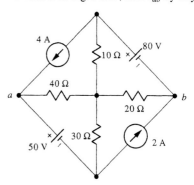

Figure P2.55

Section 2.5: Controlled Sources

2.56.

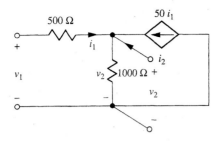

Figure P2.56

The circuit shown in Fig. P2.56 models a common transistor circuit.
(a) Find the input impedance if the output is open-circuited ($i_2 = 0$).
(b) Find the voltage gain, $A_v = v_2/v_1$ with the output open-circuited ($i_2 = 0$).
(c) Find the output impedance of the circuit. *Hint:* The results of part (b) give you the output Thévenin voltage. Find the Norton (short-circuit) current and find the output impedance from Eq. (2.22).

2.57.

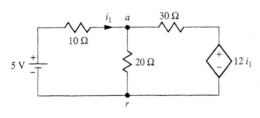

Figure P2.57

Using nodal analysis, find v_a in Fig. P2.57.

2.58.

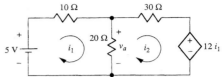

Figure P2.58

Rework P2.57 using loop currents. Fig. P2.58 defines loop-current variables.

2.59.

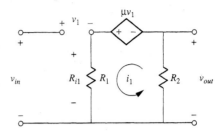

Figure P2.59

The circuit in Fig. P2.59 has a voltage-controlled voltage source.
(a) Find the voltage gain, $A_v = v_{out}/v_{in}$.
(b) Find the output impedance.

2.60.

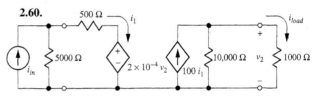

Figure P2.60

Find the current gain of the amplifier with non-negligible reverse voltage gain. The circuit is given in Fig. P2.60.

Answers to Odd-Numbered Problems

2.1. 0.714 A.
2.3. −2.29 A.
2.5. (a) −12 V; (b) +3 W.
2.7. 0 A.
2.9. (a) 3.33 V, 5 kΩ; (b) 1.25 V; (c) 5 kΩ; (d) 1.10 kΩ.
2.11. (a) −0.250 A, 2.67 Ω; (b) −0.0625 A.
2.13. (a) 1.50 Ω; (b) 39.0 W.
2.15. (a) 3.60 V, 2.40 Ω, 0.667 A; (b) 8.40 Ω.
2.17. 5.44 A.
2.19. −0.150 A.
2.21. 30 V.
2.23. 0.301 A.
2.25.

V	I	R_L
0	0.15 A	0
10 V	0.0333 A	300 Ω
12.9 V	0	∞

2.27. $R/100$.

2.29. (a) $v_a\left(\frac{1}{7}+\frac{1}{15}\right) + v_b\left(-\frac{1}{7}\right) = -2$
$v_a\left(-\frac{1}{7}\right) + v_b\left(\frac{1}{7}+\frac{1}{10}\right) = +2$
(b) $v_{ar} = -6.56$ V, $v_{br} = +4.38$ V, $v_{ab} = -10.9$ V
(c) $v'_{ar} = -13.1$ V, $v'_{br} = 4.38$ V, $v'_{ab} = -17.5$ V.

2.31. (a) $v_a = +5$ V, $V_c = -10$ V; (b) 0.0455 A.

2.33. $v_a\left(\frac{1}{10}+\frac{1}{6}+\frac{1}{4}\right) + v_b\left(-\frac{1}{4}\right) = -\frac{12}{10}$
$v_a\left(-\frac{1}{4}\right) + v_b\left(\frac{1}{4}+\frac{1}{10}\right) = 2$.

2.35. (a) 0 A; (b) 32 V, 10 Ω; (c) 90 Ω.

2.37. 0.159 A down on left, +1.19 A down on right, −1.35 A downward in middle.

2.39. 0.714 A.

2.41. −1.71 A.

2.43. $i_1(R_1+R_2+R_6) + i_2(R_1+R_2) + i_3(R_6) = +V_s$
$i_1(R_1+R_2) + i_2(R_1+R_2+R_3+R_5) + i_3(R_2-R_5) = +V_s$
$i_1(R_6+R_2) + i_2(R_2-R_5) + i_3(R_2+R_6+R_5+R_4) = 0$.

2.45. 29.5 Ω.

2.47. 634 W, 2366 W.

2.49. (a) $2R$; (b) $R_{in} = 2R$ into series resistor, and R into parallel resistor.

2.51. 768 W.

2.53. −10 V.

2.55. 129.5 V.

2.57. 3.134 V.

2.59. (a) $-\dfrac{\mu R_2}{R_2 + (\mu+1)R_1}$; (b) $R_2 \| (\mu+1)R_1$.

CHAPTER 3

The Dynamics of Circuits

Theory of Inductors and Capacitors

First-Order Transient Response of RL and RC Circuits

Advanced Techniques

Chapter Summary

Glossary

Problems

objectives

1. To understand how to identify the amplitude, frequency, and phase of a sinusoidal function
2. To understand the phasor concept and be able to find the sinusoidal steady-state response of a circuit, given the differential equation
3. To understand how to use phasors and impedance to determine the sinusoidal steady-state response of a circuit
4. To understand the effect of varying frequency on series and parallel *RL*, *RC*, and *RLC* circuits
5. To understand admittance and its advantages in analyzing parallel impedances

Alternating-current circuits comprise the vast majority of all power distribution and power consumption circuits. This chapter introduces the methods by which ac circuits are analyzed.

3.1 THEORY OF INDUCTORS AND CAPACITORS

Time and Energy

Statics and dynamics. In mechanics, dynamics usually follows statics. Statics deals with the distribution of forces in a structure; time is not a factor. Dynamics deals with the exchange of energy between components in a system, and time is an important factor because energy cannot be exchanged except as a time process.

In our study of electrical circuits, we have not yet considered time as an important quantity. Our dc circuits involve only KVL, KCL, and Ohm's laws, and none of these equations has time as a factor. As a result, energy flows smoothly from sources to loads. In a true dynamics problem, rates of energy flow vary with time.

With this chapter, we begin the study of electrical circuits in which rates of energy exchanged between circuit components are important. We begin by introducing the two circuit components that store energy in electrical circuits. The presence of inductors or capacitors in an electrical circuit suggests a true dynamics problem. We first identify the two types of energy that may be stored in a circuit.

electric energy, magnetic energy

Electric energy and magnetic energy storage. To understand what we mean by magnetic energy and electric energy, let us recall from Chap. 1 that there are two types of forces between electric charges: electrostatic and magnetic. Here we wish to emphasize energy. From mechanics, you recall that when a displacement is made against a force, work is done (mechanical energy is exchanged). Similarly, if we displace a charge in the presence of a motional (magnetic) force, magnetic energy is exchanged. To store much magnetic energy, we must bring many moving charges close together, which is what an *inductor* does. On the other hand, when we move charges in the presence of positional (electrostatic) forces, electric energy is exchanged. To store electric energy, we must separate charges, yet keep them close together, which is what a *capacitor* does.

inductor, capacitor

Analogy between mechanical and electric energy. Magnetic and electric energy are the two forms of electrical energy arising from the two types of electrical force. In a mechanical system, we also have two types of force and two types of energy. Forces that depend on position, as in a spring or a gravitational field, store *potential energy*. Forces that depend on changes in velocity lead to *kinetic energy*. Potential energy and kinetic energy in a mechanical system are analogous to electric energy and magnetic energy in an electrical system. We must emphasize, however, that this is only an analogy. Magnetic energy is not kinetic energy; it is merely analogous to kinetic energy.

LEARNING OBJECTIVE 1.

To understand the properties of inductors and to be able to find the inductor voltage from the inductor current and vice versa

Inductor Basics

Physical inductor. Figure 3.1(a) shows a coil of wire, which acts as an inductor. When current flows in the wire, moving charges are close together, magnetic forces are large, and magnetic energy is stored.

Circuit-theory definition of an inductor. Figure 3.1(b) shows the circuit symbol for an inductor. The equation describing the voltage–current characteristics of an

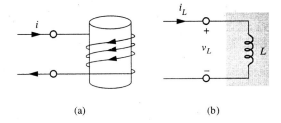

Figure 3.1 (a) A coil of wire gets moving charges close together and acts as an inductor. (b) A circuit model for an inductor.

inductor is based on Faraday's law of induction and may be stated in differential or integral form:

$$v_L(t) = + \frac{d}{dt} L i_L(t) \tag{3.1}$$

and

$$i_L(t) = i_L(0) + \frac{1}{L} \int_0^t v_L(t') dt' \tag{3.2}$$

The circuit symbol and the accompanying equations together define a circuit-theory model for an ideal inductor. Normally, the inductance, L, is considered a constant and brought outside the derivative in Eq. (3.1). The inductance of a coil may be measured or calculated from knowledge of coil geometry and material properties. Note in Fig. 3.1(b) that v_L and i_L form a load set.

Analogy with Newton's law. Equation (3.1) is analogous to Newton's second law in Fig. 3.2. Just as changes in the velocity of a mass require a force, or vice versa, changes in the current through an inductor produce a voltage, or vice versa. If the current is increased (positive di_L/dt), the inductor physical voltage opposes the change. If the current is decreased (negative di_L/dt), the inductor acts momentarily as a source polarized to keep the current going. Thus, the physical polarity of the inductor voltage will tend to keep the current constant, just as a mass will tend to maintain constant velocity. Equation (3.2) is analogous to determining the velocity by integrating the acceleration, which is proportional to an exciting force. We note that only changes in current can be determined from the voltage across the inductor, just as only changes in velocity can be computed from the force acting on a mass. To determine the current fully, we need to know the inductance, the voltage as a function of time, and the initial current, $i_L(0)$.

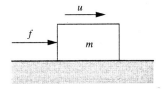

$$f = \frac{d}{dt}(mu)$$

Figure 3.2 An inductor is analogous to mass.

EXAMPLE 3.1 Battery-switch-inductor

Find the current for $t > 0$ in the circuit shown in Fig. 3.3 if the switch closes at $t = 0$.

SOLUTION:
With the switch open, the current in the circuit must be zero. With the switch closed, $v_L = V$. We may find the current for $t > 0$ by applying Eq. (3.2) with $v_L = V$:

$$i_L(t) = i_L(0) + \frac{1}{L}\int_0^t v_L(t')dt' = 0 + \frac{1}{L}\int_0^t V\,dt' = \frac{V}{L} \times t \tag{3.3}$$

Thus, the current increases linearly, as shown in Fig. 3.4. This circuit is analogous to a mass being accelerated by a constant force; theoretically the velocity will continue to increase so long as the force is applied.[1]

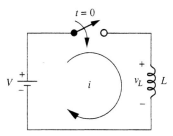

Figure 3.3 The inductor will integrate the voltage to give the current.

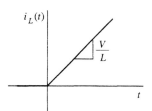

Figure 3.4 The current increases linearly because the inductor voltage is constant.

WHAT IF? What if the switch is opened?[2]

IDEA 2 · Conservation of Energy

Stored magnetic energy in an inductor. We pointed out that a load set of voltage and current is used to define the equation of an inductor. In Eq. (1.37), we compute the energy transferred by integrating the power. We thus can calculate the energy stored in the inductor by integrating its input power, $+v_L i_L$.

$$p_L = \frac{dW_m}{dt} = +v_L i_L = \left(L\frac{di_L}{dt}\right)i_L = \frac{d}{dt}\left(\frac{1}{2}Li_L^2\right) \tag{3.4}$$

$$W_m = \int p_L\,dt = \int \frac{d}{dt}\left(\frac{1}{2}Li_L^2\right)dt = \frac{1}{2}Li_L^2 \tag{3.5}$$

where p_L is the power into the inductor and W_m is the stored magnetic energy in the in-

---
[1] Until relativity effects must be considered.
[2] This is a tricky question and will be discussed later in the chapter. For now, suffice it to say that a spark would occur at the switch as it was opened.

henry,
millihenry,
microhenry

Units of inductance. The inductance, L, depends on coil dimensions and the number of turns of wire in the coil. The inductance also depends on the material (if any) located near the coil. In particular, the magnetic properties of iron increase greatly the inductance of a coil wound on an iron core. The units of inductance are volt-seconds per ampere, but to honor Joseph Henry (1797–1878), we use the name *henry* (H) for this unit; *millihenries* (10^{-3} H, or mH) and *microhenries* (10^{-6} H, or µH) are also in common use.

EXAMPLE 3.2 Stored magnetic-energy calculation

Use the definition of an inductor to compute the voltage across, the power into, and the energy stored in a 2-H inductor with the current shown in Fig. 3.5.

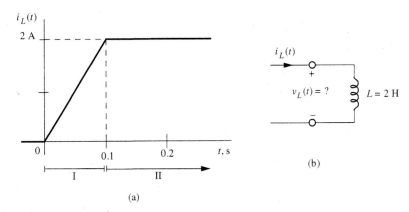

Figure 3.5 A changing current will produce a voltage.

SOLUTION:
Because the current function is piecewise continuous, we cannot represent it by a single mathematical formula, but rather must represent it separately in its several regions. During interval I, $0 < t < 0.1$ s, the slope is constant at 20 A/s; thus, during this period, the inductor voltage will be 2 H × (+20 A/s), or +40 V. During interval II, 0.1 s $< t < \infty$, the slope is zero and hence the voltage will be zero also.

When we compute the power into the inductor as the product of v_L and i_L, we note that the power is positive during interval I and zero during II. During interval I, $i_L = +20t$ and hence:

$$p_L(t) = +v_L i_L = \frac{d}{dt}(+2 \times 20t) \times 20t = +800t \text{ W} \tag{3.6}$$

where we have used Eq. (3.1) to calculate the inductor voltage. The magnetic energy stored in the inductor can be computed by integrating the input power, but the easier way uses Eq. (3.5), with the results shown in Fig. 3.6. During interval I, when the current is increasing in magnitude, we must supply power to the inductor to increase the stored magnetic energy.

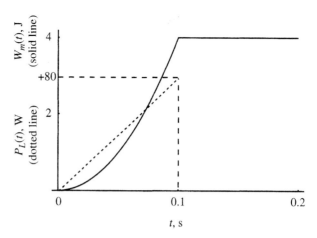

Figure 3.6 The power curve (dotted line) is the slope of the stored energy.

$$W_m(t) = \frac{1}{2} Li_L^2 = \frac{1}{2} \times 2(20t)^2 = 400t^2 \text{ J} \tag{3.7}$$

During interval II, the current is constant and hence no energy is exchanged between inductor and source: The system "coasts forever" like a mass in constant motion.

WHAT IF? What if the times in Fig. 3.5 were doubled such that the current increased from 0 to 2 A in 0.2 seconds, etc.? Which of the following would not change in magnitude: inductor voltage, inductor power, final inductor stored energy?[3]

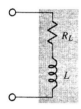

Figure 3.7 The circuit model for a coil of wire must contain a resistance to account for loss.

 Conservation of Energy

 Equivalent Circuits

Figure 3.8 Inductors in series add like resistors in series.

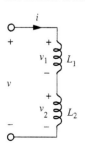

Model for a real inductor. The wire in a physical inductor would have resistance and would not store all the energy delivered to it. Some of the energy would heat the wire and be lost to the electrical circuit. This energy is not lost to the entire physical system because it appears as thermal energy. The equivalent circuit model for a real coil of wire, shown in Fig. 3.7, accounts for loss as well as the storage of magnetic energy.

An inductor that uses a ferromagnetic material such as iron to increase the magnetic energy storage will have losses and possibly some nonlinearities from the iron. For such inductors, the simple linear model in Fig. 3.7 may be inadequate.

Inductors in series and parallel. Figure 3.8 shows two inductors in series, which we may replace by a single equivalent inductor. KVL yields

$$v = v_1 + v_2 = \frac{dL_1 i}{dt} + \frac{dL_2 i}{dt} = \frac{d(L_1 + L_2)i}{dt} = \frac{dL_{eq} i}{dt} \tag{3.8}$$

where $L_{eq} = L_1 + L_2$ (series inductors). Thus, inductors in series add like resistors in series. Similarly, one can show that inductors in parallel add like resistors in parallel,

---

[3] The inductor voltage and power depend on rates of change, but the final stored energy depends only on the current and would not change in magnitude.

$$L_{eq} = L_1 \| L_2 = \frac{1}{1/L_1 + 1/L_2} \quad \text{(parallel inductors)} \tag{3.9}$$

Capacitor Basics

Physical capacitors. Electric (electrostatic) forces arise from interactions between separated charges, and the associated energy is called electric energy. To store electric energy, we must separate charges, as in Fig. 3.9.

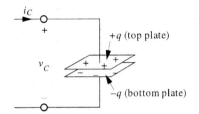

Figure 3.9 Structure having capacitance.

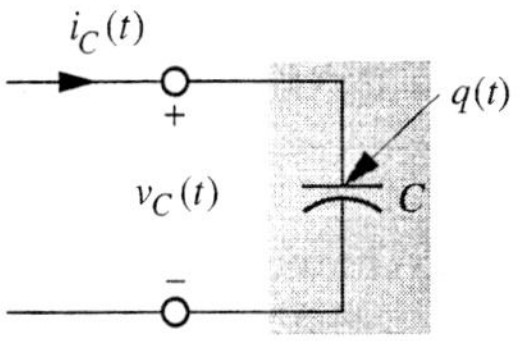

Figure 3.10 Circuit symbol for capacitance.

LEARNING OBJECTIVE 2.

To understand the properties of capacitors and to be able to find the capacitor voltage from the capacitor current and vice versa

Circuit-theory definition of a capacitor. Figure 3.10 shows the circuit symbol for a capacitance. Capacitance is defined as the constant relating charge and voltage in a structure that supports a charge separation. If q is the charge on the + side of the capacitor and the voltage is v_C, as shown in Fig. 3.9, the capacitance, C, is defined as

$$q = Cv_C \tag{3.10}$$

For every charge arriving at the + side of the capacitor, a charge of like sign will depart from the − side and the structure as a whole will remain charge-neutral. Thus, KCL will be obeyed because charge flowing into the + terminal side is matched by charge flowing out of the − terminal, as for a resistor. If current is positive into the + terminal, positive charge will accumulate there and will be increasing in proportion to the current. From the definition of current, we can relate the charge in Eq. (3.10) and current as follows:

$$i_C = \frac{dq(t)}{dt} \tag{3.11}$$

where $q(t)$ is the charge in the + side of the capacitor. Thus, we may define the relationship between current and voltage for a capacitor as

$$i_C = \frac{d}{dt} Cv_C(t) = C \frac{dv_C(t)}{dt} \tag{3.12}$$

where the last form of Eq. (3.12) is valid if the capacitance is constant. Note that we have used a load set in Fig. 3.10 for the voltage and current variables.

farad, microfarad, nanofarad, picofarad

Units of capacitance. The unit of capacitance is the coulomb/volt, but to honor Michael Faraday (1791–1867), we use the name *farad* (F). Realistic capacitor values come small and usually are specified in microfarads (10^{-6} F, or μF), nanofarads (10^{-9} F, or nF), or picofarads (10^{-12} F, or pF). When a capacitor is constructed from parallel

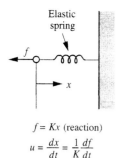

Figure 3.11
Mechanical analog for capacitance.

plates, as in Fig. 3.9, the capacitance depends on the area, separation, and material (if any) between the plates.

Mechanical analog of capacitance. The mechanical analog of a capacitor is a spring, as shown in Fig. 3.11. The analog of velocity is current and thus displacement of the spring is analogous to charge. The voltage across a capacitor is analogous to the force produced by the spring. Comparison of the second equation in Fig. 3.11 with Eq. (3.12) shows that capacitance corresponds to the inverse of the stiffness constant, K, and thus capacitance is analogous to the compliance of a spring.

EXAMPLE 3.3 Equation of RC circuit

Derive the equation for the current in Fig. 3.12(a). Assume $v_C(0) = 0$.

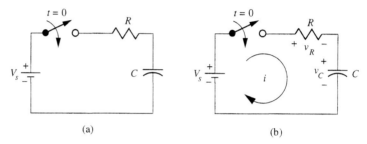

Figure 3.12 (a) Closing the switch will cause a momentary current. (b) The same circuit with variables defined.

SOLUTION:
Figure 3.12(b) defines the capacitor and resistor voltages in a load-set convention. With the switch closed, KVL is

$$-V_s + v_R(t) + v_C(t) = 0 \tag{3.13}$$

We may express both voltages in terms of the current if we differentiate Eq. (3.13):

$$0 + \frac{d}{dt}Ri(t) + \frac{dv_C(t)}{dt} = R\frac{di(t)}{dt} + \frac{i(t)}{C} = 0 \tag{3.14}$$

The last form in Eq. (3.14) is readily manipulated to the form

$$\frac{di}{i} = -\frac{1}{RC}dt \Rightarrow \ln i = -\frac{t}{RC} + \ln A \Rightarrow i(t) = Ae^{-t/RC} \tag{3.15}$$

where A is a constant. The current will charge the capacitor until the capacitor voltage is equal to the battery voltage. Equation (3.16) equates the final charge on C to the integral of the current.

$$q(\infty) - q(0) = CV_s = \int_0^\infty i(t)dt = \left.\frac{Ae^{-t/RC}}{-1/RC}\right|_0^\infty = A(RC) \qquad (3.16)$$

Because $v_C(0) = 0$, $q(0) = 0$, and the final charge is

$$q(\infty) = CV_s = A(RC) \qquad (3.17)$$

Hence

$$i(t) = \frac{V_s}{R} e^{-t/RC} \qquad (3.18)$$

WHAT IF? What if the switch is opened after being closed a long time? What would happen to the capacitor voltage?[4]

Integral i–v equation. Equation (3.12) is useful in determining the current through a capacitor, given the voltage as a function of time. If we know the current and wish to determine the voltage, we must integrate. Let us consider that we know the capacitor voltage at some time, say, $t = 0$, and wish to determine the voltage at a later time, t. We can integrate Eq. (3.12) from 0 to t, with the result

$$v_C(t) = v_C(0) + \frac{1}{C} \int_0^t i_C(t')dt' \qquad (3.19)$$

We have used t' as the dummy variable of the integration process because t is one limit of the integral.

EXAMPLE 3.4 **Capacitor discharge and charge**

As an example of the use of Eq. (3.19), consider a capacitor with current, as shown in Fig. 3.13.

Here we know the voltage at the beginning time, $v_C(0) = -2$ V, and the current through the capacitor. We wish to compute the voltage for all $t > 0$. We will use Eq. (3.19), noting that we must divide the integration into several regions due to the piecewise nature of the current function. During interval I, $0 < t < 1$ ms, the current is zero; hence, the voltage does not change. During interval II, the slope of the current is 0.1 A/1 ms = 100 A/s; hence, the current is

---

[4] No current could flow after the switch opened, and according to Eq. (3.12), the voltage would remain constant. A real capacitor would discharge over time due to leakage current.

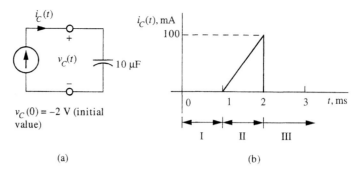

$v_C(0) = -2$ V (initial value)

(a)   (b)

Figure 3.13 (a) The capacitor has an initial voltage. (b) Current into the capacitor for positive time.

$$i_C(t) = 100(t - 1\text{ ms}) \text{ A}, \quad 1\text{ ms} < t < 2\text{ ms} \tag{3.20}$$

Thus, during interval II, the voltage will be

$$v_C(t) = -2.0 + \frac{1}{10\mu\text{F}} \int_{1\text{ ms}}^{t} 100(t' - 1\text{ms})dt' \Big|_{\text{II: } 1 < t < 2\text{ ms}}$$

$$= -2.0 + \frac{10^{+5} \times 100(t - 1\text{ms})^2}{2} \text{ V} \tag{3.21}$$

which indicates a parabolic increase, as shown in Fig. 3.14. The final value at $t = 2$ ms is $+3$ V, which is easily checked from the area under the current curve:

$$\Delta v_C = \frac{\Delta q}{C} = \frac{\text{area under current curve}}{C} = \frac{\frac{1}{2} \times 1\text{ms} \times 100 \text{ mA}}{10\ \mu\text{F}} = +5 \text{ V} \tag{3.22}$$

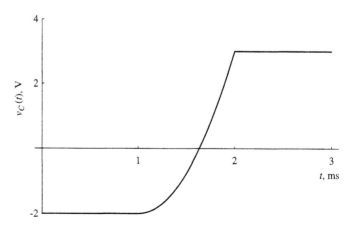

Figure 3.14 Capacitor voltage for positive time.

After the current pulse, the capacitor voltage remains constant because the current is zero. This

> constant voltage would hold for an ideal capacitor; a physical capacitor would discharge eventually due to leakage current.
>
> **WHAT IF?** What if the current were constant at 100 mA for 1 ms $< t <$ 2 ms? What then would be the final value?[5]

Mechanical analogy. The process of integrating current through a capacitor to determine voltage is analogous to integrating velocity to determine change in force in a spring problem. The only difference is that the value of the capacitance scales the integral of the current, whereas the scale factor depends on K, the spring constant, in the mechanical analog.

Stored energy in a capacitor. The charge separation in a capacitor stores electric energy. This energy is analogous to potential energy stored in a stressed spring. We may derive the stored electric energy in a capacitor by integrating the power into the capacitor,

$$W_e = \int p_C \, dt = \int v_C \times C \frac{dv_C}{dt} dt = \int d\left(\frac{1}{2} C v_C^2\right) = \frac{1}{2} C v_C^2 = \frac{1}{2} \frac{q^2}{C} \quad (3.23)$$

where p_C is the power into the capacitor and W_e is the stored electric energy in the capacitor. The constant of integration must be zero because the uncharged capacitor stores no energy. We see from Eq. (3.23) that the stored energy depends uniquely on the voltage, or the charge, and the capacitance. For example, if we take a 10-μF capacitor and connect it briefly to a 12.6-V battery, the capacitor will receive $0.5 \times 10^{-5}(12.6)^2 = 7.94 \times 10^{-4}$ J from the battery. Although this is not much energy, it would take only about 1 μs to charge the capacitor. Hence, the rate of energy flow would be rather high, about 800 W.

IDEA 6: Equivalent Circuits

Capacitors in series and parallel. Figure 3.15 shows two capacitors in series. KVL yields

$$v(t) = v_1 + v_2 = \frac{1}{C_1} \int_0^t i \, dt + \frac{1}{C_2} \int_0^t i \, dt = \frac{1}{C_{eq}} \int_0^t i \, dt \quad (3.24)$$

where

$$C_{eq} = C_1 \| C_2 = \frac{1}{1/C_1 + 1/C_2} \quad \text{(series capacitors)}$$

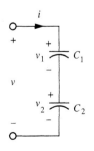

Figure 3.15
Capacitors in series add like resistors in parallel: $C_{eq} = C_1 \| C_2$.

Thus, capacitors connected in series combine numerically like resistors in parallel. Similarly, we can easily show that capacitors in parallel add like resistors in series, that is,

$$C_{eq} = C_1 + C_2 \quad \text{(parallel capacitors)} \quad (3.25)$$

Mechanical analog for resistance. A mechanical analog for a resistance is frictional loss of a special type. Voltage is analogous to force, and current is analogous to

---
[5] +8 V.

TABLE 3.1 Summary of Mechanical and Electrical Analogs	
Mechanical	Electrical
Force	Voltage
Velocity	Current
Displacement	Charge
Mass	Inductance
Spring compliance	Capacitance
Shock absorber	Resistance

velocity. Voltage/current, or resistance, would thus imply a force that is proportional to velocity. We experience such a force when we try to move underwater. A common mechanical component having this property would be an automotive shock absorber. Table 3.1 summarizes the analogies between mechanical and electrical quantities.

Check Your Understanding

1. An inductor has a stored energy of 5 J and an inductance of 0.1 H. What is the current through the inductor?
2. If an ideal 10-V battery were connected for 1 s to an ideal 1-H inductor, how much energy would be given to the inductor?
3. A capacitor has a stored energy of 500 µJ and a capacitance of 0.15 µF. What is the voltage across this capacitor?
4. What is the mechanical analog of inductance?

Answers. (**1**) 10 A; (**2**) 50.0 J; (**3**) 81.6 V; (**4**) mass.

3.2 FIRST-ORDER TRANSIENT RESPONSE OF *RL* AND *RC* CIRCUITS

transients, first-order transient

First-order transients. In the remainder of this chapter, we show how to analyze an important class of circuit problems. Typical situations are shown in Fig. 3.16. These circuits are characterized by having the following:

- A single energy storage element, one capacitor or inductor
- Loss, represented by one or more resistors
- Constant sources
- A switch[6] that either opens or closes at a known time, usually $t = 0$, causing a sudden change in the circuit
- Only linear components.

In all cases, the circuit will have one dc state before the switch action and another dc state long after the switch action. Consequently, the state of the circuit goes through a transition. Because this transition lasts for a brief period of time, these problems are often called *transient* problems. We consider primarily circuits with one energy storage element, which lead to *first-order transients*.

---
[6] The pulse in Fig. 3.16(c) can be considered a voltage switched ON, then OFF.

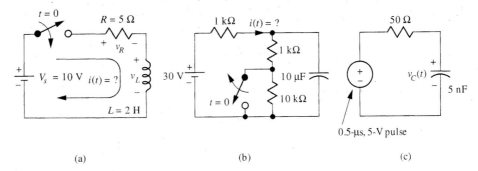

Figure 3.16 Each circuit has one energy storage element, a constant source, and a sudden change.

These problems are important because they represent switching something on or off, which is often a critical period in the operation of a device. Also included in this class of problems are digital signals, such as a computer uses in processing information. Such digital signals can often be represented as a dc source being turned ON and OFF.

system

A *system* consists of several components that work together to accomplish some purpose. Thus, an electric circuit is a system. The circuits we will consider are linear, so our results are typical of linear systems. Thus, in this section, we use some of the vocabulary associated with linear systems. Chapter 5 deals more fully with system theory.

Classical Differential Equation Solution

Deriving the differential equation. Our approach will be first to analyze a representative problem using the techniques of differential equations (DEs). Then we will develop a much simpler, more physical method that can also be applied to electrical, mechanical, thermal, or chemical problems that fall into the class described before. We now analyze the circuit of Fig. 3.16(a). With the switch open, the full 10 V appears across the switch because there is no current to cause a voltage across the resistor or the inductor. After the instant at which the switch is closed, KVL for the circuit is

$$-V_s + v_R(t) + v_L(t) = 0 \tag{3.26}$$

Equation (3.26) becomes a DE when we express v_R and v_L in terms of the unknown current, $i(t)$.

$$L\frac{di(t)}{dt} + Ri(t) = V_s, \qquad t > 0 \tag{3.27}$$

steady-state or forced response

Form of the solution. Equation (3.27) is a linear DE with constant coefficients and a constant forcing term on the right side. The solution of a DE of this type usually proceeds by separating the unknown solution into two parts. In mathematics terminology, the parts are called the homogeneous part and the particular integral. In engineering terminology, the particular integral, the part of the solution identified with the voltage term on the right side of Eq. (3.27), is usually called the *forced response* or the *steady-state response*. In this problem, and all problems we solve in this chapter, this response must be constant because the forcing function is constant in time.

homogeneous solution, natural or transient response

The *homogeneous solution* is determined by the left side of the equation, with the right side equal to zero. This part of the solution, also called in engineering terminology the *natural* or *transient response*, satisfies the equation with the forcing function set to zero. A linear DE with constant coefficients is always satisfied by a function of the form $e^{\alpha t}$, where α is an unknown constant with the dimensions of time^{-1}. The general solution to Eq. (3.27), therefore, must be of the form

$$i(t) = \underbrace{A}_{\text{forced response}} + \underbrace{Be^{\alpha t}}_{\text{natural response}} \tag{3.28}$$

where A, B, and α are unknown constants to be determined from Eq. (3.27) and the initial conditions of the circuit.

Determining the unknown constants. We can determine A and α by substituting Eq. (3.28) back into Eq. (3.27).

$$LB(\alpha)e^{\alpha t} + R(A + Be^{\alpha t}) = V_s \tag{3.29}$$

The coefficient of the exponential term must vanish if the equation is valid for all times.

$$B(L\alpha + R)e^{\alpha t} + AR = V_s \Rightarrow \alpha = -\frac{R}{L} = -2.5 \text{ s}^{-1} \tag{3.30}$$

time constant

In engineering contexts, it is customary to express the exponential term in Eq. (3.28) in the form $e^{-t/\tau}$, where

$$\tau = -\frac{1}{\alpha} = \frac{L}{R} = \frac{2}{5} = 0.4 \text{ seconds} \tag{3.31}$$

for the circuit in Fig. 3.16(a). The *time constant*, τ, is a characteristic time for the transient; we explore its significance shortly. Setting the coefficient of the exponential term to zero in Eq. (3.30) leads also to the value of A, the steady-state response.

$$AR = V_s \Rightarrow A = \frac{V_s}{R} = \frac{10}{5} = 2 \text{ A} \tag{3.32}$$

Initial condition. To determine B, we must consider the initial condition. The initial condition for this, and for all such systems, arises from consideration of energy. Time is required for energy to be exchanged and hence processes involving energy carry the system from one state to another, particularly when, as here, a sudden change occurs.

With the switch open, there is no current and no stored magnetic energy in the inductor. The closing of the switch will allow the inductor to store energy, but at the instant after the switch is closed, the stored energy must still be zero. Zero energy implies zero current because the stored energy in an inductor, Eq. (3.5), is $\frac{1}{2}Li^2$; hence, $i(0^+)$ is zero, where 0^+ denotes the instant after the switch is closed. This condition leads to the value for B:

$$0 = A + Be^{-0/\tau} \Rightarrow B = -A = -2 \text{ amperes} \tag{3.33}$$

The final solution is given in Eq. (3.34) and plotted in Fig. 3.17.

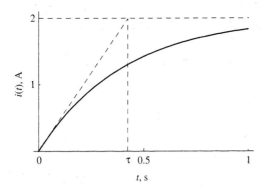

Figure 3.17 Current for circuit in Fig. 3.16(a).

$$i(t) = \frac{V_s}{R} - \frac{V_s}{R}e^{-t/\tau} = 2 - 2e^{-t/0.4 \text{ s}} \text{ A} \tag{3.34}$$

Physical interpretation of the solution. The mathematical solution is complete, but we wish to examine carefully the physical interpretation of the response, for we will develop our simpler method from a physical understanding of this type of problem.

The current is zero before the switch is closed and approaches V_s/R as a final value. This final value does not depend on the value of the inductor, but would be the current if no inductor were in the circuit. The inductor cannot matter in the end because we have a dc excitation in the circuit and eventually the current must reach a constant value. The constant current renders the inductor to act as a short circuit because an inductor will exert itself only when its current is changing.

Without the inductor, the current would change instantaneously from zero to V_s/R when the switch is closed. The inductor effects a smooth transition between the initial and final values of the current. The smooth change moderated by the inductor takes place over a *characteristic time* of τ, the *time constant*.

Note from Fig. 3.17 that the current initially increases with a rate as if to arrive at the final value in one time constant. This time constant for the *RL* circuit, $\tau = L/R$, can be understood from energy considerations. The inductor will require little energy if L is small or R is large; hence, the transition will be rapid for small L/R. But if the inductor is large or R is small, much energy will eventually be stored in the inductor; hence, the transition period will be much longer for large L/R.

Summary. We see that the response consists of a transition between two constant states, an initial state and a final state. The energy storage element, the inductor, effects a smooth transition between these states. Because of the form of the DE for this class of problems, the transition between states is exponential and is characterized by a time constant. The stored energy in the inductor carries the state of the circuit across the instant of sudden change and leads directly or indirectly to the initial condition.

A Simpler Method

We will now rework this problem using a more direct method. The goal is to write the circuit response directly based on physical understanding. There are four steps.

> **LEARNING OBJECTIVE 3.**
>
> To understand how to analyze first-order transients by the initial-value/final-value method

1. **Find the time constant.** We already know the time constant for this type of problem, $\tau = L/R$, but in general, we can determine the time constant for a first-order system by putting the DE into the special form of Eq. (3.35). We divide by the coefficient of the linear term, so that x appears in the equation without a multiplier.

$$\tau \frac{dx}{dt} + x = \text{constant} = x_\infty \qquad (3.35)$$

In Eq. (3.35), x refers to the physical quantity being determined. It would represent a voltage or current in a circuit problem, but in other physical systems, x might represent a temperature, a velocity, or something else. When we put the DE in this form, τ will always be the characteristic time, the time constant. To put Eq. (3.27) into this form, we have to divide by R to see that $\tau = L/R$ by this method.

Because the purpose of this method is to avoid DEs, we hesitate to suggest that you must write the DE to get started. Usually, you will know the equation of the time constant for the circuit or system from previous experience. For an RL circuit, for example, the time constant is always L/R. Because we solve circuits having only one energy storage element with this method, we have only one L. In Sec. 3.3, we show how to handle circuits with more than one resistor. Returning to the problem we are solving, we now know our time constant: $L/R = 0.4$ s for the circuit in Fig. 3.16(a).

2. **Find the initial condition.** The initial condition (x_0 in general, i_0 in this case) always follows from consideration of the energy condition of the system at the beginning of the transient. In this case, we argued earlier that because the inductor had no energy before the switch was closed, it must have no energy the instant after the switch is closed. Hence, the initial current is zero ($i_0 = 0$). Later, we discuss generally how to determine initial values in circuits.

3. **Find the final value.** The final value (x_∞ in general, i_∞ in this case) follows from the steady-state solution of the system. Earlier, we argued that the inductor approaches a short circuit. This is true because the final state of the circuit is a dc state, and the inductor has no voltage across it for a constant current. Thus, application of Ohm's law to the circuit in Fig. 3.16(a) shows that the final current must be $i_\infty = 2$ A.

If the DE has been written and put into the form of Eq. (3.35), then the final value is the constant on the right-hand side of the DE. In the final state, time derivatives must vanish and Eq. (3.35) reduces to $x = x_\infty$.

4. **Substitute the time constant and the initial and final values into a standard formula.**

$$x(t) = x_\infty + (x_0 - x_\infty)e^{-t/\tau} \qquad (3.36)$$

Equation (3.36) is the solution for all first-order systems having dc (constant) excitation. In our case, the x_0 and x_∞ are currents of known numerical value, so our solution is

$$i(t) = \frac{V_s}{R} + \left(0 - \frac{V_s}{R}\right)e^{-t/\tau}$$

$$= 2 + (0 - 2)e^{-t/0.4} \text{ A} \qquad (3.37)$$

As before, we see that as $t \to \infty$, $i(t) \to V/R = 2$ A, and initially $i(0) = 0$ because $e^0 = 1$. Equation (3.37) describes a *rising* exponential because the current begins at zero and increases to a final value.

Figure 3.18 shows a generalized plot of Eq. (3.36). The curve starts at x_0 and asymptotes to x_∞. Its initial slope is such as to move from x_0 to x_∞ in one time constant, but it only reaches $(1 - e^{-1})$ or 63% of the way during the first time constant. You might note that we have shown a discontinuity at the origin. This can happen in general, although it does not happen in our present example. We show a *falling* exponential in Fig. 3.18 because the curve decreases to a final value. The sign of the exponential determines whether the curve rises or falls.

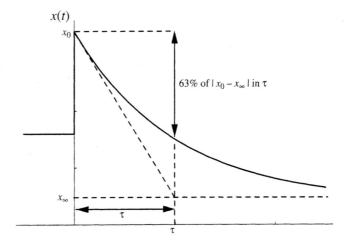

Figure 3.18 Generalized response in Eq. (3.36).

Summary. Our direct method consists of determining three constants and substituting these into a standard formula, Eq. (3.36). The time constant can be determined from the DE, but usually the formula for τ is known from prior experience. The other two constants are the initial value of the unknown, which is determined from energy considerations, and the final value, which is determined from the dc (static) solution. The energy storage element effects a smooth transition between the initial and final values and influences the initial value through energy considerations.

EXAMPLE 3.5 Same circuit, find the voltage

Determine the voltage across the inductor in Fig. 3.16(a).

SOLUTION:
We already know the time constant. The initial condition $v_L(0^+)$ follows indirectly from energy

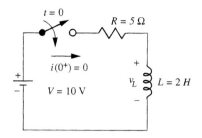

Figure 3.19 Because $i(0^+) = 0$, $v_L = V_S$.

considerations. The initial energy in the inductor is zero, which implies zero current. Consider the state of the circuit in Fig. 3.16(a) at the instant after the switch is closed, as shown in Fig. 3.19. Because of the inductor, there can be no current at this initial instant; hence there can be no voltage across the resistor. If there is no voltage across the resistor and no voltage across the switch, the full 10 V of the battery must appear across the inductor. Hence, the initial value of the inductor voltage is 10 V.

The final value of the inductor voltage must be zero because we have a dc source and eventually the current becomes constant. Consequently, di/dt approaches zero, and the inductor voltage must also approach zero. We now know the time constant, the initial value, and the final value, and therefore we can write the full solution with our standard form,

$$v_L(t) = 0 + (10 - 0)\, e^{-t/0.4} = 10 e^{-t/0.4} \text{ V} \tag{3.38}$$

The plot of Eq. (3.38) is shown in Fig. 3.20. Notice that we have shown a discontinuity in the inductor voltage at the origin. Before the switch closes, the inductor has no voltage. When the switch closes, the current begins to increase, but the inductor voltage instantly jumps to 10 V to oppose that increase. The initial rate of increase of the current is limited to

$$10 = 2 \left.\frac{di}{dt}\right|_{t=0^+} \Rightarrow \frac{di}{dt} = \frac{10}{2} = 5 \,\frac{\text{A}}{\text{s}} \tag{3.39}$$

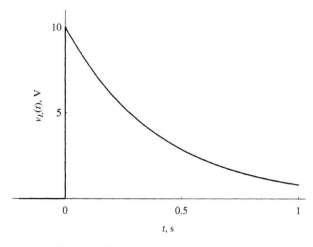

Figure 3.20 Inductor voltage in Fig 3.16(a).

As the current increases, the voltage across the resistor increases accordingly, and the voltage across the inductor decreases. Finally, the current increases to a level where it is limited by the resistor. Thus, the inductor dominates the beginning and the resistor dominates the end of the transient.

WHAT IF? What if we want the time when the stored energy in the inductor is one-half its final value?[7]

RC Circuits

We have already written the DE for the RC circuit shown in Fig. 3.12(a) in Eq. (3.14). There we solved the problem by direct mathematical analysis. Here we will merely extract from that work the time constant for an RC circuit. We may put the last form of Eq. (3.14) into the required form for Eq. (3.35) by multiplying by C. The result is

$$RC \frac{di(t)}{dt} + i(t) = 0 \qquad (3.40)$$

and hence the time constant is

$$\tau = RC \qquad (3.41)$$

This will be the time constant for *all* RC circuits whether the resistor and capacitor are in series or parallel.

Armed with this result, we will now apply our direct method to the RC circuit shown in Fig. 3.12(a) with $R = 1000\ \Omega$, $C = 1\ \mu F$, and $V_s = 100\ V$, which we repeat in Fig. 3.21. We will determine the voltage across the resistor, $v_R(t)$, assuming zero initial energy in the capacitor.

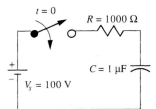

Figure 3.21 Figure 3.12(a) repeated with $C = 1\ \mu F$ and $R = 1000\ \Omega$.

Time constant. From Eq. (3.41), the time constant for this circuit $\tau = RC = 1000 \times 10^{-6} = 1$ ms.

Find the initial condition. The initial condition emerges from consideration of energy. The electric energy stored in a capacitor is given in Eq. (3.23) as $\frac{1}{2}Cv_C^2$. The capacitor is unenergized at $t = 0^-$, the instant before the switch is closed. Hence it remains unenergized at $t = 0^+$, the instant after the switch is closed, because finite time

---

[7] 0.491 s.

is required for energy to be stored in the capacitor. At $t = 0^+$, therefore, KVL around the loop is

$$-100 + v_R(0^+) + \underbrace{v_C(0^+)}_{0} = 0 \quad \Rightarrow \quad v_R(0^+) = 100 \text{ V} \tag{3.42}$$

Find the final value. The final value of the resistor voltage follows from the requirement that the voltage and current eventually become constant. Because dv_C/dt is zero, i is zero, and v_R must therefore be zero as $t \to \infty$.

Substitute the time constant and the initial and final values into the standard form. Now that we know the time constant, the initial value, and the final value for v_R, we substitute into the standard form of Eq. (3.36).

$$v_R(t) = 0 + (V_s - 0)e^{-t/\tau}$$
$$= 100 e^{-t/1 \text{ ms}} \text{ V} \tag{3.43}$$

Equation (3.43) is plotted in Fig. 3.22. As the current flows, charge accumulates and builds up voltage across the capacitor. The voltage across the resistor must diminish with time and eventually vanish. When the full voltage of the battery appears across the capacitor, the current stops.

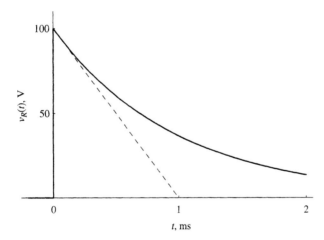

Figure 3.22 Resistor voltage in Fig 3.21.

Summary. We have developed a direct method for analyzing first-order transient circuits involving one resistor and one inductor or capacitor. The method consists of determining the time constant, initial value, and final value of the unknown, and substituting into a standard formula. In the next section, we extend and generalize this method.

3.3 ADVANCED TECHNIQUES

Equivalent Circuits

Circuits with Multiple Resistors

Thévenin equivalent circuit. Circuits with one inductor or capacitor but more than one resistor can be analyzed with the concept of equivalent circuits. The circuit in

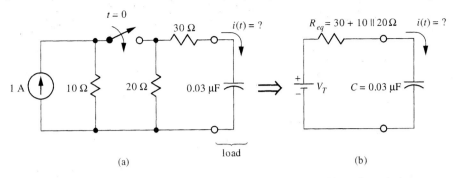

Figure 3.23 (a) The three resistors can be combined using a Thévenin equivalent circuit; (b) the time constant is easily identified in this equivalent circuit.

Fig. 3.23(a), for example, can be reduced to the equivalent circuit in Fig. 3.23(b). From this equivalent circuit, clearly the time constant is $R_{eq}C$, where R_{eq} is the output impedance of the circuit with all sources turned OFF and the switch closed. Thus, R_{eq} is the output impedance of the circuit to the capacitor as a load. In this case, R_{eq} is $30 + 20 \parallel 10 = 36.7 \, \Omega$, and thus τ is $36.7 \times 0.03 \times 10^{-6}$ or $1.1 \, \mu s$.

We may or may not be interested in the Thévenin equivalent circuit as a means for calculating the initial or final values of the unknown. That is a separate problem to be approached by the most efficient method. It is the *concept* of the Thévenin circuit, specifically the concept of the output impedance, or impedance level, that leads to the time constant for a transient circuit with multiple resistors. In general, the relevant resistance for the time constant is the output resistance presented to the energy storage element as a load.

EXAMPLE 3.6 Three R's and a C

Find the time constant for the circuit of Fig. 3.16(b) for $t > 0$.

SOLUTION:
Here we clearly should not replace the circuit external to the capacitor with a Thévenin equivalent circuit, for then we would eliminate the unknown, $i(t)$. We invoke the concept of the equivalent circuit only to calculate the equivalent resistance seen by the capacitor with the switch open, $(1 \, k\Omega + 10 \, k\Omega) \parallel 1 \, k\Omega$ or $917 \, \Omega$. Thus, $\tau = 917 \times 10 \, \mu F = 9.17 \, ms$. We continue this example in what follows.

WHAT IF? What would be the time constant if the switch were closed at $t = 0$?[8]

initial and final values

Initial and Final Values

What we mean by "initial" and "final." We now look generally at how to calculate the initial and final values. By *initial* we mean the instant after a change occurs

---
[8] $\tau = 5 \, ms$.

in the circuit, usually a switch closing or opening. This does not have to be the time origin, although often we define $t = 0$ as the time of switch action. By *final* we mean the steady-state condition of the circuit, its state after a large period of time. The final state may be hypothetical because the circuit may never reach that state due to subsequent switch action. We may first close a switch and then open it before the circuit reaches the final state. The circuit cannot anticipate the second switch action and hence reacts to the first switch action as if it *would* reach steady state. In this section, we develop guidelines for determining the final and initial states of the circuit.

Determining final values. Final values arise out of the eventual steady state of the circuit. All time derivatives must eventually vanish. Consequently, the current through capacitors and the voltage across inductors must approach zero, as suggested in

$$v_C \to \text{constant} \Rightarrow i_C = C\frac{dv_C}{dt} \to 0 \quad \text{as } t \to \infty$$
$$i_L \to \text{constant} \Rightarrow v_L = L\frac{di_L}{dt} \to 0 \quad \text{as } t \to \infty$$
(3.44)

Thus, capacitors act as open circuits and inductors act as short circuits in establishing final values, as shown in Figs. 3.24 and 3.25, respectively.

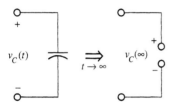

Figure 3.24 The capacitor behaves as an open circuit as time becomes large.

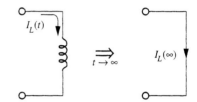

Figure 3.25 The inductor behaves as a short circuit as time becomes large.

Determining initial values. Initial values always follow from energy considerations. The fundamental principle is that the stored electric energy in a capacitor and the stored magnetic energy in an inductor must be continuous. From continuity of energy, we conclude that the voltage across a capacitor and the current through an inductor must be continuous functions of time. Thus, we always calculate capacitor voltage or inductor current before the switch is thrown and then carry this value over to the moment after the switch is thrown. From these known values of the capacitor voltage or inductor current, required circuit unknowns can be calculated.

A model for an energized capacitor at the instant after the switch action is shown in Fig. 3.26. For that first instant, the charged capacitor acts like a voltage source because the voltage across the capacitor cannot change instantaneously. As current flows through the capacitor, its voltage will change, and hence the capacitor acts as a battery *only* at that first instant. Similarly, a current source models an energized inductor at the instant after switch action, as shown in Fig. 3.27. Although these models are valid only for the first instant of the new regime, they suffice for the calculation of the initial values of the circuit unknowns of interest.

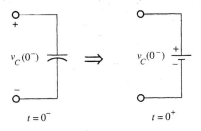

Figure 3.26 A battery models the initial voltage of the capacitor.

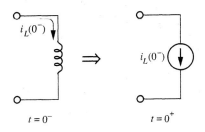

Figure 3.27 A current source models the initial current of the inductor.

EXAMPLE 3.7 Three R's and an L

Find the voltage across the inductor in Fig. 3.28 after the switch is opened at $t = 0$.

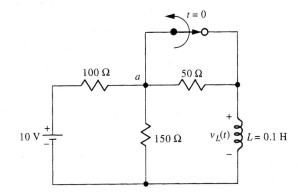

Figure 3.28 After being closed a long time, the switch is opened.

SOLUTION:

The switch is opened after having been closed for a time long enough to establish steady state throughout the circuit. We determine the voltage across the inductor after the switch is opened. We first calculate the inductor current at the instant before the switch is opened. We note that with the switch closed and with the circuit in steady state, the 50-Ω resistor is short-circuited by the switch and the 150-Ω resistor is shorted by the switch and the dormant inductor. Thus, the current from the battery is 10 V/100 Ω or 0.1 A and this current flows through the switch and inductor.

This information does not give us the initial value of $v_L(t)$ directly, but it leads indirectly to the initial value through the analysis of the circuit in Fig. 3.29. We have replaced the inductor by a current source and eliminated the switch because it is now an open circuit. As you can see, we have labeled the circuit to suggest solution by nodal analysis. You may confirm that v_a is zero at $t = 0^+$. From this information, we can calculate v_L from KVL around the right-hand loop, with the result that $v_L(0^+) = -5$ V.

The time constant is L/R_{eq}, where R_{eq} is the equivalent resistance seen by the inductor with the switch open. Thus, R_{eq} is $50 + 150 \parallel 100$ or $110 \, \Omega$, and the time constant is $0.1/110$ or 909 μs. Because the final value of the inductor voltage is zero, the full solution is

$$v_L(t) = 0 + (-5 - 0)e^{-t/909 \, \mu s} = -5e^{-t/909 \, \mu s} \text{ V} \tag{3.45}$$

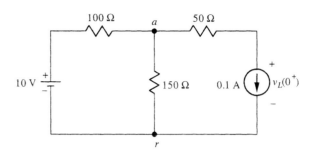

Figure 3.29 Equivalent circuit at $t = 0^+$.

WHAT IF? What if, after being open a long time, the switch is closed at $t = 0$? What is v_L for that case?[9]

Summary. An energized capacitor acts as a voltage source in the calculation of initial values. A special case is an unenergized capacitor, which acts as a short circuit. In the calculation of final values, a capacitor acts as an open circuit.

An energized inductor acts as a current source in the calculation of initial values. A special case is an unenergized inductor, which acts as an open circuit. For final values, an inductor acts as a short circuit. We can solve for initial and final values by replacing capacitors and inductors by these models.

The initial and final values thus are derived from the analysis of a circuit containing only resistors and sources. These principles are valid for circuits containing multiple inductors and capacitors, which lead to higher-order DEs that cannot be solved by our four-step procedure. We introduce the transient analysis of such circuits in what follows, but deal fully with them in Chapter 5.

EXAMPLE 3.8 **Three R's and a C (continued)**

Determine $i(t)$ for the circuit of Fig. 3.16(b).

SOLUTION:
We have already determined the time constant to be $\tau = RC = 9.17$ ms. To find the initial value, we must determine the voltage across the capacitor at $t = 0^+$. At the instant before the opening of the switch, the circuit will be in steady state; indeed, this is the final state from previous actions that established the circuit. Thus, the capacitor will act as an open circuit, as shown in Fig. 3.30.

The voltage across C is 15 V because the 30 V of the source divides equally between the two 1-kΩ resistors, the 10-kΩ resistor being shorted by the switch. Consequently, the circuit that must be analyzed for the initial value of $i(t)$ is shown in Fig. 3.31. Notice that $i(0^+)$ can

---

[9] $v_L = 2.727 e^{-\tau/1.67 \text{ ms}}$.

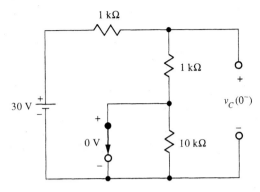

Figure 3.30 The capacitor acts as an open circuit in the steady state before the switch is opened.

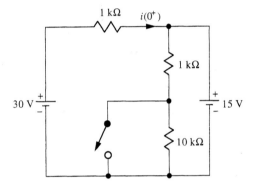

Figure 3.31 The capacitor now acts as a battery in the initial-value calculation.

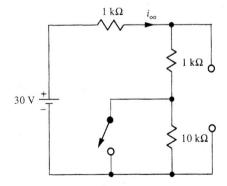

Figure 3.32 Equivalent circuit for final-value calculation.

be determined by writing KVL around the outer loop containing the two sources, with the result $i(0^+) = 15$ mA.

The final value of $i(t)$ can be established from the equivalent circuit in Fig. 3.32, where the capacitor is again treated as an open circuit. The value of i_∞ is easily derived from this series circuit: $i_\infty = 30 \text{ V}/12 \text{ k}\Omega = 2.5$ mA. We now have determined the time constant, the initial value, and the final value; hence, the solution follows from Eq. (3.36).

$$i = 2.5 + (15 - 2.5)e^{-t/9.17 \text{ ms}} \text{ mA} \tag{3.46}$$

WHAT IF? What would be $i(t)$ if the switch were closed after being open a long time?[10]

Pulse Problem

Pulse circuits. We now consider the circuit in Fig. 3.16(c) as detailed in Fig. 3.33. Although this problem appears to differ significantly from the others, we can analyze it with the same techniques. The pulse is treated as a dc voltage that is turned ON, then OFF before steady state is reached.

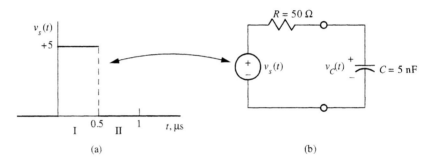

Figure 3.33 (a) Voltage pulse; (b) the capacitor is uncharged when the pulse begins.

Charging transient. We assume that the capacitor is initially unenergized. When the source voltage jumps to +5 V, current will flow, and the capacitor will begin charging toward +5 V. Bear in mind that the circuit cannot anticipate the end of the pulse, but will respond as if a 5-V battery were attached. Thus, the transient proceeds toward a final value, even though that value will never be attained. During interval I, the time constant will be $RC = 0.25$ μs, the initial value will be zero, and the final value would be +5 V. Thus, the capacitor voltage during interval I will be

$$v_C(t) = 5 + (0 - 5)e^{-t/0.25 \text{ μs}} = 5(1 - e^{-t/0.25 \text{ μs}}) \text{ V} \tag{3.47}$$

---

[10] $i(t) = 15 + (2.5 - 15)e^{-t/5 \text{ ms}}$ mA.

Discharging transient. When the end of the pulse comes along 0.5 μs after the leading edge (the beginning), the capacitor now has a voltage across it. Because the pulse width is twice the time constant, the value of the capacitor voltage at the beginning of the second transient would be $5(1 - e^{-2}) = 4.32$ V. This becomes the initial value for the transient during interval II. During this interval, the voltage source is OFF and thus is equivalent to a short circuit. The situation in interval II is therefore that shown in Fig. 3.34. We know the initial voltage, the final value is clearly zero, and the time constant is unchanged. Hence the capacitor voltage during interval II is

$$v_C(t) = 0 + (4.32 - 0)e^{-t'/0.25\,\mu s}, \qquad t' > 0$$
$$= 4.32 \exp\left(-\frac{t - 0.5\mu s}{0.25\mu s}\right) \text{V}, \qquad t > 0.5\mu s \qquad (3.48)$$

where t' is time counted from the beginning of interval II. For example, at $t = 1$ μs, we calculate the voltage to be

$$v_C(1\,\mu s) = 4.32\, e^{-(1.0-0.5)/0.25} = 0.585 \text{ V} \qquad (3.49)$$

Putting the two parts of the transient together, we plot the results in Fig. 3.35.

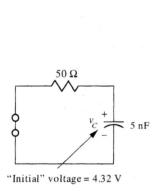

Figure 3.34 During interval II, the "initial" voltage is 4.32 V.

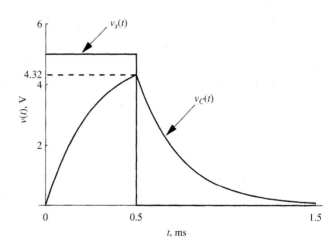

Figure 3.35 The capacitor voltage is a distorted version of the input pulse.

Comparing a pulse with a battery–switch. It is interesting to contrast the pulse problem with that shown in Fig. 3.36. Here we simulate the pulse with a battery and a switch that closes for 0.5 μs. The charging part of the transient will be the same as for the pulse problem, but when the switch is opened, there is no discharge path for the capacitor. Hence, the capacitor will charge up to 4.32 V in both cases, but with the battery–switch, no discharge will occur.[11]

---

[11] Except discharge that is due to leakage. This discharge will take place over a long period of time compared with the time constant if the physical circuit is well represented by the circuit model.

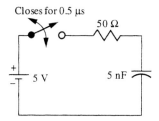

Figure 3.36 The battery switch has the same Thévenin voltage but different output impedance from the pulse source.

Impedance Level

We can understand the difference between the pulse source and the battery-switch source by considering the Thévenin equivalent circuits for each. The Thévenin voltage source is the same for both, but the impedance level differs. With the switch closed, the output impedance is 50 Ω for both cases, but with the switch open, the output impedance becomes infinite (an open circuit) for the battery-switch source.

A puzzle. While we are considering battery-switch sources, let us consider what would happen in the circuit of Fig. 3.16(a) if we opened the switch after the current is established. Energy considerations require that the current in the inductor continue after the switch is opened. But no current can flow through an open switch. What will happen?

This question can be answered on two levels. At the theoretical level, we must outlaw this situation. We have created our dilemma by violating the definitions of the circuit elements we are using. Strictly speaking, we can no more open the switch on an energized inductor than we can short circuit an ideal voltage source—the definitions of these elements are contradictory. On the theoretical level, opening the switch on an inductor is like setting $1 = 0$ in mathematics—it is nonsense.

But this answer does not fully satisfy us, does it? There are, after all, real inductors and real switches. What happens when we perform the experiment? If you try it, you will witness a spark when you open a switch connected in series with an energized inductor. The voltage across the inductor, and hence the voltage across the switch, rises instantaneously to a high value, such that the air between the switch contacts becomes ionized. The ionized air provides a resistive path for deenergizing the inductor. Indeed, this is the principle behind the conventional automotive ignition system: the coil is the inductor, the points are the switch, and you know where the spark occurs.

Higher-Order Transients

Introduction. This chapter, after introducing inductors and capacitors, has focused on first-order transients. We have analyzed circuits containing one energy-storage element by a method that built the solution out of the time constant, the initial and final values, and a standard formula.

Our method fails, however, when a circuit has two or more independent[12] energy-storage elements. In this concluding section, we give a brief introduction to such higher-order transients, and in Chap. 5 we develop a powerful method for analyzing this type of circuit.

---

[12] Two or more capacitors or inductors in series or parallel act as a single energy-storage element.

Two capacitors. Figure 3.37 shows a circuit with two capacitors, two resistors, two loops, and two nodes. We draw the circuit to indicate battery-switch input and an output voltage across C_2. The capacitors are initially uncharged (zero volts); we will find output voltage, v_2.

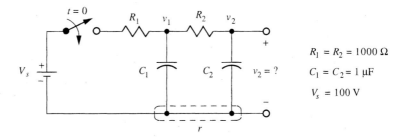

Figure 3.37 Two independent capacitors will lead to a second-order DE.

Deriving the DE. We mentioned loops and nodes because we must choose how to analyze the circuit. We have chosen nodal analysis and labeled the circuit accordingly. Using the standard procedure, augmented by the definition of a capacitor, Eq. (3.12), we write KCL for node 1 as

$$\frac{v_1 - V_s}{R_1} + C_1 \frac{dv_1}{dt} + \frac{v_1 - v_2}{R_2} = 0 \tag{3.50}$$

where the middle term is the current through C_1 to the reference node. Similarly, KCL for node 2 is

$$\frac{v_2 - v_1}{R_2} + C_2 \frac{dv_2}{dt} = 0 \tag{3.51}$$

We develop the DE for v_2 by eliminating v_1 between Eqs. (3.50) and (3.51), which yields

$$C_1 C_2 \frac{d^2 v_2}{dt^2} + \left(\frac{C_2}{R_1} + \frac{C_1 + C_2}{R_2}\right) \frac{dv_2}{dt} + \frac{v_2}{R_1 R_2} = \frac{V_s}{R_1 R_2} \tag{3.52}$$

Form of the solution. We anticipate a solution of the form[13]

$$v_2(t) = A + Be^{\alpha t} \tag{3.53}$$

and generally we proceed as in the classical DE solution on page 129. However, the algebra gets out of hand quickly if we retain the symbolic notation, so we will use numbers from here on. We substitute $V_s = 100$ V, $R_1 = R_2 = 1000\ \Omega$, and $C_1 = C_2 = 1\ \mu$F, and calculate the coefficients as

---

[13] Because the equation is second order, we will find two values of α that satisfy the circuit conditions and add a second exponential term.

$$10^{-12} \frac{d^2v_2}{dt^2} + 3 \times 10^{-9} \frac{dv_2}{dt} + 10^{-6} v_2 = 10^{-4} \tag{3.54}$$

Substitution of Eq. (3.53) into Eq. (3.54) yields

$$[(10^{-12}) \alpha^2 + (3 \times 10^{-9}) \alpha + 10^{-6}] Be^{\alpha t} + 10^{-6} A = 10^{-4} \tag{3.55}$$

which can be valid at all times only if

$$(10^{-12}) \alpha^2 + (3 \times 10^{-9}) \alpha + 10^{-6} = 0 \quad \text{and} \quad 10^{-6} A = 10^{-4} \tag{3.56}$$

The quadratic equation yields two roots, $\alpha_1 = -382$ and $\alpha_2 = -2618$, and the linear equation yields $A = 100$. Because we have two α's, we split the B in Eq. (3.53) into two constants, and write the solution in the form

$$v_2(t) = 100 + B_1 e^{-382t} + B_2 e^{-2618t} \tag{3.57}$$

Before proceeding with the analysis, we need to relate Eq. (3.57) to our earlier results. We should not be surprised to learn that this second-order circuit has two time constants:

$$\tau_1 = \frac{1}{382} = 2.62 \text{ ms} \quad \text{and} \quad \tau_2 = \frac{1}{2618} = 382 \text{ μs} \tag{3.58}$$

Nor should we be surprised, after examining Fig. 3.37, to find the constant term is 100 V, because that would be the voltage across C_2 if we replaced both capacitors with open circuits.

Initial conditions. To determine B_1 and B_2, we need to consider the initial conditions for the circuit. When the switch closes at $t = 0$, both capacitors will have zero volts and will act as short circuits. The voltage across C_2 cannot change instantaneously, so the initial condition on $v_2(t)$ is

$$\underbrace{v_2(0^+)}_{0} = 100 + B_1 e^0 + B_2 e^0 = 0 \Rightarrow B_1 + B_2 = -100 \tag{3.59}$$

Furthermore, the current through R_1 at $t = 0^+$ will all pass through C_1, because that path is a short circuit compared with the parallel path through R_2 and C_2. Thus, the initial current through C_2 will be zero, and the resulting initial condition on $v_2(t)$ will be

$$\left. \frac{dv_2}{dt} \right|_{t=0^+} = 0 \Rightarrow B_1(+\alpha_1) e^{\alpha_1 t} + B_2(+\alpha_2) e^{\alpha_2 t} = 0 \tag{3.60}$$

Simultaneous solution of Eqs. (3.59) and (3.60) yields $B_1 = -117.1$ V and $B_2 = +17.1$ V; hence, the output voltage of the circuit in Fig. 3.37 is

$$v_2(t) = 100 - 117.1 e^{-t/2.62 \text{ ms}} + 17.1 e^{-t/382 \text{ μs}} \text{ V} \tag{3.61}$$

which is plotted in Fig. 3.38. The response is dominated by the longer time constant, with the shorter time constant showing its effect in keeping the slope zero at $t = 0^+$.

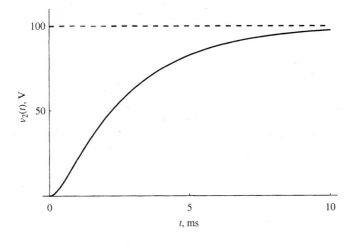

Figure 3.38 The second-order response looks like first-order except for the zero slope at the origin.

Summary. Although the analysis of the two-capacitor circuit in Fig. 3.37 was accomplished through a combination of classical DE and circuit-theory methods, nothing new emerged. We cannot use the initial-value/final-value method, but we still have time constants and the response looks similar to those we have seen in first-order circuits. As you will soon see, we must use capacitors *and* inductors to get something really different.

An *RLC* Circuit

Deriving the DE. Figure 3.39 shows a series *RLC* circuit with a battery-switch input and an output of the voltage across the inductor. Because there is one loop, we will develop the DE through loop-current analysis. With the switch closed, the KVL equation following the loop current is

$$-V_s + Ri + v_C(0) + \frac{1}{C}\int_0^t i(t')dt' + \underbrace{L\frac{di}{dt}}_{v_L(t) = v_{out}(t)} = 0 \tag{3.62}$$

We have identified the last term as the output voltage because we wish to derive a DE for $v_L(t)$. We differentiate Eq. (3.62) once to get rid of the integral:

$$0 + R\frac{di}{dt} + 0 + \frac{i}{C} + \frac{dv_L}{dt} = 0 \tag{3.63}$$

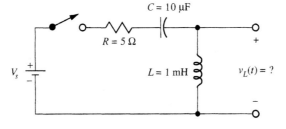

Figure 3.39 The series *RLC* circuit can exhibit oscillations.

This differentiation gets rid of the source voltage, but it will reappear when we consider initial conditions. The first term in Eq. (3.63) can easily be expressed in terms of v_L, but we must differentiate again to get rid of the current term:

$$\frac{R}{L}\frac{dv_L}{dt} + \frac{v_L}{LC} + \frac{d^2v_L}{dt^2} = 0 \tag{3.64}$$

where we have in Eqs. (3.63) and (3.64) used $di/dt = v_L/L$ repeatedly to convert variables from current to output voltage. We will consider solutions of Eq. (3.64) under two conditions: no loss ($R = 0$) and small loss ($R =$ small).

No loss ($R = 0$) solution. Setting $R = 0$ in Eq. (3.64) eliminates the first derivative term and leaves

$$\frac{d^2v_L}{dt^2} + \frac{v_L}{LC} = 0 \tag{3.65}$$

Equation (3.65) is a homogenous second-order linear DE, and has two solutions, $\cos(\omega_0 t)$ and $\sin(\omega_0 t)$, where

$$\omega_0^2 = \frac{1}{LC} = 10^8 \tag{3.66}$$

Thus, the general solution of Eq. (3.65) is

$$v_L(t) = A\cos(\omega_0 t) + B\sin(\omega_0 t)$$
$$= A\cos(10^4 t) + B\sin(10^4 t) \tag{3.67}$$

Initial conditions. We assume that the capacitor is initially uncharged. Examination of Fig. 3.39 reveals that the full battery voltage must appear across the inductor at $t = 0^+$ because the current must be zero; hence, the voltage across both the resistor and the capacitor will be zero. This conclusion could also be deduced from Eq. (3.62) at $t = 0^+$. Thus, $v_L(0^+) = V_s$. Equation (3.67) thus requires that $A = V_s$. Similarly, we can reason from Eq. (3.63) at $t = 0^+$ that

$$\left.\frac{dv_L}{dt}\right|_{t=0^+} = -\frac{R}{L}v_L(0^+) = -\frac{R}{L}V_s = 0 \quad \text{for } R = 0 \tag{3.68}$$

The requirement in Eq. (3.68) forces the sine term in Eq. (3.67) to be zero, and the output response to the switch closure is, therefore,

$$v_L(t) = V_s \cos(\omega_0 t) = 100\cos(10^4 t) \tag{3.69}$$

which is shown in Fig. 3.40.

Interpretation. We may understand this response physically in terms of the mechanical analog, which is shown in Fig. 3.41. The closing of the switch corresponds to the sudden application of a force to the spring–mass system, and the output voltage corresponds to the acceleration of the mass. The mass will respond immediately to the application of the force and will resonate with the spring in a lossless oscillation. Thus,

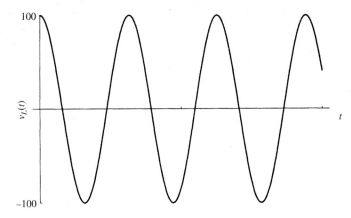

Figure 3.40 Without loss, the response is a pure oscillation.

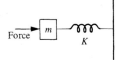

Figure 3.41
Mechanical analog to circuit in Fig. 3.39 with $R = 0$.

we expect the simple harmonic motion we see in Fig. 3.40. In the electric circuit, the "momentum" is associated with the current in the inductor, and the spring action associated with the capacitor.

Low-loss response ($R = 5\;\Omega$). We now discuss the response with small loss. The mathematics to derive the response is beyond our present ambition, especially since we deal with this problem thoroughly in Chap. 5. The response is

$$v_L(t) = e^{-2500t}[100\cos(9682t) - 25.8\sin(9682t)] \tag{3.70}$$

which is plotted in Fig. 3.42. You will note the similarity to the response for no loss, but of course the response dies out due to the loss in the resistor. With loss, the circuit will eventually reach steady state, which means zero voltage across the inductor.

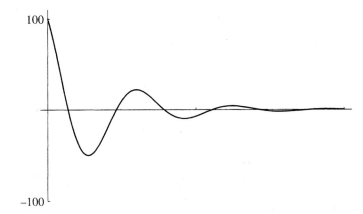

Figure 3.42 With an inductor and a capacitor in the circuit, an oscillation is possible. This oscillation shows the effect of the losses in the resistor.

Check Your Understanding

1. In the differential equation, $5\,dx/dt + 2x = 3$, what is the time constant? What is the final value for x?
2. In a first-order transient, if the initial current in a capacitor is 5 mA, what is the current one time constant later in time?

3. In using a Thévenin equivalent circuit to determine the equivalent resistance in an *RL* transient problem, what must be considered the load?

4. During a switch action, which is constant: voltage, current, stored energy, or power?

5. In a simple circuit containing a resistor and an energy-storage element (*L* or *C*), when the value of the resistor was doubled, the transient lasted twice as long. What was the other element?

6. As $t \to \infty$, what circuit element approaches a short circuit?

7. Will a capacitor allow an instantaneous change in its current or voltage?

8. An *RC* circuit has a time constant of 10 ms. The capacitor is replaced by a 50-mH inductor and the time constant changes to 8 ms. What was *C*?

9. In a simple transient, the initial energy in an inductor is 10 mJ and the final energy is 0. What is the energy after two time constants?

Answers. (1) 2.5 s, 1.5; (2) 1.84 mA; (3) the inductor; (4) stored energy; (5) a capacitor; (6) an inductor; (7) its current; (8) 1600 µF; (9) 0.183 mJ.

CHAPTER SUMMARY

Chapter 3 introduces inductors and capacitors as circuit elements that store magnetic and electric energy, respectively. Time processes enter into circuit analysis, and circuit behavior is described by differential equations. We focus on first-order systems: one inductor or capacitor in a circuit containing one or more resistors and switches.

Objective 1. To understand the properties of inductors, and be able to find the inductor voltage from inductor current and vice versa. Inductors store magnetic energy by getting currents close together. The inductor voltage is equal to the time derivative of the current times a constant, the inductance. An inductor tries to keep its current constant.

Objective 2. To understand the properties of capacitors, and be able to find the capacitor voltage from capacitor current and vice versa. Capacitors store electric energy by getting charges close together. The capacitor current is equal to the time derivative of its voltage times a constant, the capacitance. Capacitor voltage is temporarily constant when sudden changes occur on a circuit.

Objective 3. To understand how to analyze first-order transients by the initial-value/final-value method. When switches are opened or closed in a dc circuit containing a single energy-storage element, a first-order transient carries the circuit from one condition to another over a transient period of time. We develop a method for analyzing a first-order transient based upon the initial and final values of the electrical voltage or current of interest, plus the time constant of the circuit.

The chapter ends with a brief introduction to second-order circuits. First-, second-, and higher-order circuits are studied in detail in Chap. 5, where powerful methods of analysis are developed. In Chap. 4, first-order transients play a small role, but the chapter emphasizes steady-state response of circuits with alternating-current sources.

GLOSSARY

Capacitance, capacitor, p. 123, the capacity of a circuit element to store electric energy.

Final value, p. 138, the steady-state condition of the circuit, its state after a large period of time.

Forced response, p. 129, the response of a system or electrical circuit to a periodic source, including a dc source. Also called **steady-state response**.

Inductor, inductance, p. 118, the capacity of a circuit element to store magnetic energy, normally associated with a coil of wire.

Initial value, p. 138, the value of a circuit variable the instant after a change occurs in the circuit, usually a switch closing or opening.

Natural response, p. 130, the response of a system or electrical circuit stimulated by external force but not dependent on the form of that force; a response that arises out of internal energy exchanges. Also called **transient response**.

Steady-state response, p. 129, the response of a system or electrical circuit to a periodic source, including a dc source. Also called **forced response**.

System, p. 129, several components that work together to accomplish some purpose.

Time constant, p. 132, a characteristic time for a transient, associated with the natural response of a system or electrical circuit. For a dc transient, changes are 63% complete in one time constant.

Transient, p. 128, the transition of a circuit from one steady state to another, usually associated with a switch action.

Transient response, p. 130, the response of a system or electrical circuit, stimulated by an external force but not dependent on the form of that force, that arises out of internal energy exchanges. Also called **natural response**.

PROBLEMS

Section 3.1: Theory of Inductors and Capacitors

3.1. A 50-mH inductor has a current of $i_s(t) = 0$, $t < 0$, and $i_s(t) = 150t^3$ A, $t > 0$. Calculate the voltage across the inductor, $v_L(t)$, with the polarity shown in Fig. P3.1.

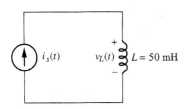

Figure P3.1

3.2. A 6.3-V battery is connected to an ideal 0.3-H inductor, as shown in Fig. P3.2.

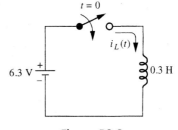

Figure P3.2

(a) Calculate the current as a function of time after the switch is closed.

(b) At what time does the stored energy in the inductor reach 10 J? Verify the stored energy by integrating the input power (product of v_L and i_L) from $t = 0$ to the time you calculate.

3.3. (a) Find the equivalent inductance of a 1-H inductor in parallel with a 2-H inductor.
(b) Find the ratio of the stored energy in the inductors, W_{m2}/W_{m1}, if they initially have zero energy.
(c) Repeat if the inductors are in series.

3.4. If an ideal 4.5-V battery were connected to an ideal 1-H inductor, how much energy would be given to the inductor in the first second?

3.5. Figure P3.5 shows a 20-μF capacitor that has a voltage

$$v_s(t) = 0, \quad t < 0$$
$$v_s(t) = 10^4 t^2, \quad 0 < t < 0.1 \text{ s}$$
$$v_s(t) = 100, \quad t > 0.1 \text{ s}$$

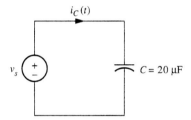

Figure P3.5

(a) Find the current, $i_C(t)$.
(b) At what time is the energy in the capacitor 50 mJ?
(c) At what time is the power into the capacitor 1 watt?

3.6. A 0.1-μF capacitor is charged with a 1-μs pulse of current, as shown in Fig. P3.6. Find the voltage across the capacitor, with the polarity shown, as a function of time.

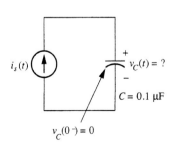

Figure P3.6

3.7. To move the spot of a CRT smoothly across the screen, the voltage across a pair of deflection plates must be increased in a linear fashion, as shown in Fig. P3.7. If the capacitance of the plates is 1 pF, find the resulting current through the capacitor.

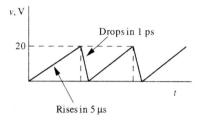

Figure P3.7

3.8. (a) Find the equivalent capacitance of a 1-μF capacitor in series with a 5-μF capacitor.
(b) Find the ratio of the stored energy in the capacitors, W_{e2}/W_{e1}, if they initially have zero energy.
(c) Repeat parts (a) and (b) if the capacitors are in parallel.

3.9. An inductor and a capacitor are placed in series with a current source whose current increases with time, as shown in Fig. P3.9 (b). The circuit is shown in Fig. P3.9 (a).

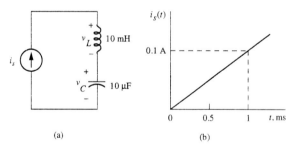

Figure P3.9

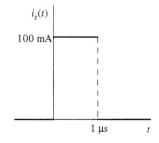

(a) Find the voltage across the inductor, $v_L(t)$.
(b) Assuming no initial charge on the capacitor, find its voltage, $v_C(t)$.
(c) Calculate the instant when the stored energy in the capacitor first exceeds that in the inductor.

3.10. For the circuit shown in Fig. P3.10, the voltage across the resistor is 0 V for negative time and $v_R = t$ V for positive time. The capacitor is initially unenergized.

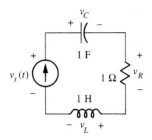

Figure P3.10

(a) What is the stored energy in the inductor at $t = 2$ s?
(b) What is the voltage across the capacitor at the same time?
(c) What is the voltage across the source at the same time?

3.11. For the series RLC circuit in Fig. P3.11, the voltage across the inductor is shown.
(a) Determine and sketch the voltages across the resistor and the capacitor. Assume $v_C = v_R = 0$ at $t = 0$.
(b) What is the value of v_s at $t = 3$ ms?

3.12. In the circuit shown in Fig. P3.12, the initial voltage on the capacitor is 10 V, with the + at the bottom, and the initial current in the inductor is zero. The current into the capacitor is 5 A dc for $t > 0$, as shown.
(a) Find the source voltage.
(b) Determine the *two* times after $t = 0$ when the power out of the voltage source is zero.

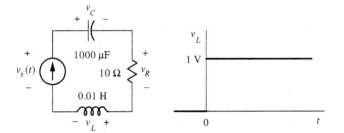

Figure P3.11

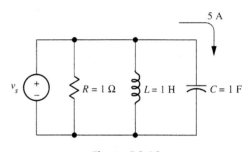

Figure P3.12

Section 3.2: First-Order Transient Response of *RL* and *RC* Circuits

3.13. The solution of a first-order DE with constant coefficients and a constant term on the right side is $x(t) = 5 - 3e^{-t/1\text{ms}}$.
(a) Plot this solution. What is the value at the origin and as time becomes large?
(b) What is the DE that this solution satisfies? What initial condition? Verify by substitution that your DE is correct.

3.14. Verify that the standard solution, Eq. (3.36), satisfies Eq. (3.35) and has the required values at $t = 0$ and t very large.

3.15. A system is described by the DE

$$\frac{1}{3}(dx/dt) + 5x = 10$$

(a) What is the time constant for this system?

(b) What would be the final value, x_∞?
(c) If $x_0 = -2$, find and sketch $x(t)$ for $t > 0$.

3.16. The circuit shown in Fig. P3.16 has two capacitors, one of which is charged, and hence would appear to

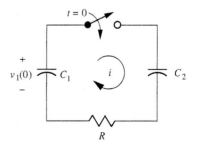

Figure P3.16

be second-order. Derive the DE for $i(t)$ and show that it has the form of Eq. (3.35). What is the time constant?

3.17. Figure P3.17 shows a switch that is changed from a to b at $t = 0$. Assume that the switch has been in position a for a long time before the switch action. Find $v_L(t)$, the voltage across the inductor with the polarity shown.

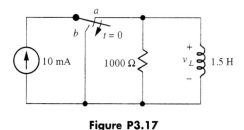

Figure P3.17

3.18. A current source is placed in series with a resistor and inductor, as shown in Fig. P3.18. During this period, the switch is open. Then the switch is closed, and the circuit is separated into two independent loops that share a common short circuit but do not interact.

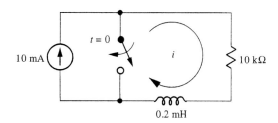

Figure P3.18

(a) Calculate the current in the right-hand loop after the switch closes at $t = 0$.
(b) Plot the current in the switch, referenced downward.

3.19. For the circuit shown in Fig. P3.19, assume that the switch has been in position a for a long time and then is changed to position b.
(a) Find and sketch $i(t)$.
(b) Calculate by integration the energy lost in the 100-Ω resistor during positive time. Confirm that all the energy stored initially in the capacitor is accounted for by the loss in the resistor.

3.20. The circuit shown in Fig. P3.20 has equilibrium established with the switch closed, then the switch is

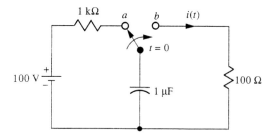

Figure P3.19

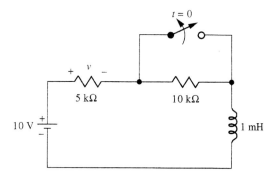

Figure P3.20

opened. Determine the voltage across the 5-kΩ resistor after the switch is opened.

3.21. After being closed a long time, the switch in Fig. P3.21 is opened at $t = 0$.
(a) Find the voltage across the switch as a function of time for $t > 0$.
(b) In what period of time is half the initial stored energy in the capacitor lost to the resistor?

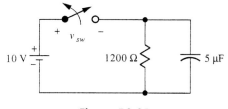

Figure P3.21

3.22. The switch in Fig. P3.22 is in position a for negative time, moved to b at $t = 0$, and to c at $t = 10$ ms. Sketch the voltage across the capacitor for $0 < t < 30$ ms. At what time should the switch be switched to c for no "transient"?

3.23. The switch in Fig. P3.23 is closed at $t = 0$. Find the time at which the power into the inductor is maximum.

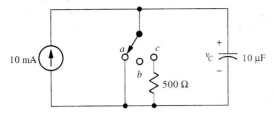

Figure P3.22

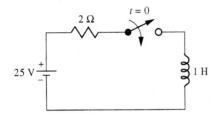

Figure P3.23

3.24. A capacitor is charged to 100 V and disconnected. Leakage reduces its voltage to 35 V in 45 min. Estimate the additional time required for the voltage to drop to 8 V. *Hint:* The leakage is represented by a resistance in parallel with the capacitance.

3.25. For the circuit shown in Fig. P3.25, the switch is open a long time, closed for 1 second, and then opened again for a long time.
  (a) Determine and sketch the current in the inductor.
  (b) How much energy comes out of the battery in this process?

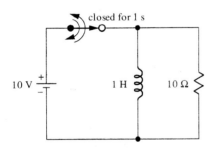

Figure P3.25

3.26. The switch in Fig. P3.26 is closed at $t = 0$. Find di/dt at the instant of time when the stored energy in the inductor is 0.5 µJ.

3.27. For the circuit in Fig. P3.27, the original energy stored in the capacitor is 500 J and the switch is open. The switch is then closed for 1 second and opened again after 1 second. Find the final energy stored in the capacitor.

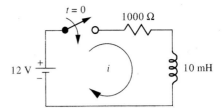

Figure P3.26

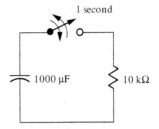

Figure P3.27

3.28. The 1000-µF capacitor in Fig. P3.28 is charged to 10 V and the switch is closed at $t = 0$. The circuit must meet the requirement that the current exceeds 250 mA for at least 10 ms. Find R to meet this requirement. (The answer is not unique.)

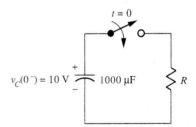

Figure P3.28

3.29. For the circuit shown in Fig. P3.29, the switch is open a long time, then closed at $t = 0$. It is required that the capacitor voltage equals or exceeds 0.75 V at $t = 10$ ms. Find R to satisfy this criterion. (The answer is not unique.)

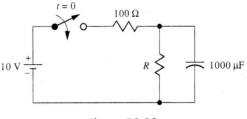

Figure P3.29

Section 3.3: Advanced Techniques

3.30. For the circuit in Fig. P3.30 find the following:
(a) τ.
(b) i_0.
(c) i_∞.
(d) $i(t)$.

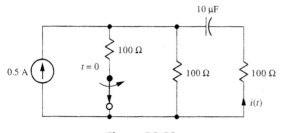

Figure P3.30

3.31. After being closed a long time, the switch in Fig. P3.31 is opened at $t = 0$. Find the following:
(a) The time constant.
(b) The initial value of $i(t)$.
(c) The final value of $i(t)$.
(d) The time function, $i(t)$.

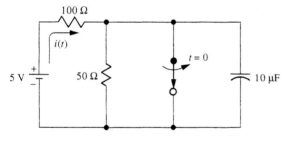

Figure P3.31

3.32. After being closed for a long time, the switch in Fig. P3.32 is opened at $t = 0$. Find the following:
(a) The voltage across the capacitor with + at top as a function of time.

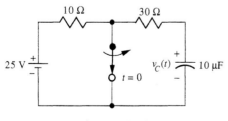

Figure P3.32

(b) The integral of the current during the period $0 < t < \infty$.
(c) The total energy given to the circuit by the battery during the period $0 < t < \infty$.

3.33. For the circuit shown in Fig. P3.33, the switch has been open a long time and is closed at $t = 0$.

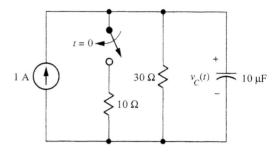

Figure P3.33

(a) What is the time constant of the circuit with the switch closed?
(b) Determine the capacitor voltage with the polarity shown and sketch.

3.34. For the circuit in Fig. P3.34, find $v(t)$ for $t > 0$ after the switch is closed.

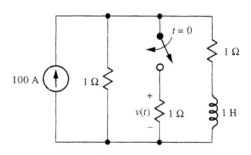

Figure P3.34

3.35. For the circuit shown in Fig. P3.35, the switch is closed at $t = 0$ after being open a long time. Find the following:
(a) The voltage across the 10-Ω resistor for $t > 0$.
(b) The stored energy in the capacitor in the steady-state condition.

3.36. After being open a long time, the switch in Fig. P3.36 closes at $t = 0$. Find the power out of the 6-V source at $t = 10$ ms.

3.37. For the circuit shown in Fig. P3.37, the switch has been closed for a long time and then is opened at

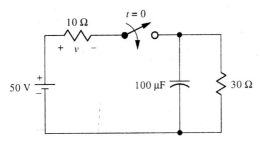

Figure P3.35

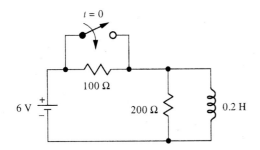

Figure P3.36

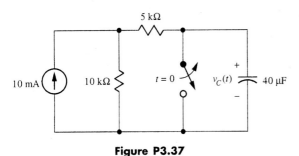

Figure P3.37

$t = 0$. Determine the voltage across the capacitor with the polarity shown.

3.38. For the circuit shown in Fig. P3.38, the switch is open a long time, closed for 5 ms, and then opened again. Find and sketch $i(t)$.

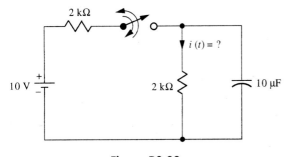

Figure P3.38

3.39. After being open a long time, the switch in Fig. P3.39 is closed at $t = 0$ for 1 ms, then opened again. Determine the voltage across the 125-Ω resistor for $t > 0$.

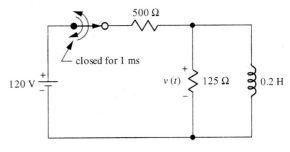

Figure P3.39

3.40. For the circuit shown in Fig. P3.40, the switch has been closed for a long time and then is opened at $t = 0$. Determine the current through the inductor with the reference direction shown.

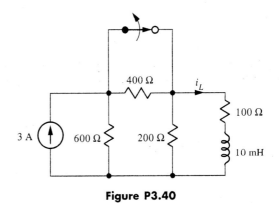

Figure P3.40

3.41. The parallel RL circuit shown in Fig. P3.41 is excited by the current pulse as shown. Calculate and sketch the resulting voltage, v_L. Assume that the inductor is initially unenergized.

3.42. A pulse can be modeled as two sources, one switching the voltage on and the other switching it off, as shown in Fig. P3.42. Note that the two sources add to zero except during the period $0 < t < t_1$. The voltage across the capacitor can be calculated by superposition. Use this model to rework the problem in Fig. 3.33. Specifically, solve for the voltage across the capacitor at $t = 1.0$ μs and verify the result given on page 129, where $V = 5$ V, $R = 50$ Ω, and $C = 5$ nF.

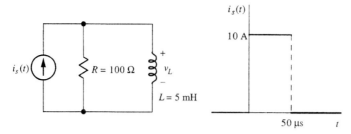

Figure P3.41

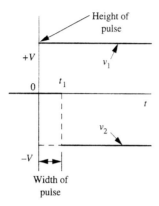

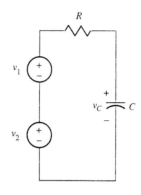

Figure P3.42

3.43. The circuit in Fig. P3.43 is the same as in Fig. 3.37, except that the output is taken across R_2 instead of C_2.
  (a) Derive the differential equation for v_2. [Should be very similar to Eq. (3.52).]
  (b) Determine the initial conditions for the output voltage and its derivative, assuming the capacitors initially uncharged.
  (c) Find $v_2(t)$.

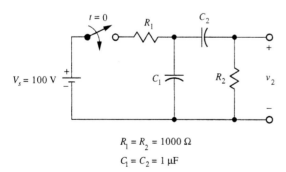

$R_1 = R_2 = 1000\ \Omega$
$C_1 = C_2 = 1\ \mu F$

Figure P3.43

3.44. The RLC circuit in Fig. P3.44 is similar to that in Fig. 3.39, except the inductor and capacitor are in parallel rather than in series.

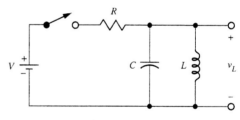

Figure P3.44

  (a) Derive the DE for $v_L(t)$.
  (b) Find the initial values for the $v_L(t)$ and its derivative.

3.45. The circuit in Fig. P3.45 has loop currents defined. Derive the DE for $i_2(t)$.

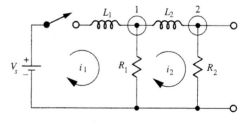

Figure P3.45

General Problems

3.46. When the switch in the RC circuit of Fig. P3.46 is closed, for a period of time energy flows from the battery into the circuit. Once the current stops, the energy flow ceases.

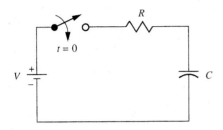

Figure P3.46

(a) Calculate by integration the total energy given to the circuit by the battery for $t > 0$. Show that this is Vq, where q is the charge on the capacitor.

(b) Show by direct calculation that one-half this total energy is stored in the capacitor and the other half is lost to the circuit as losses in the resistor.

3.47. The switch in Fig. P3.47 is in position a for a long time, then switched to b for a period of time Δt, and then switched to position c. Find Δt such that the voltage across the 50-Ω resistor is 60 V at 100 μs after the switch was switched to position b, that is, 100 μs $-\Delta t$ after the switch was put in position c. *Hint:* The equation is nonlinear and must be solved by numerical methods.

3.48. A microwave oven will boil water in 3 minutes, starting with water at 75°F. If the initial heating rate

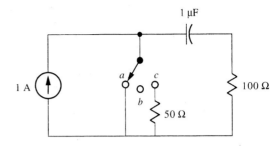

Figure P3.47 The switch is in position b for Δt and then put in position c.

is 55°/minute, estimate the temperature to which the water would heat were it not for the phase change at 212°F. *Hint:* This fits the model for a first-order transient. The unknown is the final value of the temperature (which it never reaches).

3.49. The circuit in Fig. P3.49 will operate as a variable delay in operating the alarm. The alarm operates when its input current exceeds 100 μA. Find the range of R such that the delay is between 0.1 and 1 second.

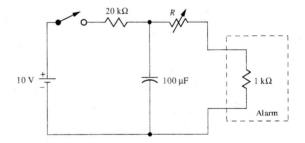

Figure P3.49

Answers to Odd-Numbered Problems

3.1. $22.5t^2$, $t > 0$.

3.3. (a) 2/3 H; (b) 1/2; (c) 2/1.

3.5. (a) $0.4t$ for $0 < t < 0.1$ s, 0 elsewhere; (b) 0.0841 s; (c) 0.0630 s.

3.7. 4 μA when increasing, -20 A when decreasing.

3.9. (a) 1 V; (b) $10^7 t^2/2$, (c) 0.632 ms.

3.11. (a) $i = 100t$, $v_R = 1000t$, $v_C = 10^5 t^2/2$; (b) $v_s = 4.45$ V at 3 ms.

3.13. (a) Starts at 2 and approaches 5 for large time with 1 ms time constant; (b) $10^{-3} dx/dt + x = 5$ with $x(0) = 2$.

3.15. (a) 1/15 s; (b) +2; (c) $x(t) = 2 - 4e^{-15t}$.

3.17. $v_L(t) = -10e^{-t/1.5 \text{ ms}}$ V.

3.19. (a) $i(t) = 1e^{-t/100 \text{ μs}}$ A; (b) 5 mJ.

3.21. (a) $10(1-e^{-t/6 \text{ ms}})$ V; (b) 2.08 ms.

3.23. 0.347 s.

3.25. (a) $i_L(t) = 10$ A for $0 < t < 1$, $i_L(t) = 10e^{-(t-1)/0.1 \text{ s}}$ A, $t > 1$ s; (b) 60 J.

3.27. 409 J.

3.29. (a) $R > 19.5$ Ω works.

3.31. (a) 0.333 ms; (b) 50 mA; (c) 33.3 mA; (d) $i(t) = 33.3 + (50 - 33.3)e^{-t/0.333 \text{ ms}}$ mA.

3.33. (a) 75 μs; (b) $v_C(t) = 7.5 + (30 - 7.5)e^{-t/75 \text{ μs}}$ V.

3.35. (a) $v(t) = 12.5 + (50 - 12.5)e^{-t/0.75 \text{ ms}}$ V; (b) 70.3 mJ.

3.37. $v_C(t) = 100(1 - e^{-t/0.6 \text{ s}})$ V.

3.39. $v_L(t) = 24e^{-t/2 \text{ ms}}$ V for $t < 1$ ms and $-11.8e^{-(t-1 \text{ ms})/1.6 \text{ ms}}$ V for $t > 1$ ms.

3.41. $v_L(t) = 1000e^{-t/50 \text{ μs}}$ V for $t < 50$ μs and $-632e^{-(t-50 \text{ μs})/50 \text{ μs}}$ V for $t > 50$ μs.

3.43. (a) Same as Eq. (3.52) except right-hand side $= 0$;

(b) $v_2(0^+) = 0$, $\left.\dfrac{dv_2}{dt}\right|_{t=0^+} = 10^5$ V/s;

(c) $v_2(t) = 44.7e^{-382t} - 44.7e^{-2620t}$ V.

3.45. $L_1 L_2 \dfrac{d^2 i_2(t)}{dt^2} + [R_1 L_2 + (R_1 + R_2)L_1]\dfrac{di_2(t)}{dt} + R_1 R_2 i_2(t) = R_1 V_s$.

3.47. 72.2 μs.

3.49. $3.36 \text{ k}\Omega < R_L < 33.6 \text{ k}\Omega$.

CHAPTER 4

The Analysis of AC Circuits

Introduction to Alternating Current (AC)

Representing Sinusoids with Phasors

Impedance: Representing the Circuit in the Frequency Domain

Phasor Diagrams for *RL*, *RC*, and *RLC* Circuits

Chapter Summary

Glossary

Problems

objectives

1. To understand how to identify the amplitude, frequency, and phase of a sinusoidal function
2. To understand the phasor concept and be able to find the sinusoidal steady-state response of a circuit, given the differential equation
3. To understand how to use phasors and impedance to determine the sinusoidal steady-state response of a circuit
4. To understand the effect of varying frequency on series and parallel RL, RC, and RLC circuits
5. To understand admittance and its advantages in analyzing parallel impedances

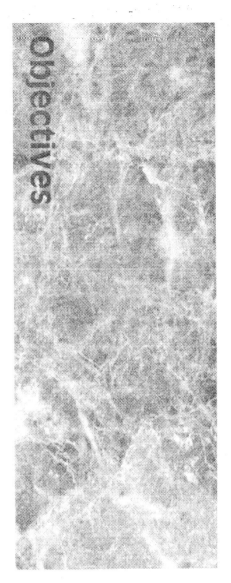

Alternating-current circuits comprise the vast majority of all power distribution and power consumption circuits. This chapter introduces the methods by which ac circuits are analyzed.

4.1 INTRODUCTION TO ALTERNATING CURRENT (AC)

Importance of AC

Historical perspective. Thomas A. Edison was a clever, determined inventor whose activities excited the public imagination toward the practical uses of electricity. He pioneered in, among other things, the generation and distribution of electric power for lighting. But Edison was committed to direct current (dc). The power plants built by the Edison Electric Lighting Company produced dc.

Edison had many young inventors and scientists working for him. His was, in fact, the first industrial research laboratory. One of these underlings was Nikola Tesla, a young engineer from Croatia. Tesla appears to have been the first person to recognize the possibilities of ac and he is credited with inventing the ac induction motor. But his efforts to convince Boss Edison of the benefits of ac were in vain, and Tesla eventually quit.

Tesla went to a rival company and battle was pitched: Was it to be dc or ac? Nasty ads were placed in the newspapers by Edison's group, claiming that ac was unsafe. From our perspective, these warnings of the dangers of ac seem odd, but at the time they created a serious debate.

Edison was wrong and Tesla was right. Today, the vast majority of all electric power is generated, distributed, and consumed in the form of ac power. Before World War II, "electrical engineering" meant ac generators, motors, transformers, transmission lines, and the like. The ac power industry currently employs a mature technology and is a vital part of modern civilization.

In this chapter, you will learn how to analyze ac circuits and, more broadly, how electrical engineers think about ac waveforms. The techniques of ac analysis, once mastered, are powerful for solving problems and stimulating insight. We begin by acquainting you with the sinusoidal waveform.

Sinusoids

> **LEARNING OBJECTIVE 1.**
> To understand how to identify the amplitude, frequency, and phase of a sinusoidal function

Physical model for a sinusoid. Most of us were introduced to sines and cosines through the study of triangles: sine equals opposite over hypotenuse, and that sort of thing. Later we learned that circular motion leads to sine and cosine functions. Figure 4.1 shows a crank. The horizontal projection of the crank is the length times the cosine of the angle ϕ. If the rotation speed of the crank is uniform, the horizontal projection becomes a sinusoidal function of time. We may describe this waveform mathematically as a sine function or a cosine function, but we will simply call it a *sinusoid*, or sinusoidal waveform.

sinusoid

Mathematical form for a sinusoid. Electrical engineers have adopted the cosine function as the standard mathematical form for sinusoidal waveforms. Figure 4.2 shows the peak value, period, and phase of a sinusoidal waveform. The corresponding mathematical form is

$$v(t) = V_p \cos(\omega t + \theta) \tag{4.1}$$

period, event frequency

The peak value of the voltage is V_p. The sinusoid repeats with a *period*, T, which determines the frequency of the sinusoid. The *event frequency*, namely, the number of cycles during a period of time, is the reciprocal of the period:

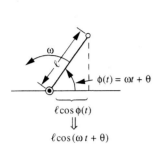

Figure 4.1 The projection of the rotating crank is a sinusoid.

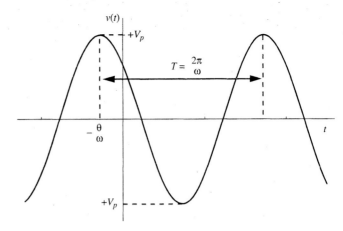

Figure 4.2 The sinusoidal function is defined by its peak value, its phase, and its period (or frequency).

$$f = \frac{1}{T} \text{ hertz} \tag{4.2}$$

For example, if the period were 0.02 s, the event frequency would be 1/0.02, or 50 cycles per second. Because the unit "cycles per second" is awkward to say, most people tend to shorten it to "cycles," which is misleading to the novice and offensive to the purist. To honor Heinrich Hertz (1857–1894), the unit hertz, abbreviated Hz, has replaced cycles per second as the common unit for event frequency. So 50 hertz means 50 cycles/second.

angular frequency When most people talk about a frequency, they mean the event frequency. When we write mathematical expressions for sinusoidal waveforms, however, we more often deal with the proper mathematical measure, the angular frequency, ω (Greek lowercase omega), in radians per second. Because there are 2π radians in a full circle (a cycle), the relationship between *angular frequency*, event frequency, f, and period, T, is

$$\omega = 2\pi f = \frac{2\pi}{T} \frac{\text{radians}}{\text{second}} \tag{4.3}$$

The scientific dimensions of frequency are reciprocal seconds. The numerator of Eq. (4.3) represents an angle and is therefore dimensionless. Just as we retain radians or degrees to remind us which measure of an angle we are using, so we need to state the units for frequency to make explicit which frequency we mean, f or ω.

Units of phase. The phase of the sinusoid, θ (Greek lowercase theta), is what permits the waveform in Fig. 4.2 to represent a general sinusoid. We have drawn the curve for a phase of $\theta = +50°$, but we can shift the position of the sinusoid by varying the phase. The phase is related to the time origin when we use a mathematical description, but in an ac problem, what matters are the relative phases of the various sinusoidal voltages and currents.

Here you must tolerate one of the traditional inconsistencies of electrical engineers. The mathematical unit for ωt is radians and hence the correct unit for phase should also be radians. For example, we may wish to compute the time when the voltage reaches its positive peak. The cosine is maximum for zero angle, so the peak occurs when the total angle is zero:

$$\omega t_{peak} + \theta = 0 \Rightarrow t_{peak} = -\frac{\theta}{\omega} = -\frac{\theta}{2\pi} T \qquad (4.4)$$

When we solve an equation like Eq. (4.4), we must use radian measure for the phase, as the last form of Eq. (4.4) suggests. But electrical engineers usually speak of phase in degrees, as we did before (+50°). Thus, for $\theta = 50°$,

$$t_{peak} = -\frac{50 \times (\pi/180)}{2\pi} \times T \qquad (4.5)$$

where $\pi/180$ converts 50° to radians. This inconsistency is tolerable because only relative phase is what matters in most situations. Electrical engineers, like most people, still are more comfortable thinking about and sketching angles in degree measure. Probably you also think best in degree measure, so we will usually express phase in degrees unless radians are required by the mathematics.

EXAMPLE 4.1 **Sinusoids**

A sinusoidal voltage is 20 V peak to peak, is 5 ms between peak and trough, and at $t = 0$ is -3.6 V and decreasing. Find $v(t)$, and the value of the sinusoid at $t = 12$ ms.

SOLUTION:
Figure 4.3 shows the sinusoid. The amplitude is clearly 10 V. The period is the time between peaks, 10 ms; so the event frequency is

$$f = \frac{1}{T} = \frac{1}{10 \text{ ms}} = 100 \text{ Hz} \qquad (4.6)$$

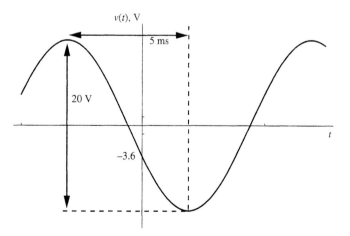

Figure 4.3 Sinusoid described by peak-to-peak voltage, half-period, and value with trend at the origin.

and the radian frequency is

$$\omega = 2\pi f = 200\pi \text{ radians/second} \tag{4.7}$$

Therefore, the sinusoid is of the form

$$v(t) = 10\cos(200\pi t + \theta) \text{ V} \tag{4.8}$$

For $v(0) = -3.6$ V, we require $10\cos(\theta) = -3.6$. The calculator gives $111.1°$ for $\cos^{-1}(-0.36)$, but we should realize that this is the principal value and $360° - 111.1° = 248.9°$ is also a possibility.

We decide between the two possible phases by choosing the one that gives a negative derivative. A simple way to determine the sign of the derivative is use the calculator to evaluate $\cos(\theta + 1°)$, which is equivalent to evaluating the function at a time slightly after the origin in time. The calculator gives $\cos(111.1° + 1°) = -0.376$, but $\cos(248.9° + 1°) = -0.344$. Hence, $111.1°$ is the correct phase because it gives a decreasing function at the origin. Therefore,

$$v(t) = 10\cos(200\pi t + 111.1°) \text{ V} \tag{4.9}$$

The value at 12 ms would be

$$v(0.012) = 10\cos(200\pi \times 0.012 + 111.1°) \text{ V} \tag{4.10}$$

where the first term is in radians: $200\pi \times 0.012 = 7.54$ radians $(= 432.0°)$. Thus, the value at 12 ms would be

$$v(0.012) = 10\cos(432.0° + 111.1°) = -9.985 \text{ V} \tag{4.11}$$

WHAT IF? What if the voltage were $+3.6$ V at $t = 0$ and increasing? What then would be the phase?[1]

Some familiar frequencies. Frequency is a familiar concept. When we speak of an engine speed as 4000 rpm, for example, we are indicating a frequency of 66.7 Hz, and each of the eight spark plugs would be firing with a frequency of 33.3 Hz. The power system frequency is 60 Hz in this country, although 50 Hz is used in much of the world, and 400 Hz is used in airborne and some naval applications.[2] When the radio announcer tells you that you are "tuned to 1200 on your radio dial," she is giving her station frequency, 1200 kHz (kilohertz) or 1.2×10^6 Hz. The FM stations broadcast at frequencies of about 100 MHz (megahertz) or about 10^8 Hz. The UHF TV band extends to about 800 MHz, and communication satellites relay signals at about 5 GHz (gigahertz), or 5×10^9 Hz. The highest frequencies currently used for radio signals are about 300 GHz, 3×10^{11} Hz. We find infrared, optical, and X-ray radiation at even higher frequencies.

Frequency has great importance in electrical engineering, particularly in communication systems. In this section, we have introduced the concept of frequency,

---

[1] The phase would be $-68.9°$.

[2] Because motors and transformers are smaller and lighter at the higher frequency.

and in the next section, we show how ac waveforms are represented for ac circuit analysis.

AC Circuit Problem

The differential equation (DE). Figure 4.4 shows the ac circuit we will analyze. The ac source has a frequency of 60 Hz, or 120π rad/s, and is connected by a switch that closes at $t = 0$. We wish to solve for the current, $i(t)$. After the switch is closed, we write KVL as

$$-v_s(t) + v_R(t) + v_L(t) = 0 \tag{4.12}$$

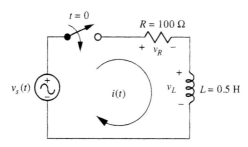

$v_s = V_p \cos(\omega t + \theta)$
$V_p = 100$ V, $f = 60$ Hz, $\theta = 30°$

Figure 4.4 Solve for $i(t)$ for positive time. The switch closes at $t = 0$.

Conservation of Energy

We may introduce $i(t)$ through Ohm's law and the definition of inductance, with the result

$$L\frac{di}{dt} + Ri = V_p \cos(\omega t + \theta) \tag{4.13}$$

Equation (4.13) is a linear first-order DE, but the technique we used in Chap. 3 does not fit because we do not have a dc term on the right side.

Form of the solution. The character of the response, however, is similar to our earlier results. Here also there will be a transition from the state of the circuit before the switch was closed to the state with the closed switch. Here also the transition period is expressed in terms of the L/R time constant of the circuit. In fact, the form of the solution is

$$i(t) = Ae^{-t/\tau} + i_{ss}(t), \quad \text{with } \tau = \frac{L}{R} \tag{4.14}$$

where A is an unknown constant, τ the time constant, and $i_{ss}(t)$ the particular integral, or steady-state current. The steady-state current, $i_{ss}(t)$, is no longer a constant, but results from a dynamic equilibrium between source, resistor, and inductor.

Looking ahead. Methods for determining the steady-state response are the focus of this chapter. Equation (4.13) can be integrated directly with the aid of an integrating factor. This approach fails, however, when we try it on more complicated circuits and, besides, our goal is to learn how electrical engineers analyze ac problems. No electrical engineer would integrate this equation directly to find the steady-state solution. We will lead you down the traditional path—so please be patient. Once we arrive, you will be amazed how easily we can solve ac circuit problems.

Check Your Understanding

1. What is the frequency in hertz of the ticking of a pendulum clock if it ticks twice every second?
2. If eight cycles of a sinusoidal waveform take 2 ms, what is the angular frequency, ω?
3. What is the value of $\cos(100t + 30°)$ at $t = 10$ ms?
4. What is the time between positive peaks for the sinusoidal waveform in the previous question?

Answers. (1) 2 Hz; (2) 25,100 rad/s; (3) 0.0472; (4) 6.28×10^{-2} s.

4.2 REPRESENTING SINUSOIDS WITH PHASORS

Sinusoids and Linear Systems

Sinusoid in, sinusoid out. An important idea is suggested in Fig. 4.5. Here we have represented the circuit as a linear system, linear because the equations of R and L are linear equations and a "system" because the circuit is an interconnection of such elements. (We could also have capacitors in our circuit, but that would complicate this first effort.) Think of the voltage source as an input to this system (the circuit in Fig. 4.4) and the current $i(t)$ as an output of the system.

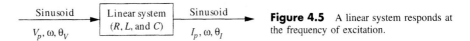

Figure 4.5 A linear system responds at the frequency of excitation.

The important idea is the following: If the input is a sinusoid, the output is also a sinusoid at the same frequency. This assertion can be justified through examination of Eq. (4.13) and reflection on the properties of sinusoids. The sinusoidal steady-state solution of Eq. (4.13) must be a function that, when differentiated and added to itself, will result in a sinusoid of frequency ω. The only mathematical function that qualifies is a sinusoid of the same frequency because the "shape" of the sinusoidal function is invariant to linear operations such as addition, differentiation, and integration.

A mechanical analogy. To get a feeling for this, think about a system of springs and masses. If you shake such a system with a certain frequency,[3] the entire system will shake at the same frequency. In other words, no new frequencies are generated in the system; the only frequency that exists is the one you applied externally. Therefore, if we apply a voltage at 60 Hz to the circuit in Fig. 4.4, the current will respond at 60 Hz.

New unknowns. The input voltage, being a sinusoid, is completely described by three numbers: the amplitude ($V_p = 100$ V), the frequency ($\omega = 120\pi$), and the phase

---

[3] The ac input is an electrical shaking.

($\theta_V = 30°$). The output current must also be a sinusoid, and hence can also be described by an amplitude ($I_p = ?$), frequency ($\omega = 120\pi$), and phase ($\theta_I = ?$). Because the output frequency is known, only the amplitude and phase need to be derived to determine the steady-state current. Our object, therefore, is to develop an efficient method for finding the amplitude and phase of the output; thus, I_p and θ_I become our new unknowns. We now develop a mathematical model suited to finding the unknown amplitude and phase. The first step is to represent the amplitude and phase of a sinusoid by a complex number.

A mathematical model. We model a sinusoid as a rotating point in the complex plane. An initial difficulty at this stage is that many readers are not up to speed in the complex number system without review. So now we must follow a detour in this development to refresh your knowledge of the complex plane. If you do not need the detour, skip ahead to "Phasor Idea," page 177.

Mathematics of the Complex Plane

Most of us were introduced to complex numbers through the study of quadratic equations. We discovered that the solution to certain equations like $x^2 + 4 = 0$ required the introduction of a new type of number:

$$x = \pm \sqrt{-4} = \pm i2 \tag{4.15}$$

imaginary number, complex number

The new numbers are called *imaginary* and the symbol i is chosen (by mathematicians) to identify these new numbers, like "−" is used to identify negative numbers. The connotation of the word "imaginary" is unfortunate, because these new numbers are no less the product of mathematical imagination than "real" numbers. The solutions of other quadratic equations are combinations of the real and imaginary numbers, such as $2 \pm i\sqrt{2}$. These combinations are called *complex numbers*. Complex numbers are perhaps well named because the rules for manipulating them are more complicated than those for real numbers, but on the whole, the complex numbers represent a reasonable and useful extension of our number system.

We expect that you already have some skill in dealing with complex numbers. However, we will review the properties of complex numbers, because we will need many of these properties to understand fully why a rotating point in the complex plane is an ingenious tool for analyzing ac circuits. We now list and illustrate some of the important properties of complex numbers.

Complex plane. All numbers can be represented as points in a complex plane, such as we show in Fig. 4.6. The horizontal axis represents real numbers, and the vertical axis represents imaginary numbers. You will note that we have changed to j instead of i to indicate imaginary numbers. This is customary among electrical engineers to avoid confusion between imaginary numbers and currents, which have traditionally been symbolized by i. In Fig. 4.6, we show two complex numbers, z_1 and z_2, each having real and imaginary components. In general, we can state that a complex number has a real and imaginary part,

$$z = x + jy \tag{4.16}$$

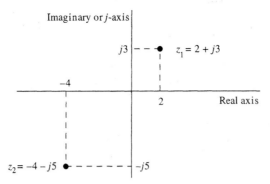

Figure 4.6 A complex number can be represented by a point in the complex plane.

In what follows, we give the principal algebraic and geometric properties of complex numbers, for we need these properties to represent sinusoids as rotating points in the complex plane.

Addition, subtraction, multiplication, and equality of complex numbers. A good first approximation to the algebra of complex numbers is to use ordinary algebra plus two additions: (1) Keep real and imaginary parts separate and (2) treat j^2 as -1. For example, when we add complex numbers, we add the real parts and the imaginary parts separately:

$$z_1 + z_2 = (2 + j3) + (-4 - j5) = (2 - 4) + j(3 - 5) = -2 - j2 \quad (4.17)$$

and similarly for subtraction. Thus, complex numbers add and subtract like vectors in a plane. This is one of the few properties common between complex numbers and vectors.

Similarly, multiplication of z_1 and z_2 is

$$\begin{aligned} z_1 z_2 &= (2 + j3)(-4 - j5) \\ &= 2(-4) + 2(-j5) + j3(-4) + j3(-j5) \\ &= -8 - j^2 15 - j10 - j12 \\ &= -8 + 15 + j(-10 - 12) \\ &= 7 - j22 \end{aligned} \quad (4.18)$$

EXAMPLE 4.2 **The square root of a complex number**

Find the square root of $-1 + j2$.

SOLUTION:
We have two unknowns, the real and the imaginary parts of the root, so we can proceed as follows:

$$(x + jy)^2 = -1 + j2$$
$$x^2 + 2x(jy) + (jy)^2 = -1 + j2$$
$$x^2 - y^2 + j(2xy) = -1 + j2$$
$$x^2 - y^2 = -1 \quad \text{and} \quad 2xy = 2 \quad (4.19)$$

Notice that we treated j^2 as -1, and we separated real and imaginary parts. Thus, an equation involving complex numbers is equivalent to two ordinary equations, one for the real part and one for the imaginary part. We can continue the problem by solving simultaneously for x and y, but we will stop here. As we will soon see, there is a better way to find the roots of complex numbers.

WHAT IF? What if you eliminate y from Eq. (4.19) and got the equation $x^4 + x^2 - 1 = 0$? Because this equation has four solutions and we expect only two roots for the square root of $-1 + j2$, how do we pick the right values of x?[4]

Division, conjugation, and absolute values. Division requires a trick to get the results into standard form:

$$\frac{z_2}{z_1} = \frac{-4 - j5}{2 + j3} = \frac{-4 - j5}{2 + j3} \times \frac{2 - j3}{2 - j3}$$

$$= \frac{(-4)(2) + (-j5)(-j3) + (-4)(-j3) + (-j5)(2)}{(2)^2 - (j3)^2}$$

$$= \frac{-8 - 15 + j(+12 - 10)}{4 + 9} = -\frac{23}{13} + j\frac{2}{13} \quad (4.20)$$

The first form we wrote to the right of the first equal sign is considered nonstandard because it contains a complex number in the denominator. To force the denominator to be real, we multiply top and bottom by the *complex conjugate* of z, which is the same complex number except that the sign of the imaginary part is changed. This causes the cross term in the product in the denominator to drop out and thus forces the denominator to be real and positive. Meanwhile, the numerator requires lots of careful work, but the rules are simple: Separate real and imaginary parts and let $j^2 = -1$. The final form is now considered standard because at a glance we can identify the real and imaginary parts of the quotient.

complex conjugate

If these tedious manipulations of complex numbers discourage you, take heart—there is a better way to multiply and divide complex numbers. Before we present this better way, however, we must look more closely at the complex conjugate. In Fig. 4.7, we show the complex conjugate of z_1, denoted by z_1^*, as the mirror image of z_1. We have already shown that the product of a complex number with its conjugate is a real and positive number:

$$zz^* = (x + jy)(x - jy) = x^2 - (jy)^2 = x^2 + y^2 \quad (4.21)$$

---

[4] The roots of the biquadratic are $x^2 = 0.618$ and -1.618. Because x must be a real number in Eqs. (4.19), we have only $x = \pm\sqrt{0.618} = \pm 0.786$.

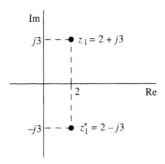
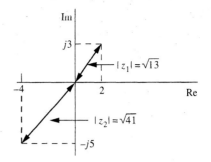

Figure 4.7 The complex conjugate of z_1 is z_1^*.

Figure 4.8 The magnitude of a complex number is its distance from the origin.

absolute value

This suggests the definition of the *absolute value* (or the magnitude) of a complex number:

$$|z| = \sqrt{zz^*} = \sqrt{x^2 + y^2} \qquad (4.22)$$

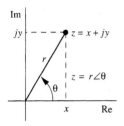

Figure 4.9 Complex number in rectangular and polar forms.

The absolute value of a complex number thus is the Pythagorean sum of the real and imaginary parts. Geometrically, we identify this with the distance from the origin to the point representing the complex number, as shown in Fig. 4.8.

Polar form. Figure 4.9 shows a general complex number. We have shown that z, a point in the complex plane, can be located either by its x, y location or by a distance and an angle. We have used r for the distance; clearly, r is $|z|$, the absolute value of z. We call the x, y form of z the *rectangular* form and we call the r, θ form the *polar* form. We can symbolize the two forms as

$$z = x + jy = r \angle \theta = |z| \angle \theta \qquad (4.23)$$

rectangular form, polar form

where the symbol $\angle$ is read "at an angle of." The right triangle yields simple transformations between *rectangular* and *polar* form:

$$
\begin{array}{ccc}
x, y & \Longleftrightarrow & r, \theta \\
x = r \cos \theta & & r = \sqrt{x^2 + y^2} \\
& \Longleftrightarrow & \\
y = r \sin \theta & & \theta = \tan^{-1} \dfrac{y}{x}
\end{array}
\qquad (4.24)
$$

Most engineering calculators have built-in functions for this transformation. Presumably, you have used this feature in working with vectors.[5]

At this stage of our review, the polar form represents only a notation. If we wanted to multiply or take the square root of a complex number, we would have to do it with the rectangular form. But the polar form is closely related to the exponential form, and because of this relationship, the polar form proves to be extremely useful, not merely as

---

[5] You will save yourself lots of time and grief by learning to use this feature.

a notation but also as a computational aid. All this arises out of Euler's theorem, an amazing relationship between the exponential function and angles in the complex plane.

Euler's theorem. The Swiss mathematician Euler discovered an important property of complex numbers. He began with the series expansion for the function e^x:

$$e^x = 1 + x + \frac{x^2}{2!} + \frac{x^3}{3!} + \cdots \qquad (4.25)$$

and made a series expansion for e^x when x is imaginary, $x = j\theta$.

$$e^{j\theta} = 1 + (j\theta) + \frac{(j\theta)^2}{2!} + \frac{(j\theta)^3}{3!} + \cdots \qquad (4.26)$$

Euler simplified with $(j\theta)^2 = -\theta^2$, $(j\theta)^3 = -j\theta^3$, ..., and followed the rule of grouping together real and imaginary parts. The results were

$$e^{j\theta} = \left(1 - \frac{\theta^2}{2!} + \frac{\theta^4}{4!} - \cdots\right) + j\left(\theta - \frac{\theta^3}{3!} + \frac{\theta^5}{5!} - \cdots\right) \qquad (4.27)$$

The series in parentheses Euler identified as the expansions for cosine and sine. We would speculate that he wrote something like

$$e^{j\theta} = \cos\theta + j\sin\theta \quad (?) \qquad (4.28)$$

The question mark is not there because of some suspicion about the mathematics—the algebra is correct—but it is there because Euler must have wondered what this could mean. Look, for example, at the strange combination of mathematical symbols on the left side of Eq. (4.28). A special number from differential calculus (e), the square root of -1, and the ratio of the arc to the radius of a circle (θ)—what can these have to do with each other? And on the right side, we find the ratios of the legs to the hypotenuse of a right triangle. What can these have to do with e and j?

Interpretation. The right side of Eq. (4.28) is interpreted in Fig. 4.10. Because the Pythagorean sum of cosine and sine is unity, the right side of Eq. (4.28) must be a point in the complex plane one unit from the origin located at an angle θ counterclockwise from the positive real axis, as shown. Euler's theorem requires that this point also be expressed by the function $e^{j\theta}$, as indicated by Eq. (4.28).

The best way to get acquainted with a new mathematical relationship often is to try it on some familiar specific cases to see how it works. Let us try Euler's formula on the

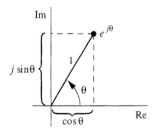

Figure 4.10 Implication of Eq. (4.28).

identity $\sqrt{-1} = j1$. We will express the left side of this equation with Euler's formula and follow the rules of exponents:

$$\sqrt{-1} = \sqrt{e^{j\pi}} = (e^{j\pi})^{1/2} = e^{j\pi/2}$$
$$= \cos\frac{\pi}{2} + j\sin\frac{\pi}{2}$$
$$= 0 + j1 \qquad (4.29)$$

The first substitution follows because -1 in the complex plane is one unit from the origin at an angle π from the positive real axis. The second change replaces the square root symbol by the one-half power, and the third change follows an ordinary rule of exponents. The form on the second line is a direct application of Euler's formula, and the third line gives the cosine and sine of $\pi/2$ radians. We conclude that Euler's formula leads to the correct answer for this special case. If we tried it on other powers and roots, it would work there also.

exponential form

Exponential form. Comparison of Figs. 4.9 and 4.10 reveals that a general complex number can be expressed as

$$z = re^{j\theta} \qquad (4.30)$$

where $r = |z|$. Clearly, this is closely related to the polar form, Eq. (4.23), and we have

$$z = r\angle\theta = re^{j\theta} = \underbrace{r\cos\theta}_{x} + j\underbrace{r\sin\theta}_{y} \qquad (4.31)$$

In Eq. (4.31), we normally express the angle of the polar form in degrees, whereas in the exponential form, the angle *must* be expressed in radians for the Euler formula to be valid. This inconsistency arises from our traditions but does not diminish the fact that both forms give the same information, namely, that the complex number falls in the complex plane at a radius r and an angle θ counterclockwise from the positive real axis. We stress that the polar form is merely a compact notation, whereas the exponential form is a legitimate mathematical function.

Multiplication, division, and conjugation revisited. The laws of exponents reveal that the magnitude of the product of two complex numbers is the product of the magnitudes of the numbers and the angle of the product is the sum of their angles.

$$(z_1)(z_2) = (r_1 e^{j\theta_1})(r_2 e^{j\theta_2}) = r_1 r_2 e^{j(\theta_1 + \theta_2)} \qquad (4.32)$$

Although the proof rests on the exponential form, the results are more easily expressed in polar form:

$$(r_1 \angle \theta_1)(r_2 \angle \theta_2) = r_1 r_2 \angle (\theta_1 + \theta_2) \qquad (4.33)$$

Hence, to multiply two complex numbers, multiply their magnitudes and add their angles. In a similar way, we can show that dividing two complex numbers requires dividing the magnitudes and subtracting the angles:

$$\frac{r_2 \angle \theta_2}{r_1 \angle \theta_1} = \frac{r_2}{r_1} \angle (\theta_2 - \theta_1) \tag{4.34}$$

EXAMPLE 4.3 Multiplication and division of complex numbers

Use polar form to confirm Eqs. (4.18) and (4.20).

SOLUTION:

First, we convert z_1 and z_2 to polar form using Eq. (4.24):[6]

$$z_1 = 2 + j3 = 3.61 \angle 56.3°$$

$$z_2 = -4 - j5 = 6.40 \angle -128.7° \tag{4.35}$$

We multiply using Eq. (4.33):

$$z_1 z_2 = (3.61)(6.40) \angle (53.6° - 128.7°)$$
$$= 23.1 \angle -72.5° = 7 - j22 \tag{4.36}$$

We divide using Eq. (4.34):

$$\frac{z_2}{z_1} = \frac{6.40}{3.61} \angle [-128.7° - (56.3°)]$$

$$= 1.78 \angle (-185.0°) = -1.77 + j0.154 \tag{4.37}$$

WHAT IF? What if you express the complex conjugate in polar form? You see how the polar form gets the same results directly.[7]

Summary of how to calculate with complex numbers. When we add and subtract complex numbers, the rectangular form must be used. This is true because we add and subtract real and imaginary parts separately. When we multiply (or divide) complex numbers, the polar form should be used because the magnitudes multiply (or divide) and the angles add (or subtract).

For the analysis of ac circuits, we must master the manipulation of complex numbers. Our techniques require frequent changes between rectangular and polar form. In the days of slide rules, these conversions were a bother, but now they require a simple keystroke on the calculator. If you currently do not know about those keys, locate the manual on your calculator and look them up. This review of complex numbers is vital to your gaining skill in solving ac circuit problems. These operations, like the vector

---

[6] Actually, I used the button on my calculator marked $\to P$.

[7] If $z = r \angle \theta$, then $z^* = r \angle (-\theta) = r \cos \theta + jr \sin(-\theta) = r \cos \theta - jr \sin \theta$.

manipulations required in statics, are simple in principle, but care is required to produce correct answers consistently.

Some examples. Here are two additional examples of complex number manipulation, which give practice and show some new principles.

$$(2 - j6)(5\angle + 30°)^* = (6.32\angle - 71.6°)(5\angle - 30°)$$
$$= 31.6\angle - 101.6°$$
$$= -6.34 - j31.0 \qquad (4.38)$$

Note that complex conjugation merely changes the sign of the angle.

Find z, where $z^3 = -1 + j0.5$.

$$z^3 = -1 + j0.5 = 1.12 \angle 153.4° = 1.12\, e^{j(153.4)(\pi/180)}$$

$$z = (1.12 e^{j2.68})^{1/3} = \sqrt[3]{1.12}\, e^{j2.68/3} \qquad (4.39)$$
$$= \text{also } \sqrt[3]{1.12}\, e^{j(2.68+2\pi)/3} \text{ and } \sqrt[3]{1.12}\, e^{j(2.68+4\pi)/3}$$
$$= 1.04\angle 51.1°, 1.04\angle 171.1°, 1.04\angle 291.1°$$

To find all three cube roots, we must add 2π and 4π (one and two additional rotations) to the angle before dividing by 3.

Phasor Idea

Expressing a sinusoid with a complex number. Now that we have reviewed the mathematics of the complex plane, we will return to the main quest, that of finding the steady-state solution of the ac circuit problem described in Fig. 4.4 and Eq. (4.13). Specifically, we wish to find a way to represent sinusoidal functions, such as that appearing in Eq. (4.1), repeated here:

$$v_s(t) = V_p \cos(\omega t + \theta), \quad V_p = 100\text{ V}, \quad \omega = 2\pi \times 60, \quad \theta = 30° \qquad (4.40)$$

Euler's formula, Eq. (4.28), allows us to express a sinusoidal waveform with the function

$$v_s(t) = V_p \cos(\omega t + \theta) = \text{Re}\{V_p e^{j(\omega t + \theta)}\} \qquad (4.41)$$

where Re, the "real part of," is the complex-plane notation for the projection on the real axis. Equation (4.41) follows from Eq. (4.42), which expands Euler's formula:

$$V_p e^{j(\omega t + \theta)} = V_p[\cos(\omega t + \theta) + j\sin(\omega t + \theta)]$$
$$= \underbrace{V_p \cos(\omega t + \theta)}_{\text{real part}} + j\underbrace{V_p \sin(\omega t + \theta)}_{\text{imaginary part}} \qquad (4.42)$$

Rotating point in the complex plane. The left side of Eq. (4.42) has great importance in this development and deserves more attention. We rearrange it as

$$V_p e^{j(\omega t + \theta)} = V_p(e^{j\omega t})(e^{j\theta}) = (V_p e^{j\theta})e^{j\omega t}$$
$$= (V_p \angle \theta)e^{j\omega t} = (100 \angle 30°)e^{j120\pi t} \qquad (4.43)$$

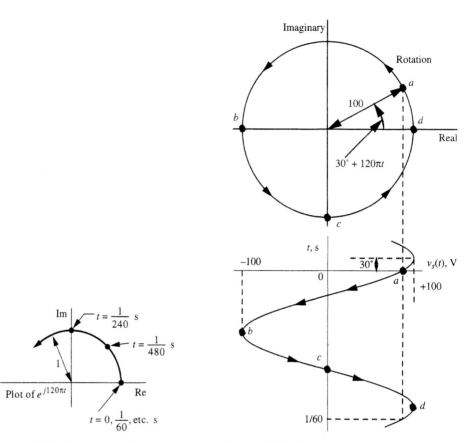

Figure 4.11 Rotating point in the complex plane.

Figure 4.12 The correspondence between the rotating point in the complex plane and the sinusoidal waveform.

The first line results from some juggling of terms and the rules for exponents. The second line emphasizes that this term consists of a complex constant, $V_p \angle \theta = 100 \angle 30°$, multiplied by a time function, $e^{j\omega t}$, which describes the rotation of a point in the complex plane. If we follow its progress as time increases, as shown in Fig. 4.11, we find it to be a point at unit distance from the origin rotating around the origin, starting at the real axis at $t = 0$ and passing through that point 60 times every second.

The time function $e^{j\omega t}$ describes a point rotating in the complex plane at an angular frequency of ω radians per second. When we multiply this rotation function by the complex number $V_p \angle \theta = 100 \angle 30°$, we move the point out to a magnitude of $V_p = 100$ and counterclockwise by an angle of $\theta = 30°$ at $t = 0$, as shown in Fig. 4.12 (point *a*). The rotation begins from that point and rotates around a circle of radius 100, 60 times a second. The projection on the real axis is shown below the complex plane. Clearly, we have the desired sinusoid—correct amplitude, phase, and frequency. Figure 4.12 is a graphical interpretation of Eq. (4.41), illustrated for the values appropriate for Eq. (4.40).

Summary. We have shown that a sinusoidal function can be represented by

$$v_s(t) = V_p \cos(\omega t + \theta) = \text{Re}\{\underline{\mathbf{V}}_s e^{j\omega t}\} \tag{4.44}$$

where $\underline{\mathbf{V}}_s = V_p \angle \theta = 100 \angle 30°$ is a complex number containing the amplitude and phase of the sinusoid. We use the uppercase bold, underlined symbols to indicate a complex number that does not vary with time. Thus, we can represent a sinusoid by a rotating point in the complex plane. To see why this is useful, we must return to the differential equation we are solving.

Back to the Circuit Problem

Using complex numbers to solve the DE. Equation (4.45) is our DE,

$$L\frac{di}{dt} + Ri = V_p \cos(\omega t + \theta) = \text{Re}\{V_p \angle \theta e^{j\omega t}\}$$

$$= \text{Re}\{\underline{\mathbf{V}}_s e^{j\omega t}\} \tag{4.45}$$

We have introduced the complex number form for the sinusoidal voltage source.

The new unknown. Recall that we know something about the unknown current, $i(t)$. Earlier we argued that the current had to be a sinusoid of the same frequency as the input voltage. Only the amplitude and phase of the current must be determined. Using our scheme for representing sinusoids by complex numbers, we know that $i(t)$ may be represented in the form

$$i(t) = \text{Re}\{\underline{\mathbf{I}} e^{j\omega t}\}, \qquad \text{where } \underline{\mathbf{I}} = I_p \angle \theta_I \tag{4.46}$$

when I_p is the peak amplitude and θ_I the phase of the current. We have thus introduced a complex number, $\underline{\mathbf{I}}$ as the unknown; everything else in the equation is known. We have numerical values for L, R, and $\underline{\mathbf{V}}_s$, and, of course, ω is known from the input. We can now substitute Eq. (4.46) into Eq. (4.45):

$$L\frac{d}{dt}\text{Re}\{\underline{\mathbf{I}} e^{j\omega t}\} + R \times \text{Re}\{\underline{\mathbf{I}} e^{j\omega t}\} = \text{Re}\{\underline{\mathbf{V}}_s e^{j\omega t}\} \tag{4.47}$$

We wish to avoid taking unnecessary excursions into mathematical proofs, so you will have to take our word for this assertion: The derivative can be taken inside the "real part" operation:

$$\frac{d}{dt}\text{Re}\{\underline{\mathbf{I}} e^{j\omega t}\} = \text{Re}\left\{\frac{d}{dt}\underline{\mathbf{I}} e^{j\omega t}\right\} = \text{Re}\left\{\underline{\mathbf{I}}\frac{d}{dt}e^{j\omega t}\right\}$$

$$= \text{Re}\{j\omega \underline{\mathbf{I}} e^{j\omega t}\} \tag{4.48}$$

We substitute the derivative back into Eq. (4.47). Because the "real part" operation distributes, we can collect terms:

$$\text{Re}\{L(j\omega \underline{\mathbf{I}} e^{j\omega t}) + R(\underline{\mathbf{I}} e^{j\omega t})\} = \text{Re}\{\underline{\mathbf{V}}_s e^{j\omega t}\} \tag{4.49}$$

$$\text{Re}\{(j\omega L\underline{\mathbf{I}} + R\underline{\mathbf{I}} - \underline{\mathbf{V}}_s)e^{j\omega t}\} = 0 \tag{4.50}$$

Equation (4.50) emerges from the algebra; but what does it mean? Look first at the term, $j\omega L\underline{\mathbf{I}} + R\underline{\mathbf{I}} - \underline{\mathbf{V}}_s$. This represents a complex number that does not vary with time. Because it contains the complex number representing the unknown current amplitude and phase, $\underline{\mathbf{I}}$, this term in Eq. (4.50) is an unknown constant, a point somewhere in the complex plane. Of course, the $e^{j\omega t}$ term rotates this unknown constant, and Re represents the projection of the rotating point on the real axis. The right side of Eq. (4.50) requires that this projection be zero at all times; hence, the complex constant must itself be zero. If the point were not rotating, many complex constants could have zero projection,[8] but the origin is the only point that continues to have zero projection when rotated about the origin.

Conclusion. Equation (4.50) requires

$$j\omega L\underline{\mathbf{I}} + R\underline{\mathbf{I}} - \underline{\mathbf{V}}_s = 0 \qquad (4.51)$$

Equation (4.51) involves only complex constants; time is no longer a factor. We know all the constants except $\underline{\mathbf{I}}$, so we can solve for the unknown $\underline{\mathbf{I}}$:

$$\underline{\mathbf{I}} = \frac{\underline{\mathbf{V}}_s}{R + j\omega L} = \frac{100\angle 30°}{100 + j(120\pi \times 0.5)}$$

$$= \frac{100\angle 30°}{213\angle 62.1°} = 0.469\angle -32.1° \qquad (4.52)$$

Recall that the magnitude of $\underline{\mathbf{I}}$ represents the peak value of the sinusoidal steady-state current flowing in the RL circuit and the angle of $\underline{\mathbf{I}}$ represents the phase angle of the current. Hence, we may write

$$i_{ss}(t) = 0.469\cos(120\pi t - 32.1°) \text{ A} \qquad (4.53)$$

This at last is the steady-state solution to the DE in Eq. (4.13) and hence this is the steady-state current that flows in the circuit shown in Fig. 4.4 after the transient period passes. Before finding the transient part of the solution, we will summarize the argument.

Summary of the argument. Because the development of Eq. (4.51) has taken many pages and extensive math review, you may feel that you have been led through a complicated argument. Actually, there are but a few steps:

1. We wrote the DE and focused on the steady-state solution.

$$L\frac{di}{dt} + Ri = V_p \cos(\omega t + \theta) \qquad (4.54)$$

2. We then represented the sinusoidal voltage and the unknown sinusoidal current by the real parts of rotating complex constants and took the derivative. The result was

$$\text{Re}\{j\omega L\underline{\mathbf{I}}\, e^{j\omega t}\} + \text{Re}\{R\underline{\mathbf{I}}\, e^{j\omega t}\} = \text{Re}\{\underline{\mathbf{V}}_s\, e^{j\omega t}\} \qquad (4.55)$$

---

[8] All points on the imaginary axis have this property.

3. We then collected terms and reasoned that the resulting complex constant had to vanish when the equation is put into the form of Eq. (4.50). This amounts to the same thing as dropping Re and $e^{j\omega t}$ in Eq. (4.55). Note that we did not *cancel* these factors—that would be mathematically incorrect—but the result is the same as if we had canceled them.

4. The resulting equation is solved for $\underline{I}$, and the result is easily interpreted in terms of the amplitude and phase of the current in the circuit.

Complete solution. We may now determine the complete solution to Eq. (4.13) using principles developed in Chap. 3. The form of the solution is

$$i(t) = Ae^{-t/\tau} + 0.469 \cos(120\pi t - 32.1°) \qquad (4.56)$$

where $\tau = L/R = 1/200$ s and A is an unknown constant to be determined from the initial condition. The initial current must be zero, $i(0^+) = 0$, due to the inductor; hence,

$$0 = Ae^{-0/\tau} + 0.469 \cos(-32.1°) \qquad (4.57)$$

Hence, $A = -0.397$. Thus, the total solution is

$$i(t) = -0.397e^{-200t} + 0.469 \cos(120\pi t - 32.1°) \text{ A} \qquad (4.58)$$

This response is shown in Fig. 4.13. The current rapidly approaches the sinusoidal steady-state part of the solution after the switch is closed.

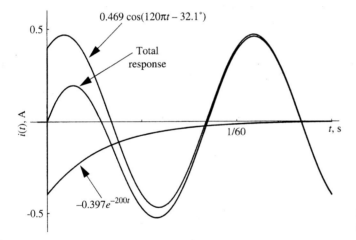

Figure 4.13 The current is the sum of the transient and steady-state responses.

phasor

Phasors. Before giving further shortcuts and simplifications, let us summarize what we have learned and introduce some vocabulary. A *phasor* is a complex number containing the amplitude and phase of a sinusoid. The phasor produces the time-function sinusoid when rotated at the proper frequency and projected on the real axis. Because it is difficult to see a single point, phasors are usually drawn as an arrow from the origin, as in Fig. 4.14. Because phasors add like vectors, they are often incorrectly referred to as vectors.

The equation relating the sinusoid $v(t)$ and the phasor $\underline{V}$ is

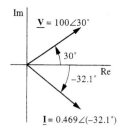

$$v(t) = \text{Re}\{\underline{V}\,e^{j\omega t}\} \tag{4.59}$$

The phasor encodes the amplitude and phase of the sinusoid in the form $\underline{V} = V_p \angle \theta$. Equation (4.59) can be differentiated to show that the phasor representing the derivative is

$$\frac{dv(t)}{dt} = \text{Re}\left\{\underline{V}\frac{de^{j\omega t}}{dt}\right\} = \text{Re}\{\underbrace{j\omega\underline{V}}_{\text{phasor for derivative}} e^{j\omega t}\} \tag{4.60}$$

Phasors can be added to represent the addition of two sinusoids of the same frequency, as is shown in the next example.

We use the expression *frequency domain* to refer to the form of the DE after it has been transformed into a complex equation. The DE is called the *time-domain* formulation of the problem, but the transformed equation belongs to the frequency domain—time is no longer a factor.

Equation (4.60) shows that the differentiation in the time domain corresponds to multiplication by $j\omega$ in the frequency domain. The phasor voltages and currents belong to the frequency domain. We move from the frequency domain to the time domain by rotating the phasors at the proper frequency and taking projections on the real axis.[9]

Figure 4.14 Voltage phasor and current phasor.

💡 **The Frequency Domain** 7

frequency domain, time domain

EXAMPLE 4.4 **Phasor addition**

Add the two sinusoids $v_1(t) = 12\cos(100t)$ and $v_2(t) = 8\cos(100t - 48°)$.

SOLUTION:
First, we find the phasors

$$v_1(t) = 12\cos(100t) \Rightarrow \underline{V}_1 = 12 \angle 0°$$

$$v_2(t) = 8\cos(100t - 48°) \Rightarrow \underline{V}_2 = 8 \angle -48° = 5.35 - j5.94 \tag{4.61}$$

The sum, $v_1(t) + v_2(t)$, will be represented by the phasor sum, $\underline{V}_1 + \underline{V}_2$, which is

$$\underline{V}_1 + \underline{V}_2 = 12 + j0 + (5.35 - j5.94) \tag{4.62}$$
$$= 17.35 - j5.94 = 18.34 \angle -18.9°$$

Thus, the sum will be $v_1(t) + v_2(t) = 18.34\cos(100t - 18.9°)$.

WHAT IF? What if it were the difference, $v_1(t) - v_2(t)$, you wanted?[10]

---

[9] That is the mathematical way. In practice, we take the magnitude and angle of the phasor and substitute into a standard cosine form, Eq. (4.1).
[10] $v_1(t) - v_2(t) = 8.92\cos(100t + 41.8°)$.

LEARNING OBJECTIVE 2.

To understand the phasor concept and be able to find the sinusoidal steady-state response of a circuit, given the differential equation

Shortening the procedure. This method for determining the sinusoidal steady state admits to additional shortcuts. If we compare Eq. (4.51) with the original DE, Eq. (4.13), we see a direct correlation between terms:

$$i(t) \Rightarrow \underline{I}, \qquad v_s(t) \Rightarrow \underline{V}_s, \qquad \frac{d}{dt} \Rightarrow j\omega \qquad (4.63)$$

We have shown that these are legitimate transformations. These suggest a shorter method:

1. Write the DE.

2. Perform the transformations shown in Eq. (4.63) on the known sinusoidal source, the unknown voltage or current, and the d/dt.

3. Solve the resulting complex equation for the unknown.

4. Interpret the results by substituting the amplitude and phase back into the standard form for a sinusoid. This amounts to the transformation

$$\underline{V} = V_p \angle \theta$$
$$\Downarrow \quad \Downarrow \quad \searrow$$
$$v(t) = V_p \cos(\omega t + \theta) \qquad (4.64)$$

EXAMPLE 4.5 Short method

Using the short method, determine the voltage in Fig. 4.15.

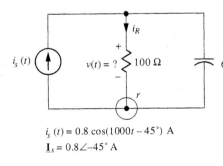

$i_s(t) = 0.8 \cos(1000t - 45°)$ A
$\underline{I}_s = 0.8\angle -45°$ A

Figure 4.15 Solve for the steady-state voltage across the parallel resistor and capacitor.

SOLUTION:
We use nodal analysis to derive the DE. At the top node, KCL is

$$\frac{v}{R} + C\frac{dv}{dt} = i_s(t) \qquad (4.65)$$

The second step is to transform this DE into a complex equation with the changes suggested by Eq. (4.63).

$$i_s \Rightarrow \underline{I}_s = 0.8\angle-45° \quad v \Rightarrow \underline{V} = ? \quad \frac{d}{dt} \Rightarrow j\omega \qquad (4.66)$$

The resulting complex equation is

$$\frac{1}{R}\underline{V} + j\omega C\underline{V} = \underline{I}_s \qquad (4.67)$$

We can solve the equation for $\underline{V}$, which represents the amplitude and phase of the unknown voltage. The results are

$$\underline{V} = \frac{\underline{I}_s}{1/R + j\omega C} = \frac{0.8\angle-45°}{1/100 + j1000 \times 6 \times 10^{-6}}$$

$$= \frac{0.8\angle-45°}{0.0117\angle 31.0°} = 68.6\angle - 76.0° \text{ V} \qquad (4.68)$$

The last step consists in writing the current in the standard form for sinusoids. From Eq. (4.64), we write

$$v(t) = 68.6\cos(1000t - 76.0°) \text{ V} \qquad (4.69)$$

You must admit that apart from the complex arithmetic, which may still be unfamiliar to you, the derivation of the DE is now the most difficult part of the solution. We will soon introduce shortcuts for simplifying that part of the problem.

EXAMPLE 4.6 **Using sine form**

In Fig. 4.16(a), we show a series RC circuit, with the voltage across the capacitor given as $5\sin(500t)$ V. Determine the voltage across the source that produces this prescribed voltage.

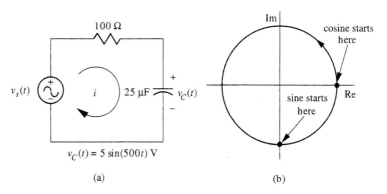

Figure 4.16 (a) The voltage across C is given; the source voltage is to be found. (b) The frequency domain equivalent of a sine function is $-j$ or $1\angle -90°$ because a rotating point starting at the bottom projects on to the real axis as a sine function.

SOLUTION:
Considering that the current is $C(dv_C/dt)$, the voltage across the resistance is $RC(dv_C/dt)$; hence, the DE for the source voltage is

$$v_s(t) = 100 \times 25 \times 10^{-6} \frac{dv_C}{dt} + v_C \qquad (4.70)$$

which transforms into the frequency domain as

$$\underline{V}_s = 25 \times 10^{-4}(j500)\underline{V}_C + \underline{V}_C = (1 + j1.25)\underline{V}_C \qquad (4.71)$$

where $\underline{V}_s$ and $\underline{V}_C$ are phasors representing the voltages across the source and capacitor, respectively. Because $v_C(t)$ is a sine (instead of a cosine) function, we must represent it as

$$v_C(t) = 5\sin(500t) \Rightarrow \underline{V}_C = -j5 \qquad (4.72)$$

We may show that sine transforms into $-j$ by using the trigonometric identity $\sin x = \cos(x - 90°)$, but a graphical demonstration is shown in Fig. 4.16(b). We have identified $-j$ with sine because that point produces a sine function when rotated and projected onto the real axis: The projection is zero at $t = 0$, but increases to unity one-quarter of a period later. Finally, we substitute Eq. (4.72) into Eq. (4.71) and transform back into the time domain.

$$\underline{V}_s = (1 + j1.25)(-j5) = 1.601 \angle 51.3° \times 5 \angle -90°$$

$$= 8.00 \angle -38.7°$$

$$v_s(t) = 8.00\cos(500t - 38.7°) \text{ V} \qquad (4.73)$$

Check Your Understanding

1. Convert $10 - j12$ and $-30 + j58$ to polar form.
2. What are the rectangular forms for $4 \angle 25°$ and $0.025 \angle -140°$?
3. Convert $13.5 e^{j0.86}$ to polar and rectangular forms.
4. What is the complex conjugate of $z = -2 + j6$?
5. Give both square roots of $1 + j1$.
6. What are the phasors representing $-6\cos(\omega t - 30°)$ and $5\sin(\omega t + 10°)$?
7. What is the time-domain sum of the two sinusoids in the previous question?
8. In transforming a DE into the frequency domain, what replaces d/dt?

Answers. **(1)** $15.6 \angle -50.2°$, $65.3 \angle 117.3°$; **(2)** $3.63 + j1.69$, $(-1.92 - j1.61) \times 10^{-2}$; **(3)** $13.5 \angle 49.3°$, $8.81 + j10.2$; **(4)** $-2 - j6$; **(5)** $1.19 \angle 22.5°$, $1.19 \angle 202.5°$; **(6)** $6 \angle 150°$, $5 \angle -80°$; **(7)** $4.74\cos(\omega t - 156°)$; **(8)** $j\omega$.

4.3 IMPEDANCE: REPRESENTING THE CIRCUIT IN THE FREQUENCY DOMAIN

The Final Shortcut

Look at the answer. Let us look again at the steady-state response of the *RL* circuit, shown again in Fig. 4.17. This is a simple circuit with *R* and *L* in series; we know the sinusoidal driving voltage and we have found the current. When expressed as phasor voltages and currents, our results were given in Eq. (4.52) as

$$\underline{\mathbf{I}} = \frac{\underline{\mathbf{V}}_s}{R + j\omega L} \tag{4.74}$$

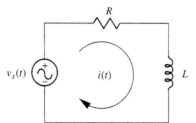

Figure 4.17 The *RL* circuit again.

We interpret Eq. (4.74) to state that the phasor current is found by dividing the phasor voltage by the sum of the resistance and $j\omega L$. Because Eq. (4.74) closely resembles Ohm's law, every part of this expression has meaning to us except the $j\omega L$ part. Clearly, this term represents the effect of the inductor in the circuit. What can it mean?

The Frequency Domain

Impedance. This question, or one similar to it, sparked long ago an idea that leads to the final shortcut in analyzing ac circuits—the idea of impedance. This idea is the following: Because we are dealing only with sinusoidal voltages and currents and because these can be represented by complex numbers, why not represent *R*'s, *L*'s, and *C*'s by complex numbers also? This suggests transforming Ohm's law and the definitions of *L* and *C* into the frequency domain and representing them by complex equations.

Impedance of an inductor. Here is how it works for the inductor: The defining equation for an inductor is

$$v_L = L \frac{di_L}{dt} \tag{4.75}$$

Because we are dealing with sinusoids, we can transform this equation into a frequency-domain equivalent:

$$v_L(t) \Rightarrow \underline{\mathbf{V}}_L, \quad i_L(t) \Rightarrow \underline{\mathbf{I}}_L, \quad \frac{d}{dt} \Rightarrow j\omega \tag{4.76}$$

The result is

$$\underline{\mathbf{V}}_L = L(j\omega \underline{\mathbf{I}}_L) = j\omega L \underline{\mathbf{I}}_L \tag{4.77}$$

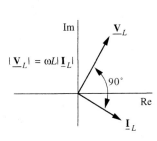

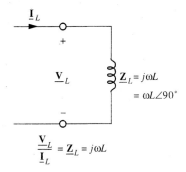

Figure 4.18 For an inductor, current lags voltage by 90°.

Figure 4.19 Representing an inductor by its impedance.

The phasor interpretation of Eq. (4.77) is shown in Fig. 4.18. Equation (4.77) gives both the magnitude and phase relationships between the phasor voltage and phasor current for an inductor. The magnitude of the phasor voltage is ωL times the magnitude of the phasor current. Because $j1 = 1 \angle 90°$, the phase of the phasor voltage leads the phase of the phasor current by 90°. We can verify this interpretation by taking the derivative in the time domain. If $i_L = I_p \cos(\omega t + \theta_I)$, then $v_L(t)$ is

$$v_L = L \frac{d}{dt} I_p \cos(\omega t + \theta_I) = -\omega L I_p \sin(\omega t + \theta_I)$$
$$= \omega L I_p \cos(\omega t + \theta_I + 90°) \tag{4.78}$$

where we have used the identity that $-\sin\phi = \cos(\phi + 90°)$. We use the word "impedance" to refer to the complex number relating the phasor voltage and phasor current of an element in an ac circuit, as shown in Fig. 4.19.

We began using the word "impedance" back in Chap. 2 in connection with output impedance and impedance level, with the promise that a precise definition would follow. We define *impedance*, $\underline{Z}$, as the phasor voltage divided by the phasor current,

impedance

$$\underline{Z} = \frac{\underline{V}}{\underline{I}} \tag{4.79}$$

Applying this definition to Eq. (4.77), we obtain the result shown in Fig. 4.19:

$$\underline{Z}_L = j\omega L \tag{4.80}$$

EXAMPLE 4.7 **Impedance of an inductor**

Find the impedance of a 10-mH inductor at a frequency of 360 Hz.

SOLUTION:
The impedance is given by Eq. (4.80):

$$\underline{Z}_L = j\omega L = j2\pi \times 360 \times 0.01 = j22.6 \ \Omega$$
$$= 22.6 \angle + 90° \ \Omega \qquad (4.81)$$

WHAT IF? What if the frequency is doubled?[11]

Kirchhoff's laws in the frequency domain. You will note that Fig. 4.19 also introduces the practice of putting the phasor voltage and current on the circuit diagram. This is legitimate and useful because, for sinusoidal voltages and currents, KVL and KCL can also be transformed directly into the frequency domain; for example,

$$-v_s(t) + v_R(t) + v_L(t) = 0 \Rightarrow -\underline{V}_s + \underline{V}_R + \underline{V}_L = 0 \qquad (4.82)$$

Conservation of Energy

Equation (4.82) gives the frequency-domain version of KVL as applied to the RL circuit. If there were nodes in the circuit, we could also write frequency-domain versions of KCL at these nodes.

Summary. We can represent all voltages and currents in an ac circuit as phasors. These phasors are complex numbers representing the amplitudes and phases of the waveforms in the sinusoidal steady state. Phasor voltages and currents obey KVL and KCL. We can also represent R and L (and C, when we get around to it) as complex numbers, which are called *impedances*. Figure 4.20 gives the impedance of R, which is Ohm's law for phasors. The impedance of a resistor is real because its voltage and current have the same phase.

Representing the circuit in the frequency domain. We can transform the entire circuit into the frequency domain, representing voltages and currents as phasors and representing the resistor and inductor as impedances. Because phasor voltages and currents obey KVL and KCL, we can use the techniques of dc circuit theory to

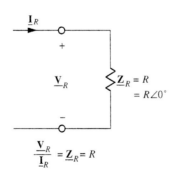

Figure 4.20 The impedance of a resistor is real because no phase shift occurs.

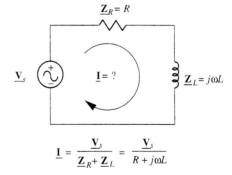

Figure 4.21 The frequency-domain version of the RL circuit.

---

[11] Then the impedance doubles in magnitude to 45.2 Ω. The angle is unchanged.

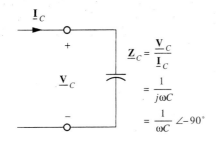

Figure 4.22 Impedance of a capacitor.

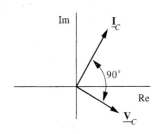

Figure 4.23 For a capacitor, the current leads the voltage by 90°.

solve the circuit, only we must now work with impedances as if they were "complex resistors."

Benefits of the impedance concept. We can now solve our original problem in the frequency domain. Figure 4.21 shows the circuit transformed into the frequency domain. Because we have a simple series circuit, we can find $\underline{I}$ by the method of equivalent resistance (actually, equivalent impedance), combining the impedances of R and L in series. If we were faced with a more complicated circuit, we might analyze the circuit using node voltages, loop currents, or a Thévenin equivalent circuit. Through the concept of impedance, all the techniques of dc circuits can be applied to ac circuit problems. The DE does not have to be written.

The impedance of a capacitor. We may use these techniques to derive the impedance of a capacitor, as symbolized in Fig. 4.22. We transform the definition of a capacitor into the frequency domain,

$$i_C = C\frac{dv_C}{dt} \Rightarrow \underline{I}_C = C(j\omega\underline{V}_C) = j\omega C\underline{V}_C \qquad (4.83)$$

Thus, the impedance of a capacitor is

$$\underline{Z}_C = \frac{\underline{V}_C}{\underline{I}_C} = \frac{1}{j\omega C} = \frac{j}{j^2\omega C} = -j\frac{1}{\omega C} = \frac{1}{\omega C}\angle -90° \qquad (4.84)$$

For a capacitor, the voltage lags the current by 90°, as shown in Fig. 4.23.

LEARNING OBJECTIVE 3.

To understand how to use phasors and impedance to determine the sinusoidal steady-state response of a circuit

Frequency–domain technique. We now illustrate the full use of the frequency domain in finding the sinusoidal steady-state response of a circuit. We will find $v(t)$ in Fig. 4.24. The steps are as follows:

1. Transform the time domain variables to phasors.

$$v_s(t) = 10\cos(2000t + 20°) \Rightarrow \underline{V}_s = 10\angle 20°$$

$$v(t) = ? \qquad\qquad \Rightarrow \underline{V} = ? \qquad (4.85)$$

2. Transform the circuit into the frequency domain as impedances.

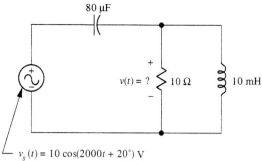
$v_s(t) = 10 \cos(2000t + 20°)$ V

Figure 4.24 The circuit in the time domain.

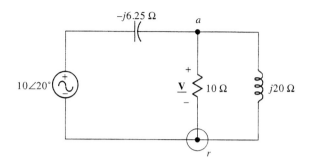

Figure 4.25 The circuit in the frequency domain. Impedances replace R, L, and C.

$$R \Rightarrow \underline{Z}_R = R = 10 \, \Omega$$

$$L \Rightarrow \underline{Z}_L = j\omega L = j2000 \times 0.01 = j20 \, \Omega$$

$$C \Rightarrow \underline{Z}_C = \frac{1}{j\omega C} = -j\frac{1}{2000 \times 80 \times 10^{-6}} = -j6.25 \, \Omega \quad (4.86)$$

The resulting circuit is shown in Fig. 4.25.

Conservation of Charge

3. Analyze the circuit by an efficient technique, treating impedances as "complex resistors." Here we have marked the circuit for nodal analysis. Kirchhoff's current law at the top node is

$$\frac{\underline{V}_a - 10 \angle 20°}{-j6.25} + \frac{\underline{V}_a - (0)}{10} + \frac{\underline{V}_a - (0)}{j20} = 0 \quad (4.87)$$

which yields

$$\underline{V}_a = \frac{10 \angle 20°/(-j6.25)}{1/-j6.25 + 1/10 + 1/j20} = 10.8 \angle 62.3° \text{ V} \quad (4.88)$$

4. Transform the resulting phasor back into the time domain.

$$10.8 \angle 62.3° \Rightarrow v(t) = 10.8 \cos(2000t + 62.3°) \text{ V} \quad (4.89)$$

EXAMPLE 4.8 **Voltage divider**

Figure 4.26(a) shows a circuit in the time domain in sinusoidal steady state. Determine the voltage across the resistor.

SOLUTION:
Figure 4.26(b) shows the circuit, source, and unknown transformed into the frequency domain. As discussed on page 184, we transform sine → $1 \angle -90°$ because the cosine function is standard. This changes the phase from $+20°$ for sine to $-70°$ for cosine. The phasor voltage

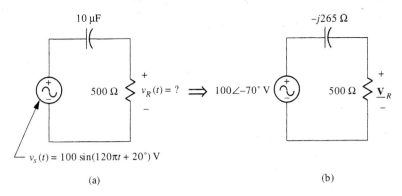

Figure 4.26 (a) Time-domain circuit; (b) frequency-domain circuit.

across the resistor can be determined from a voltage divider:

$$\underline{\mathbf{Z}}_C = \frac{1}{j120\pi \times 10 \times 10^{-6}} = -j265$$

$$\underline{\mathbf{V}}_R = 100 \angle -70° \times \frac{500}{500 - j265} = 88.3 \angle -42.1° \text{ V} \quad (4.90)$$

The time-domain voltage across the resistor is thus

$$v_R(t) = 88.3 \cos(120\pi t - 42.1°) \text{ V} \quad (4.91)$$

WHAT IF? What if the capacitor were replaced by a 1-H inductor?[12]

Summary of the frequency-domain method. We have developed an efficient method for finding the sinusoidal steady-state response of a circuit. The circuit is transformed into the frequency domain by replacing time-domain voltages and currents by phasors. The circuit elements R, L, and C are replaced by their impedances R, $j\omega L$, and $1/j\omega C$, respectively. The circuit is analyzed by the most efficient means—voltage dividers, node voltages, Thévenin equivalent circuit, etc. The resulting phasor is then transformed back into the time domain.

4.4 PHASOR DIAGRAMS FOR RL, RC, AND RLC CIRCUITS

LEARNING OBJECTIVE 4.

To understand the effect of varying frequency on series and parallel RL, RC, and RLC circuits

Varying the frequency. You may wonder why we use the grandiose word "domain" for the techniques described before. You are correct in observing that we do not yet have much of a domain, only a technique for analyzing ac steady-state circuits at a single frequency. But there really is a domain; the exploration of this idea continues as an important theme in the next chapter.

---

[12] $v_R(t) = 79.8 \cos(120\pi t - 107.0°)$.

IDEA 7: The Frequency Domain

The viewpoint in this section is that we have available a sinusoidal source, the frequency of which we vary. Here we explore the effects of frequency changes on the response of series and parallel circuits to gain insight into the properties of inductors and capacitors in ac circuits.

RL Circuits

Series RL circuit. Figure 4.27 shows a series *RL* circuit excited by a current source. We will investigate the effect of the inductor as frequency varies. If there were no inductor in the circuit, it would exhibit the same behavior at all frequencies; all voltages and currents would be in phase with each other and in a fixed ratio. With an inductor in the circuit, however, the properties of the circuit change as frequency is varied.

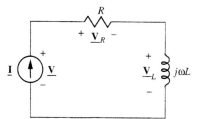

Figure 4.27 Series *RL* circuit.

The impedance of the series circuit is

$$\underline{Z} = R + j\omega L \tag{4.92}$$

The relationship between the phasor voltage and current is thus

$$\underline{V} = \underline{Z}\underline{I} = (R + j\omega L)\underline{I} \tag{4.93}$$
$$= R\underline{I} + j\omega L\underline{I} = \underline{V}_R + \underline{V}_L$$

In Eq. (4.93), we have interpreted the two terms in the impedance as relating to the voltages across the resistor, $\underline{V}_R$, and the inductor, $\underline{V}_L$. Because the current is common to both *R* and *L*, we have drawn the phasor diagram in Fig. 4.28 with $\underline{I}$ as the phase reference, that is, $\underline{I} = I_p \angle 0°$. The phase of the voltage across the resistor, $\underline{V}_R$, is the same as the phase of $\underline{I}$, but the phase of the voltage across the inductor, $\underline{V}_L$, leads by 90°. The total voltage, $\underline{V}$, the phasor sum of $\underline{V}_R$ and $\underline{V}_L$, thus leads the current by a phase angle somewhere between 0° and 90°. The phase angle by which the voltage leads the current is the geometric angle of $\underline{Z}$ in the complex plane, θ, as given in Eq. (4.94) and Fig. 4.29.

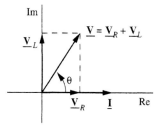

Figure 4.28 The inductor causes current to lag.

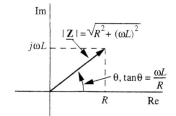

Figure 4.29 The angle of the impedance represents the phase shift.

$$\underline{Z} = R + j\omega L = |\underline{Z}| \angle \theta \qquad (4.94)$$

where

$$|\underline{Z}| = \sqrt{R^2 + (\omega L)^2}, \qquad \theta = \tan^{-1} \frac{\omega L}{R}$$

Frequency effects on the series RL circuit. Now let us examine what happens to the phasor diagram when frequency is varied, with R, L, and $\underline{I}$ held constant. Consider, first, very low frequencies. By "low," we mean those frequencies where the imaginary part of $\underline{Z}$ is small compared with the real part, that is, where $\omega L \ll R$. For low frequencies, the phase angle is small, meaning that $\underline{V}$ and $\underline{I}$ are almost in phase with each other, and the magnitude of the impedance is essentially equal to R. Thus, the inductor has little effect at low frequencies.

Zero frequency = dc. We saw in Chap. 3 that inductors become invisible at dc once their initial energy requirements are met. The foregoing discussion shows inductors to be virtually invisible at low ac frequencies. Or we can put it the other way and treat dc behavior as a limiting case of ac as ω approaches zero. The mathematics supports this approach because $e^{j\omega t} \to 1$ as $\omega \to 0$. Whichever outlook we choose, the resistance dominates the behavior of the circuit at low frequencies.

As frequency increases, the presence of the inductor is shown by an increase in the impedance of the inductor and hence an increase in the voltage across the inductor. This increases the overall voltage and also the phase difference between the total voltage and the current. When $\omega L = R$, for example, the phase difference is 45° and the voltage is increased by $\sqrt{2}$ because the magnitude of the total impedance has increased by $\sqrt{2}$ from the dc value. As the frequency goes yet higher, the inductor comes to dominate the behavior of the circuit. The phase shift approaches 90° because the impedance approaches $j\omega L$.

Mechanical analogy. To get a feeling for how the impedance of the inductor becomes large at high frequencies, imagine shaking a massive object in your hands. If you shake slowly (low frequencies), not much force is required, but as you attempt to shake faster, more force is required. Similarly, it takes more voltage to put the same current through an inductor as the frequency is increased.

Motor application. The series RL circuit in Fig. 4.27 would be an appropriate circuit model for an electric motor under steady load. The resistance would represent heat and friction losses in the motor, plus the energy converted to mechanical form and applied to a mechanical load. The inductance would represent magnetic energy storage in the motor structure. If we were to draw the phasor diagram for the motor, it would look like Fig. 4.28 except that the voltage would normally be used as the phase reference, as in Fig. 4.30.

lagging and leading current

Voltage is customarily used as the phase reference in ac power circuits because lights, motors, heaters, and so on, are parallel loads requiring a standard voltage. Because voltage is the phase reference, current is said to *lag* for an inductive load and, as we will soon see, *lead* for a capacitive load.

Resistive and reactive parts. The foregoing discussion suggests that an impedance can represent more than a physical resistor and a physical inductor in an ac circuit. The

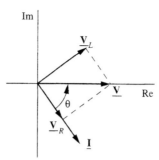

Figure 4.30 Phasor diagram redrawn with voltage as the phase reference.

real part of the impedance represents losses to the circuit, that is, energy leaving electrical form and converted to heat or some other form of energy such as mechanical work. The *imaginary part* of the impedance represents energy storage, in this instance, magnetic energy storage in a motor. Because of these broader interpretations of impedance, which we will explore more fully in the next chapter, names are given to the real and imaginary parts of the impedance. The real part of **Z** is called the *resistive part* and the imaginary part is called the *reactive part*. The reactive part is given the symbol, X, as used in

resistive, reactive, reactance

$$\mathbf{Z} = R + jX = \text{(resistance)} + j\text{ (reactance)} \tag{4.95}$$

Thus, the reactance of an inductor is $X_L = \omega L\ \Omega$, and the reactance of a capacitor is

$$X_C = -\frac{1}{\omega C} \tag{4.96}$$

Why do we need these new words? Why can't we speak in terms of inductance? One reason we have already given—that the impedance represents more than simple R and L. The other reason becomes important when frequency is varied. The reactance of a true inductor varies linearly with frequency, but the reactance of more complicated circuits or devices does not vary linearly with frequency. Reactance is a more general concept than inductance.

EXAMPLE 4.9 Reactance

A load has a voltage of $10 \cos(120\pi t + 12°)$ V and a current of $2.5 \cos(120\pi t - 37°)$ A. What is the reactance of the load?

SOLUTION:
The load impedance is

$$\mathbf{Z} = \frac{\mathbf{V}}{\mathbf{I}} = \frac{10\ \angle\ 12°}{2.5\ \angle -37°} = 4\ \angle\ 49° = 2.62 + j3.02 \tag{4.97}$$

The reactance is thus $3.02\ \Omega$.

> **WHAT IF?** What if the voltage and current at twice the frequency are $10\cos(240\pi t + 12°)$ V and $1.5\cos(240\pi t - 54.5°)$ A? Can the load be a simple coil of wire?[13]

Parallel RL load. For the parallel *RL* load shown in Fig. 4.31, voltage is convenient as a phase reference from both practical and mathematical considerations. Since both *R* and *L* are connected in parallel with an ideal voltage source, we can determine their currents independently:

$$\mathbf{I}_R = \frac{\mathbf{V}}{\mathbf{Z}_R} = \frac{\mathbf{V}}{R}, \quad \mathbf{I}_L = \frac{\mathbf{V}}{\mathbf{Z}_L} = \frac{\mathbf{V}}{j\omega L}, \quad \mathbf{I} = \mathbf{I}_R + \mathbf{I}_L = \mathbf{V}\left(\frac{1}{R} + \frac{1}{j\omega L}\right) \quad (4.98)$$

The impedance of the parallel load is calculated according to the generalized concept of impedance:

$$\mathbf{Z} = \frac{\mathbf{V}}{\mathbf{I}} = \frac{1}{1/R + 1/j\omega L} = R \| j\omega L = |\mathbf{Z}| \angle \theta \quad (4.99)$$

where $\theta = \tan^{-1}(R/\omega L)$ and $|\mathbf{Z}| = 1/\sqrt{(1/R)^2 + (-1/\omega L)^2}$.

Frequency effects on the RL parallel circuit. Figure 4.32 shows the phasor diagram for the parallel *RL* circuit. The current lags the voltage because of the inductance. Let us now consider changes in the phasor diagram as frequency is varied. At low frequencies, $\omega L \ll R$, the current in the inductor is much larger than the current in the resistor because the reactance of the inductor approaches a short circuit at dc. This would cause the phase, θ, to approach 90° (current lagging voltage).

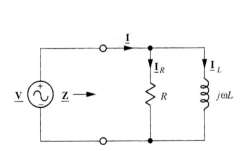

Figure 4.31 Parallel *RL* circuit.

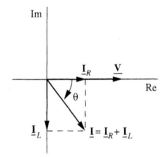

Figure 4.32 Current still lags voltage due to the inductor.

---

[13] Yes, because the impedance is $2.66 + j6.11$, the load appears to have about 2.6 Ω resistance in series with about 8.1 mH of inductance.

As frequency increases, the reactance of the inductor will increase and the resistor gains importance. At $\omega L = R$, the current phase is $-45°$, current lagging voltage, and the magnitude of the total current will be $\sqrt{2}$ times the current in the resistor. As higher frequencies are reached, the reactance of the inductor exceeds that of the resistor, and the impedance of the parallel combination approaches that of the resistor.

EXAMPLE 4.10 Parallel to series conversion

Convert $100 \| j50$ into a series form $R + jX$.

SOLUTION:
Figure 4.33(a) shows the parallel circuit. We may derive the series form by evaluating $100 \| j50$ in rectangular form:

$$100 \| j50 = \frac{1}{1/100 + 1/j50} = 44.7 \angle 63.4$$
$$= 20 + j40 \; \Omega \qquad (4.100)$$

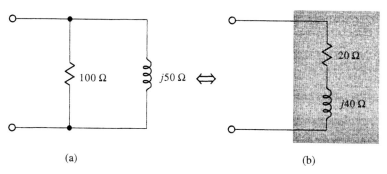

(a)   (b)

Figure 4.33 (a) A parallel circuit and (b) a series circuit can be equivalent.

Figure 4.33(b) shows the series form. The two circuits are equivalent, however, at a single frequency only; they would exhibit different characteristics if the frequency changed.

WHAT IF?   What if the frequency were doubled?[14]

RC Circuits

Series RC circuit. Figure 4.34 shows a series RC circuit. The impedance of the series combination is

---

[14] $\mathbf{Z} = 50 + j50 \; \Omega$.

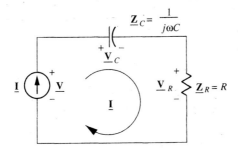

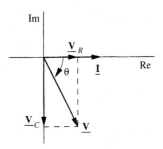

Figure 4.34 Series *RC* circuit.

Figure 4.35 Current leads voltage in the capacitive circuit.

$$\mathbf{Z} = R + \frac{1}{j\omega C} = \sqrt{R^2 + \left(\frac{-1}{\omega C}\right)^2} \angle \theta \qquad (4.101)$$

where $\theta = -\tan^{-1}(1/\omega RC)$.

Figure 4.35 shows the corresponding phasor diagram. We use the current as the phase reference, so the voltages are shown lagging. However, we would normally say the current leads voltage in a capacitive circuit because voltage would be considered the phase reference.

Frequency effects on the *RC* series circuit. In this case, we will examine the frequency behavior by starting at high frequencies. At high frequencies, the reactance of the capacitor, $-1/\omega C$, is very small and the circuit appears resistive. The phase angle is nearly zero, with the current slightly leading the voltage. As frequency is decreased, however, the reactance of the capacitor increases. This increases $\mathbf{V}_C$, which increases the total voltage and the phase angle. At $1/\omega C = R$, the phase is 45°, leading current, and the voltage has increased by $\sqrt{2}$ from its high-frequency value. At low frequencies, the reactance of the capacitor becomes very large, approaching an open circuit at dc. Thus, at low frequencies, the capacitor dominates the behavior of the series combination, and the current phase approaches 90° leading the voltage.

EXAMPLE 4.11 Series *RC*

At what frequency does a load consisting of 10 Ω in series with 0.01 µF produce a phase shift of 12.5° between voltage and current?

SOLUTION:
For a series *RC* circuit, the phase of the impedance is given by the angle in Eq. (4.101); thus,

$$12.5° = \left|-\tan^{-1}\frac{1}{\omega RC}\right| \Rightarrow \frac{1}{\omega RC} = \tan 12.5°$$

$$\omega = \frac{1}{10 \times 10^{-8} \times \tan 12.5°} = 4.51 \times 10^7 \text{ rad/s} \qquad (4.102)$$

or about 7.2 MHz.

> **WHAT IF?** What if you want the frequency for $Z_{RC} = 13\ \Omega$ in magnitude?[15]

Parallel RC circuit. Figures 4.36 and 4.37 show a parallel RC circuit and the corresponding phasor diagram. The currents in the resistor and the capacitor are independent because they are connected in parallel with an ideal voltage source.

$$\mathbf{I}_R = \frac{\mathbf{V}}{R}, \qquad \mathbf{I}_C = \frac{\mathbf{V}}{1/j\omega C} = j\omega C \mathbf{V}$$

$$\mathbf{I} = \mathbf{I}_R + \mathbf{I}_C = \mathbf{V}\left(\frac{1}{R} + j\omega C\right), \qquad \theta = \tan^{-1}(\omega RC) \tag{4.103}$$

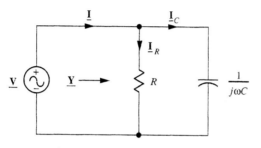

Figure 4.36 Parallel RC circuit.

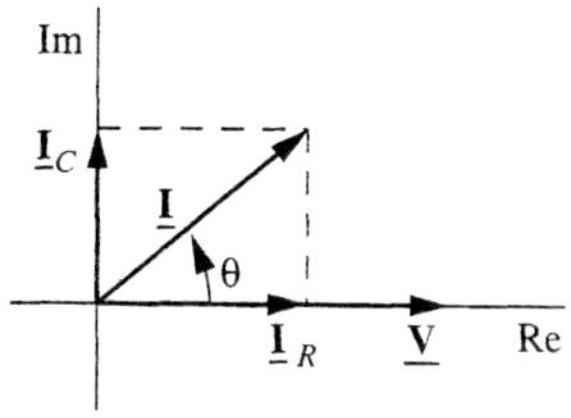

Figure 4.37 The capacitor causes the current to lead the voltage.

We can derive the impedance from Eq. (4.103), but Eq. (4.103) leads naturally to the definition of "admittance."

LEARNING OBJECTIVE 5.

To understand admittance and show its advantages in analyzing parallel impedances

Admittance. Recall from Chap. 1 that we introduced the concept of conductance, $G = 1/R$, to simplify discussion of parallel resistors. There the concept of conductance brought few practical benefits, because modern calculators handle reciprocals without difficulty, but the corresponding concept in ac circuits has considerable theoretical and practical importance. *Admittance*, $\mathbf{Y}$, is defined as

$$\mathbf{Y} = \frac{\mathbf{I}}{\mathbf{V}} = \frac{1}{\mathbf{Z}} \tag{4.104}$$

admittance

Clearly, the admittance of the parallel RC circuit in Fig. 4.36 is

$$\mathbf{Y} = \frac{\mathbf{I}}{\mathbf{V}} = \frac{1}{R} + j\omega C = G + j\omega C \tag{4.105}$$

where we have introduced the conductance of the resistor, G. Admittance is useful for dealing with parallel circuits and has importance because parallel connections are common

---

[15] $\omega = 1.2 \times 10^7$ rad/s, or 1.92 MHz.

susceptance

in practice. We have a specialized vocabulary associated with admittance. The real part of the admittance is called the *conductive part* or the *conductance*. The imaginary part of the admittance is called the *susceptive part* or the *susceptance*. Thus, we say that the conductive part of the admittance of a parallel RC circuit is G and the susceptive part is ωC. The standard symbols for the conductance and susceptance are

$$\underline{Y} = G + jB \qquad (4.106)$$

where B is a real number. For example, the conductive part of the parallel RL circuit in Fig. 4.33(a) is 0.01 S (siemens) and the susceptive part is $B = -0.02$ S. Note that the susceptance of the inductive circuit is negative.

RLC Circuits

Series RLC. Figure 4.38 shows a series RLC circuit. The impedance of the circuit is

$$\underline{Z} = R + j\omega L + \frac{1}{j\omega C} = R + j\left(\omega L - \frac{1}{\omega C}\right) = |\underline{Z}| \angle \theta$$

where

$$|\underline{Z}| = \sqrt{R^2 + \left(\omega L - \frac{1}{\omega C}\right)^2} \quad \text{and} \quad \theta = \tan^{-1}\frac{\omega L - 1/\omega C}{R} \qquad (4.107)$$

The reactive part of the impedance now combines the effects of the inductor and the capacitor. The frequency response of this reactive term gives this circuit interesting and useful characteristics. The phasor diagram is shown in Fig. 4.39. We have drawn the phasors showing the voltage across the inductor greater in magnitude than the voltage across the capacitor. This situation would be appropriate for high frequencies, where $\omega L > 1/\omega C$.

The frequency characteristics of the series RLC circuit combine those of the series RC and RL circuits. At dc, the circuit acts as an open circuit because of the capacitor. At low frequencies, the reactance of the capacitor dominates and the phase angle approaches 90°, with current leading voltage. As frequency increases, however, the inductive reactance becomes significant and at the resonant frequency grows to the point of canceling the negative reactance of the capacitor. This frequency occurs when

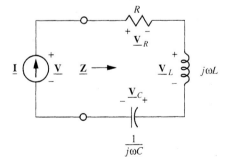

Figure 4.38 Series RLC circuit.

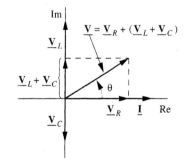

Figure 4.39 Above resonance, the inductor dominates, so current lags voltage.

$$\omega_r L = \frac{1}{\omega_r C} \quad \text{or} \quad \omega_r = 1/\sqrt{LC} \tag{4.108}$$

series resonance

and ω_r is called the frequency of *series resonance*. At resonance, the inductor and capacitor combination becomes invisible and R is the total impedance of the circuit. As the frequency increases through resonance, the phase changes from leading to lagging (current lagging voltage), and at resonance, the phase is zero. At frequencies above resonance, the inductor dominates the circuit characteristics and the phasor diagram of Fig. 4.39 shows the trend. At very high frequencies the current phase approaches 90° lagging.

This circuit, with its series resonance, is used in electronics to select one group of frequencies from a broader group. For example, this circuit can be used as part of a radio filter that selects one station for reception, rejecting all others. We discuss this circuit more fully in Chap. 6, where we discuss energy in ac circuits. We will see that resonance occurs when magnetic and electric energy requirements are equal, just as a mechanical system resonates when kinetic and potential energy requirements are balanced.

EXAMPLE 4.12 Series resonance

Calculate the width of the resonance region for the circuit shown in Fig. 4.40.

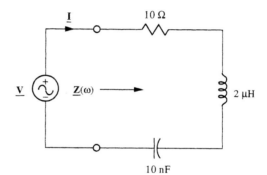

Figure 4.40 Series *RLC* circuit. We will find the resonant frequency and the width of the resonance.

SOLUTION:
The impedance is

$$\mathbf{Z}(\omega) = 10 + j\left(2 \times 10^{-6}\omega - \frac{10^8}{\omega}\right) \Omega \tag{4.109}$$

Resonance occurs when the total reactance is zero:

$$2 \times 10^{-6}\omega = \frac{10^8}{\omega} \Rightarrow \omega_r = 7.071 \times 10^6 (f_r = 1125 \text{ kHz}) \tag{4.110}$$

where ω_r (or f_r) is the resonant frequency. At the resonant frequency of 1125 kHz, the impedance of the circuit is 10 Ω resistive. Frequencies will exist below and above resonance where the reactance is equal to the resistance, and the phase will be ±45°. We will consider these frequencies to define the width of the resonance. These frequencies occur at

$$\left| 2 \times 10^{-6}\omega - \frac{10^8}{\omega} \right| = 10 \tag{4.111}$$

Equation (4.111) leads to the two quadratic equations:

$$2 \times 10^{-6}\omega - \frac{10^8}{\omega} = 10$$

and

$$2 \times 10^{-6}\omega - \frac{10^8}{\omega} = -10 \tag{4.112}$$

which yield frequencies of 1592 and 796 kHz. At the lower frequency, the circuit would be capacitive, and the angle of the impedance would be negative (leading current). At the higher frequency, the circuit would be inductive, and the angle of the impedance would be positive (lagging current). Because these frequencies are in the AM radio band, this circuit might be used as a frequency filter in an AM radio.

Parallel RLC circuit. The parallel RLC circuit, Fig. 4.41, combines the properties of the parallel RL and RC circuits. The admittance of the circuit is

$$\underline{Y} = G + j\omega C + \frac{1}{j\omega L} = G + j\left(\omega C - \frac{1}{\omega L}\right) \tag{4.113}$$

At low frequencies, the susceptance of the inductor ($-1/\omega L$) is large and dominates the admittance expression. The admittance is large (infinite at dc) and the current phase ap-

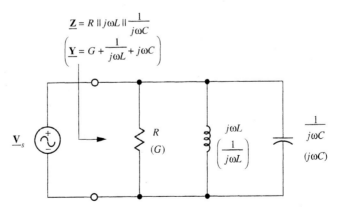

Figure 4.41 Parallel RLC circuit. Admittance (in parentheses) is natural for parallel circuits.

parallel resonance

proaches $-90°$. As frequency is increased, the inductive susceptance diminishes and the capacitive susceptance grows until they become equal. This is a *parallel resonance* and it occurs when $\omega_r C = 1/\omega_r L$, or $\omega_r = 1/\sqrt{LC}$, the same frequency as for series resonance. Thus, series and parallel resonance occur at the same frequency for the same ideal inductor and capacitor. At resonance, the admittance is pure conductance, $\mathbf{Y} = G = 1/R$. Thus, the impedance level at resonance is R.

As the frequency increases above resonance, the capacitive susceptance dominates, and as the frequency approaches very high frequencies, the admittance again becomes very large and the current phase approaches $+90°$. Thus, the admittance is minimum at resonance and becomes very large at low and high frequencies. Put differently, at low and high frequencies, the impedance is very small, approaching a short circuit, but the impedance has a maximum at the frequency of parallel resonance. This contrasts with the series case, where the impedance is minimum at resonance.

Check Your Understanding

1. What is the impedance, including units, of a 0.7-H inductor at 50 Hz?
2. What is the impedance of a 80-µF capacitor at 120 Hz?
3. A capacitor at 1 kHz has an impedance with a magnitude of 20 Ω. What is the magnitude of the impedance at 2 kHz?
4. What is the magnitude of the impedance of a 20-Ω resistor in series with a 20-mH inductor at 80 Hz?
5. What is the conductive part of $\mathbf{Y} = 0.012 \angle +12°$?
6. What is the reactive part of $\mathbf{Z} = 18 \angle -26°$?
7. If the current lags the voltage in an RL circuit, what type of circuit is it: series, parallel, or either?
8. At resonance, is the input impedance to a series RLC circuit a minimum or a maximum?

Answers. (1) $j220\ \Omega$; (2) $-j16.6\ \Omega$; (3) $10\ \Omega$; (4) $22.4\ \Omega$; (5) 11.7×10^{-3} S; (6) -7.89; (7) either; (8) minimum.

CHAPTER SUMMARY

Chapter 4 introduces methods for determining the steady-state response of circuits with sinusoidal sources. These methods are frequently required in electronics and electric power engineering. Complex numbers are used to describe voltage and current through phasors, and resistors, inductors, and capacitors are described through their impedances. The frequency responses of simple circuits are considered.

Time and frequency domains. Table 4.1 summarizes the relationships between the time domain and the frequency domain that we have developed in this chapter.

TABLE 4.1 Time-Domain and Frequency-Domain Transforms

Time Domain	Frequency Domain
Sinusoid	Phasor
Phase angle	Angle in complex plane
$V_p \cos(\omega t + \theta)$	$\underline{\mathbf{V}} = V_p \angle \theta$
DEs	Arithmetic with complex numbers
d/dt	$j\omega$
Cosine function	$1 \angle 0°$
Sine function	$-j = 1 \angle -90°$
R, L, and C	Impedances
R	$\underline{\mathbf{Z}}_R = R$
L	$\underline{\mathbf{Z}}_L = j\omega L = \omega L \angle +90°$
C	$\underline{\mathbf{Z}}_C = \dfrac{1}{j\omega C} = \dfrac{1}{\omega C} \angle -90°$

Objective 1: To understand how to identify the amplitude, frequency, and phase of a sinusoidal function. We describe the parameters and mathematical properties of the generalized sinusoid, of which the sine and cosine functions are special cases.

Objective 2: To understand the phasor concept and be able to find the sinusoidal steady-state response of a circuit, given the differential equation. After a review of the algebra and arithmetic of complex numbers, a phasor is introduced as a complex number derived from the amplitude and phase of a sinusoidal time function. Using phasors, we can transform a linear differential equation into an algebraic equation that can be solved for the phasor representing the unknown voltage and current in the circuit.

Objective 3: To understand how to use phasors and impedance to determine the sinusoidal steady-state response of a circuit. When the phasor technique is applied to the definitions of resistors, inductors, and capacitors, the equations define the impedance of these circuit elements. Impedances can be combined in series and parallel like resistors at dc; indeed, the impedance concept allows all the techniques used to analyze dc circuits to be applied to ac circuits.

Objective 4: To understand the effect of varying frequency on series and parallel *RL*, *RC*, and *RLC* circuits. The magnitude of the impedances of inductors and capacitors depend on frequency. Thus, the phase and amplitude responses of circuits containing these elements depend on frequency. We study the frequency response of series and parallel first- and second-order circuits, especially noting the resonance response in *RLC* circuits.

Objective 5: To understand how to define admittance and show its advantages in analyzing parallel impedances. Admittance is the reciprocal of impedance and is frequently used in the analysis of parallel ac circuits.

Chapter 5 generalizes the frequency domain analysis developed in this chapter. Chapter 6 continues the study of ac circuits with an emphasis on energy processes.

GLOSSARY

Absolute value (or the magnitude) of a complex number, p. 173, the Pythagorean sum of the real and imaginary parts of a complex number; its distance from the origin in the complex plane.

Admittance, p. 198, the reciprocal of impedance; the phasor current divided by phasor voltage.

Angular frequency, p. 165, 2π times the event frequency, radians/second.

Complex conjugate, p. 172, a complex number that is related to another complex number, having the same real part but having an imaginary part of the opposite sign.

Complex numbers, p. 170, the sum of a real and an imaginary number. May be represented in rectangular, polar, or exponential form.

Conductance, conductive part of the admittance, p. 199, the real part of the admittance, indicative of circuit losses.

Event frequency, p. 164, the number of cycles during a period of time; the reciprocal of the period, cycles/second = hertz.

Exponential form of a complex number, p. 175, the same as the polar form except that the angular information is expressed in Eulerian form, $e^{j\theta}$.

Frequency domain, p. 182, circuit analysis techniques and notations arising out of using frequency as the independent variable.

Imaginary numbers, p. 170, numbers that are negative when squared, indicated by i or j.

Impedance, p. 187, a complex number, the phasor voltage divided by the phasor current. The magnitude of the impedance is the ratio of the peak values of the voltage and current. The angle of the impedance is the phase of the voltage minus the phase of the current.

Lagging current, p. 193, when current phase follows voltage phase; indicative of an inductive circuit.

Parallel resonance, p. 202, occurs when a parallel-connected inductor and capacitor have identical energy requirements, leading to a maximum value of the combined impedance.

Phasor, p. 181, a complex number representing the amplitude and phase of a sinusoid.

Polar form of a complex number, p. 173, a complex number expressed as a magnitude (distance from the origin) and an angle, measured counterclockwise from the positive real axis to the location of the complex number in the complex plane.

Reactive part of the impedance, p. 194, the imaginary part of the impedance; indicative of circuit energy storage type and magnitude.

Rectangular form of a complex number, p. 173, a complex number expressed as the sum of a real and imaginary part.

Resistive part of the impedance, p. 194, the real part of the impedance; indicative of circuit losses.

Series resonance, p. 199, occurs when a series-connected inductor and capacitor have identical energy requirements, leading to a minimum value of the combined impedance.

Sinusoid, or sinusoidal waveform, p. 164, a variable that varies as the sine or cosine of time, with any phase angle.

Susceptance, susceptive part of the impedance, p. 199, the imaginary part of the admittance; indicative of circuit energy storage type and magnitude.

Time domain, p. 182, circuit analysis techniques and notations arising out of using time as the independent variable.

PROBLEMS

Section 4.1: Introduction to Alternating Current (AC)

4.1. Find the frequencies in hertz and in radians per second for the following:
  (a) The rotation of the Earth on its axis relative to the Earth–Sun line.
  (b) The rotation of a bike tire (26-in. diameter) at 20 mph.
  (c) The second hand of a watch.
  (d) A dentist's drill rotating at 200,000 rpm.
  (e) A 33.3-rpm LP phonograph record.

4.2. The sidereal day measures the Earth's rotation relative to the fixed stars and is 3 minutes, 56 seconds shorter than the mean solar day. What is the Earth's angular velocity on its axis relative to absolute space in radians/hour? Give to five-place precision.

4.3. The maximum elevation angle of the sun in Ft. Collins, Colorado, is $E(t) = 48.4° + 23.5° \cos(\omega t - \theta)$.
  (a) If t is in months, what is ω?
  (b) Estimate θ in radians if $t = 0$ on January 1.
    Hint: The summer solstice, approximately June 23, is the day when the Sun should be at its maximum northerly position. Thus, $E(t)$ should be maximum on that day.
  (c) What is the maximum elevation angle of the Sun on July 4?

4.4. A sinusoidal function is shown in Fig. P4.4. Determine the frequency, phase, and amplitude for expressing this sinusoid in the standard form:
  $i(t) = I_p \cos(\omega t + \theta)$.

4.5. A sinusoidal function is shown in Fig. P4.5.
  (a) Determine the frequency, phase, and amplitude for expressing this sinusoid in the standard form: $v(t) = V_p \cos(\omega t + \theta)$.

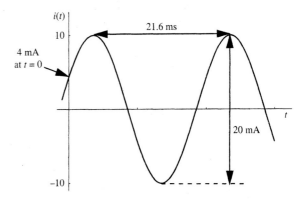

Figure P4.4

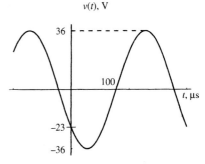

Figure P4.5

  (b) What is the first time after $t = 0$ that the voltage is at its maximum value?

4.6. Sketch the sinusoidal voltage
  $v(t) = 60 \cos(100\pi t - 120°)$ V.

4.7. (a) Sketch the sinusoidal voltage
    $v(t) = 420 \cos(1000t + 200°)$ mV.
  (b) What is the first time after $t = 0$ that this voltage is zero?

Section 4.2: Representing Sinusoids with Phasors

4.8. Given three complex numbers:
$z = 6 - j2$, $w = 7 \angle 15°$, and $u = 10e^{j2.2}$,
(a) Find $|z|$.
(b) Find u^* in rectangular form.
(c) Evaluate $z/(w - u)$ and place the result in exponential form.
(d) Solve for s (a complex number) if $z(s - w) = u$ and express s in polar form.

4.9. For the complex numbers $z_1 = -2 + j3$ and $z_2 = 1 - j6$, show the following:
(a) $|z_1 \times z_2| = |z_1| \times |z_2|$.
(b) $|z_1/z_2| = |z_1|/|z_2|$.
(c) $|z_1| + |z_2| \neq |z_1 + z_2|$.

4.10. Given that $z = x + jy$ is a general complex number, show the following:
(a) $\text{Re}\{z\} = (z + z^*)/2$.
(b) $\dfrac{1}{z} = \dfrac{x}{x^2 + y^2} - \dfrac{jy}{x^2 + y^2}$.
(c) Solve for the first time when $\text{Re}\{ze^{j\omega t}\} = 0$ if z is $2 - j6$ and $\omega = 100$.

4.11. Evaluate the following expressions:
(a) $(1.4 + j6)\, 4 \angle 18° + 6 \angle +12°$.
(b) $\dfrac{0.2 - j0.5}{13 + j7} \times \dfrac{1}{10 \angle +66°} - 10^{-3}$.
(c) $\dfrac{1}{\dfrac{1}{2 + j2} + \dfrac{1}{3 \angle +40°}}$.

4.12. (a) What is the phasor for $v(t) = 5.2 \cos(100t - 90°)$?
(b) What time function is represented by the phasor $\mathbf{I} = 6 + j9\ \mu\text{A}$ if the frequency is 400 Hz?

4.13. Two 60-Hz sinusoidal voltages are described by the phasors $\mathbf{V}_1 = 20 \angle +10°$ V and $\mathbf{V}_2 = 9 - j17$ V.
(a) Which has the larger amplitude?
(b) Find $v_1(t) - v_2(t)$ at $t = 0$.
(c) Find the first time after $t = 0$ when $v_1(t) = 0$.

4.14. (a) Sketch one cycle of the time function, $v(t)$, represented by the phasor $\mathbf{V} = e^{j\pi/3}$ mV, $f = 60$ Hz.
(b) What is $\text{Re}\{(2 + j7)e^{j\omega t}\}$ V at $t = 0$?

4.15. Use phasor techniques in the following.
(a) Find $2\cos(100t - 45°) - 3\cos(100t + 60°)$
(b) Find $50 \sin(100t) + (d/dt)\cos(100t - 30°)$.
Hint: Do not take the derivative in the time domain; replace it by $j\omega$ in the frequency domain.

(c) Use phasor techniques to evaluate the derivative of $i(t) = 20 \sin(500t)$ at $t = 2$ ms. *Hint:* Write the formula in the time domain and transform into the frequency domain, using $j\omega$ for d/dt. Then put the specific ω and t in the $e^{j\omega t}$ and take the real part.

4.16. On page 180, it was argued that, if $\text{Re}\{\mathbf{Z}e^{j\omega t}\} = \text{Re}\{\mathbf{W}e^{j\omega t}\}$ for all t, then $\mathbf{Z} = \mathbf{W}$. In effect, Re and $e^{j\omega t}$ can be dropped. Prove this by letting $\mathbf{Z} = Z_r + jZ_i$ and $\mathbf{W} = W_r + jW_i$ and evaluating the equation at the times when $\omega t = 0$ and $\omega t = \pi/2$.

4.17. Solve for $v(t)$ in the circuit shown in Fig. P4.17 using the phasor methods described in Section 4.2. You may use the short transformations in Eq. (4.63) if you wish.

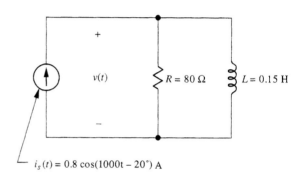

Figure P4.17

4.18. Solve for $v_C(t)$ in the circuit shown in Fig. P4.18 using the phasor methods described in Section 4.2. You may use the short transformations in Eq. (4.63) if you wish.

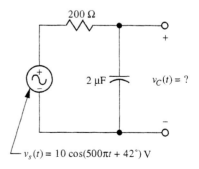

Figure P4.18

4.19. Rework Problem 4.18 for the total response if the voltage is applied with a switch closure at $t = 0$. Assume the initial voltage on the capacitor is zero.

4.20. Find the steady-state value of the voltage across the inductor in the circuit of Fig. P4.20.

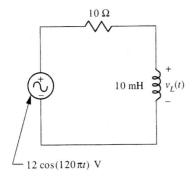

Figure P4.20

(a) Write the DE for $v_L(t)$.
(b) Transform into the frequency domain.
(c) Solve for the unknown phasor representing $v_L(t)$.
(d) Transform back into the time domain.

4.21. Find the steady-state value of the voltage across the resistance in the circuit of Fig. P4.21.
(a) Write the DE for $v_R(t)$.
(b) Transform into the frequency domain.
(c) Solve for the unknown phasor representing $v_R(t)$.
(d) Transform back into the time domain.

4.22. The mechanical system shown in Fig. P4.22 is excited by a rotating wheel that gives approximately

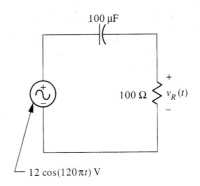

Figure P4.21

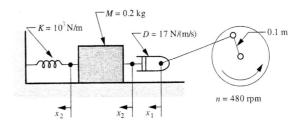

Figure P4.22

a sinusoidal displacement for x_1. The differential equation of the displacement x_2 is

$$M\frac{d^2 x_2}{dt^2} + D\frac{dx_2}{dt} + Kx_2 = D\frac{dx_1}{dt}$$

(a) Find ω.
(b) Transform the DE to the frequency domain: $x_2(t) \Rightarrow \mathbf{X}_2$.
(c) Solve for $\mathbf{X}_2$.
(d) Write $x_2(t)$.

Section 4.3: Impedance: Representing the Circuit in the Frequency Domain

4.23. (a) What is the impedance of a 5-H inductor at 5 Hz in polar form?
(b) A resistor and capacitor, connected in series, have an impedance of $20 \angle -32°$ at a frequency of 2 kHz. Find R and C.
(c) A resistor and capacitor, connected in parallel, have an impedance of $20 \angle -32°$ at 2 kHz. Find R and C.

4.24. Make a chart for resistors, capacitors, and inductors with the following columns: name, symbol, time-domain equation, frequency-domain equation, impedance in rectangular form, and impedance in polar form.

4.25. (a) What value of capacitance and what value of inductance have an impedance with a magnitude of 12 Ω at a frequency of 800 Hz?
(b) What would be the reactance of this C and this L at 1.6 kHz?
(c) What would be the impedance of this inductance and capacitor connected in series at a frequency of 1.2 kHz?

4.26. Using the frequency-domain versions of KVL and KCL, show that two impedances in series add like resistors in series, that is, $\underline{Z}_{eq} = \underline{Z}_1 + \underline{Z}_2$. Show also that two impedances in parallel add like resistors in parallel, that is,

$$\underline{Z}_{eq} = \underline{Z}_1 \| \underline{Z}_2 = \frac{1}{(1/\underline{Z}_1) + (1/\underline{Z}_2)}$$

4.27. Find the impedance in Fig. P4.27.

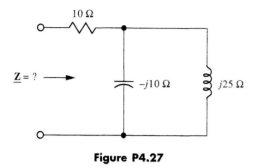

Figure P4.27

4.28. Determine the input impedance of the circuits shown in Fig. P4.28.

4.29. For the circuit shown in Fig. P4.29, determine $v(t)$ using phasor techniques. Sketch $v(t)$ in the time domain.

4.30. Use the techniques of the frequency domain to solve for $i(t)$ in the circuit shown in Fig. P4.30.
  (a) Find the frequency-domain version of the circuit, using phasors to represent sinusoidal functions, known and unknown, and impedances to represent circuit components.
  (b) Using parallel and series combinations, find $\underline{Z}_{eq}$ as seen by the voltage source.
  (c) Solve for $\underline{I}$.
  (d) Convert back to the time domain.

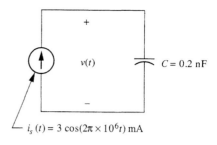

Figure P4.29

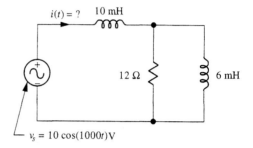

Figure P4.30

4.31. For the circuit in Fig. P4.31, find the first time after $t = 0$ when the instantaneous current is maximum, and give the maximum current.

4.32. The circuit shown in Fig. P4.32 is in sinusoidal steady state. The source voltage is shown in the graph. Sketch the capacitor voltage on the same graph.

4.33. Consider a 4-μF capacitor and a 10-Ω resistor.
  (a) They are connected in series. At what frequency in hertz is their series impedance 20 Ω in magnitude?

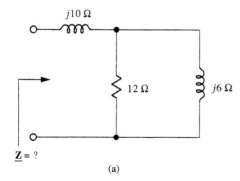

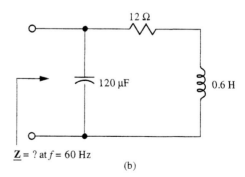

Figure P4.28

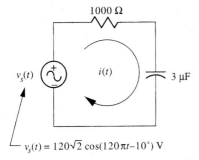

Figure P4.31

$v_s(t) = 120\sqrt{2}\cos(120\pi t - 10°)$ V

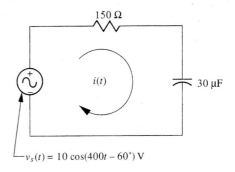

Figure P4.34

$v_s(t) = 10\cos(400t - 60°)$ V

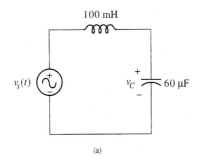

(a)

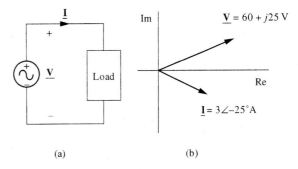

(a)   (b)

Figure P4.35

Assume the load consists of a resistor in series with a reactive component. The frequency is 60 Hz.
(a) What is the voltage at $t = 0$?
(b) What is the magnitude of the impedance?
(c) What is the resistance of the circuit?
(d) What is the reactive component (type and value)?

4.36. A 60-Hz ac source and load are connected as indicated in Fig. P4.36. The phasor voltage is $\mathbf{V} = 120 + j0$ V and the phasor current is $\mathbf{I} = 7 - j5$ A.
(a) Find the instantaneous current at $t = 0$.

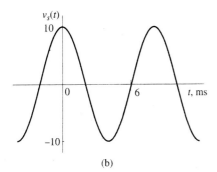

(b)

Figure P4.32

(b) If the resistor and capacitor are now placed in parallel, find the frequency at which their combined impedance is 7 Ω in magnitude.
(c) Still connected in parallel, at what frequency is the angle of the impedance $-45°$?

4.34. The circuit shown in Fig. P4.34 is in sinusoidal steady state. Determine the maximum value of the current and the first time after $t = 0$ at which the maximum current occurs.

4.35. Figure P4.35(a) shows a circuit in sinusoidal steady state, with the phasor diagram in Fig. P4.35(b).

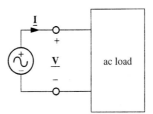

Figure P4.36

(b) What is the impedance of the ac load in rectangular form?
(c) Assuming that the load is a series resistor and inductor, find the value of the inductance.

4.37. For the circuit shown in Fig. P4.37, find the following:

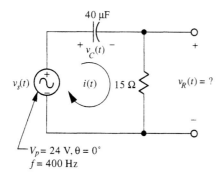

$V_p = 24$ V, $\theta = 0°$
$f = 400$ Hz

Figure P4.37

(a) Draw a phasor diagram showing $\underline{V}_s$, $\underline{I}$, $\underline{V}_R$, and $\underline{V}_C$. The voltages must be drawn to consistent scale and shown to add in accordance with the phasor KVL.
(b) Find $v_R(t)$ and sketch along with the source voltage.

4.38. Figure P4.38 shows an ac circuit in the frequency domain. The phasor voltage across the capacitor is given as $10 \angle 45°$.

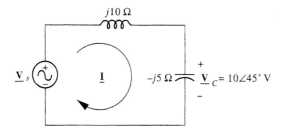

Figure P4.38

(a) Determine the phasor current in the circuit, $\underline{I}$.
(b) Determine the phasor source voltage, $\underline{V}_s$.

4.39. For the circuit shown in Fig. P4.39, find the following:
(a) At what frequency would the magnitude of the input impedance be $200\,\Omega$?
(b) What is the angle of the impedance of this frequency?
(c) What value of C should be added in series to make the circuit appear purely resistive at this frequency?
(d) What value of C should be added in parallel to make the circuit appear purely resistive at this frequency? (This is not the same answer as the previous part.)

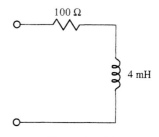

Figure P4.39

4.40. A resistor in series with a capacitor or inductor has a current of $i(t) = 1.2 \cos(1000t + 75°)$ mA and a voltage across the series combination of $v(t) = 0.8 \cos(1000t + 47°)$ V. What is the value of the resistance and the capacitor or inductor?

4.41. A resistor in series with a capacitor or inductor has a current of $i(t) = 12 \cos(1200t + 60°)$ mA and a voltage across the series combination of $v(t) = 0.8 \cos(1200t + 85°)$ V. What is the type and value of the resistor and the capacitor or inductor?

4.42. (a) Convert the circuit shown in Fig. P4.42 to the equivalent parallel circuit at $f = 1.2$ kHz.
(b) Find the impedances of the two circuits at a frequency of 1 kHz to show that the circuits are equivalent only at 1.2 kHz.

4.43. The circuit in Fig. P4.43 is to be represented by a Norton equivalent circuit. Determine $\underline{I}_N$ and $\underline{Z}_{eq}$.

4.44. (a) What is the total admittance of $\underline{Y}_1 = 1 + j6$ and $\underline{Y}_2 = 2.5 - j2.5$ connected in parallel?
(b) What is the magnitude of the input impedance of this parallel combination?

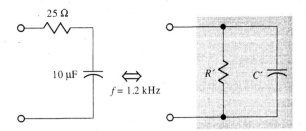

Figure P4.42

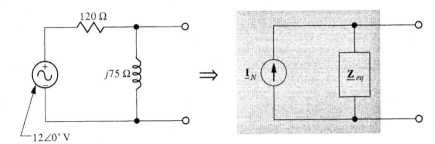

Figure P4.43

General Problems

4.45. By proper choice of X_C and X_L, the 10-Ω resistance in Fig. P4.45 can be transformed to "look" like a 50-Ω resistor at a specified frequency, as indicated in Fig. P4.45. Find X_C and X_L and, from them, C and L to accomplish this transformation at 1 kHz.

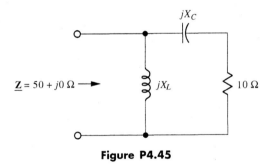

Figure P4.45

4.46. Find $v_a(t)$ in the circuit shown in Fig. P4.46 using nodal analysis.

4.47. The circuit shown in Fig. P4.47(a) is in sinusoidal steady state, with the output waveform, $v_{out}(t)$, shown in Fig. P4.47(b). On the same graph, sketch the input waveform, $v_{in}(t)$.

4.48. A voltage source with $v_s(t) = 120\sqrt{2}\,\cos(250t)$ V is connected in series with a 100-Ω resistance, a 0.2-H inductance, and a 25-μF capacitance.

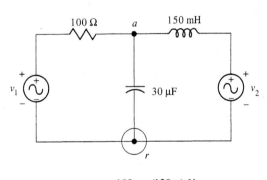

$v_1 = 100 \cos(120\pi t)$ V

$v_2 = 80 \sin(120\pi t)$ V

Figure P4.46

(a) Find the impedance of the circuit at the source frequency.
(b) Determine the sinusoidal steady-state current, $i(t)$, in the series connection.
(c) What is the first time after $t = 0$ when the voltage across the capacitance is zero?

4.49. For the circuit shown in Fig. P4.49, find the following:
(a) $v_R(t)$ for $\omega = 0$, that is, for a dc voltage of 8 V.
(b) $v_R(t)$ for $\omega = 4200$ rad/s.

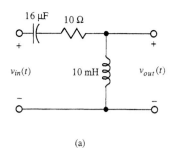

(a)

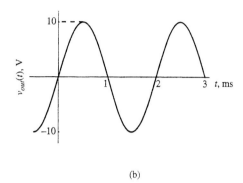

(b)

Figure P4.47

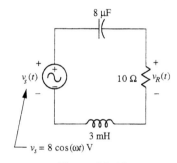

Figure P4.49

(c) The frequency in hertz for which the amplitude of $v_R(t)$ is maximum.

4.50. For the circuit in Fig. P4.50, $\underline{V} = 50 \angle 0°$ V.
  (a) Show $\underline{V}$ and $\underline{I}$ on a phasor diagram.
  (b) With $\underline{V}$ unchanged except that the frequency is doubled, show the phasor diagram. Use primed $\underline{V}'$ and $\underline{I}'$ for this case.

4.51. In Fig. P4.51, the design goal is to have $i(t)$ lead $v_s(t)$ by 55° of phase.
  (a) What is in the box: R, L, or C?
  (b) What is its numerical value?
  (c) What is the peak value of $i(t)$?
  (d) If the frequency were doubled, what would be the phase difference between $i(t)$ and $v_s(t)$? Consider phase positive if $i(t)$ leads $v_s(t)$.

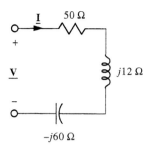

Figure P4.50

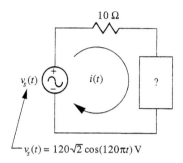

Figure P4.51

4.52. Figure P4.52(b) shows a circuit with an input and output voltage, and Fig. P4.52(a) shows the input voltage. Sketch the output voltage in the same graph.

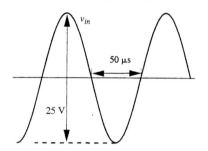

(a)

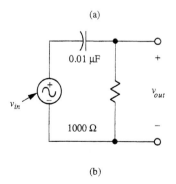

(b)

Figure P4.52

Answers to Odd-Numbered Problems

4.1. (a) 7.27×10^{-5} rad/s; (b) 4.31 Hz; (c) 0.105 rad/s; (d) 3,330 Hz; (e) 0.555 Hz.

4.3. (a) 0.524 rad/mo; (b) 3.02 rad or 173°; (c) 71.5°.

4.5. (a) $36\cos(24500t + 130°)$; (b) 164 μs.

4.7. (a)

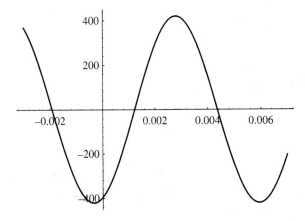

(b) 1.22 ms.

4.9. (a) 21.9 for both; (b) 0.593 for both; (c) $9.69 \neq 3.61$.

4.11. (a) $3.78 + j25.8$; (b) $(-4.48 - j1.10) \times 10^{-3}$; (c) $1.07 + j0.986$.

4.13. (a) $\underline{V}_1$; (b) 10.7 V; (c) 3.70 ms.

4.15. (a) $4.01\cos(100t - 91.2°)$; (b) $62.0\cos(100t + 36.2°)$; (c) 5400 A/s.

4.17. $56.5\cos(1000t + 8.07°)$ V.

4.19. $-8.34e^{-2500t} + 8.47\cos(500\pi t + 9.86°)$ V.

4.21. (d) $11.6\cos(120\pi t + 14.9°)$ V.

4.23. (a) $157 \angle 90°$; (b) 17.0 Ω, 7.51 μF; (c) 23.6 Ω, 2.11 μF.

4.25. (a) 16.6 μF, 2.39 mH; (b) $X_C = -6$ Ω, $X_L = 24$ Ω; (c) $j10$ Ω.

4.27. $19.4 \angle -59.0°$ Ω.

4.29. $2.39\sin(2\pi \times 10^6 t)$ V.

4.31. 0.127 A, 15.2 ms.

4.33. (a) 2300 Hz; (b) 4060 Hz; (c) 3980 Hz.

4.35. (a) 60 V; (b) 21.7 Ω; (c) 14.6 Ω; (d) inductor, 42.5 mH.

4.37. (a)

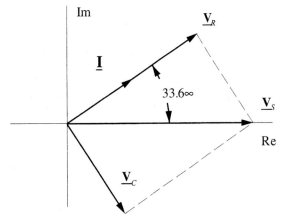

(b) $20.0\cos(800\pi t + 33.6°)$ V.

4.39. (a) 6890 Hz; (b) 60.0°; (c) 0.133 μF; (d) 0.100 μF for 400 Ω real.

4.41. 60.4-Ω resistor + 23.5 mH inductor.

4.43. $\underline{I}_N = 0.1 \angle 0°$ A, $\underline{Z}_{eq} = 33.7 + j53.9$ Ω

4.45. $X_C = -20$ Ω, 7.96 μF, $X_L = 25$ Ω, 3.98 mH.

4.47. $4.86\cos(1000\pi t - 131°)$ V.

4.49. (a) 0 V; (b) $4.03\cos(4200t + 59.8°)$ V; (c) 1030 Hz.

4.51. (a) C; (b) 186 μF; (c) 9.73 A; (d) +35.5°.

Circuit and System Analysis Using Complex Frequency

Complex Frequency

Impedance and the Transient Behavior of Linear Systems

Transient Response of *RLC* Circuits

Filters and Bode Plots

Systems

Chapter Summary

Glossary

Problems

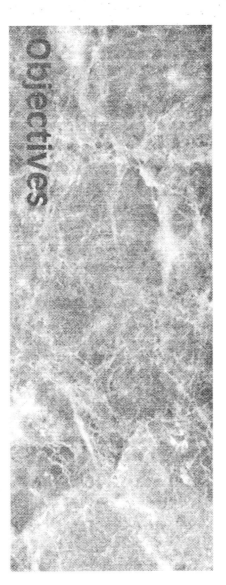

objectives

1. To recognize the class of functions that can be described by complex frequencies
2. To use generalized impedance to find the forced and natural response of a system
3. To recognize conditions for undamped, underdamped, critically damped, and overdamped responses in second-order systems
4. To understand the concept of electric filters and the use of Bode plots to describe signals and filters
5. To use system notation to describe the properties of composite systems
6. To determine the transient response of linear systems to the sudden application of a dc input

The concept of complex frequency allows further exploration of the frequency domain. Here we define complex frequency and use it to solve steady-state and transient problems of first-, second-, and higher-order. We introduce the language of linear system theory.

Introduction to Complex Frequency Techniques

system

In this chapter we introduce circuit and system analysis techniques based on complex frequency. Much of what we have learned about circuits will be integrated and illuminated by this more general approach. Such techniques have come to be associated primarily with system analysis because the language of linear system theory is based on complex frequency.

What is a system? A *system* consists of several components that together accomplish some purpose. For example, an automobile has a motor, steering mechanism, lights, padded seats, entertainment features, and more—operating together to give safe and pleasant transportation. Likewise, a stand-alone ac generator requires a control system to regulate the frequency and voltage of its output. Often systems are modeled with linear equations. Thus, electric circuits containing resistors, inductors, and capacitors are linear systems. The analysis of such linear systems has furnished a powerful language for system description that builds on our earlier study of the frequency domain. This chapter introduces system models and explores basic techniques of linear system description and analysis.

Contents of this chapter. We begin by generalizing the concept of frequency. We then introduce the language of system notation by defining the generalized impedance of electrical circuits. From this impedance we determine the natural frequencies and natural response of electrical circuits, and we then investigate the transient response of first- and second-order circuits. We introduce the system function to describe the relationship between the input and output of a linear system, especially electrical filters. We show how Bode plots, which can be derived from the system function, describe the frequency response of a linear system. We analyze the transient response of linear systems to sudden application of dc signals.

5.1 COMPLEX FREQUENCY

complex frequency

Frequency is a variable. Throughout most of Chap. 4, we held frequency constant, for example, at the frequency of the power system. By the end of the chapter, however, frequency became a variable as we explored the frequency response of series and parallel *RC*, *RL*, and *RLC* circuits. In this chapter we continue of think of frequency as a variable.

The Frequency Domain

Definition of Complex Frequency. We resume our exploration of the frequency domain through the following definition of complex frequency. A time-domain variable, for instance, a voltage, is said to have a *complex frequency* $\underline{s}$ when it can be expressed in the form given in Eq. (5.1)

$$v(t) = \text{Re}\{\underline{V} e^{\underline{s}t}\} \tag{5.1}$$

where $\underline{V}$ is a complex number, a phasor, and

$$\underline{s} = \sigma + j\omega \quad s^{-1} \tag{5.2}$$

where $\underline{s}$, σ, and ω all have units of inverse seconds, s^{-1}. Equation (5.1) is a slightly modified version of Eq. (4.44), except that frequency is now a complex number.

> **LEARNING OBJECTIVE 1.**
>
> To recognize the class of functions that can be described by complex frequencies

Functions that can be represented by complex frequencies. Before we explore the implications of complex frequency, we wish to relate Eq. (5.1) to familiar results. We have in previous chapters introduced several important functions that can be represented by complex frequencies.

When $\underline{s}$ is zero. When the complex frequency in Eq. (5.1) is zero, $\underline{s} = 0$, the voltage is a constant, or a dc voltage

$$v(t) = \text{Re}\{\underline{V}e^{0t}\} = V_{dc} \quad \text{for } \underline{s} = 0 \tag{5.3}$$

When $\underline{s}$ is real and negative. When the complex frequency $\underline{s}$ is real, $\underline{s} = \sigma$, the time-domain voltage is

$$v(t) = \text{Re}\{\underline{V}e^{\sigma t}\} = Ae^{\sigma t} \quad \text{for } \underline{s} = \sigma \tag{5.4}$$

where A is a constant. For negative σ, the voltage is a decreasing exponential function such as we encountered in first-order transient problems in Chap. 3. Specifically, the time constant would be the negative of the reciprocal of σ

$$\sigma = -\frac{1}{\tau} \quad \text{where} \quad \tau = R_{eq}C \text{ or } \frac{L}{R_{eq}} \tag{5.5}$$

EXAMPLE 5.1 **Transient problem**

Find the complex frequency representing the response of the circuit in Fig. 5.1.

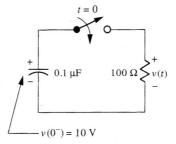

Figure 5.1 The circuit response can be described by a complex frequency.

SOLUTION:
Using the techniques from Chap. 3, we find the response to be

$$v(t) = v(0^-)e^{-t/RC} = 10e^{-10^5 t} \text{ V} \tag{5.6}$$

Comparison with Eq. (5.4) shows $\underline{s} = \sigma = -10^5 \text{ s}^{-1}$.

WHAT IF? What if the switch is open?[1]

---

[1] Then the only possible response is $\underline{s} = 0$. In effect the resistance becomes infinite, but a constant voltage and a constant current (zero current) are still possible.

When s is real and positive. If σ is positive in Eq. (5.4), the voltage is an increasing exponential, which we have not encountered before. Growing exponential functions are found in some biological systems and can occur in electric circuits as well. Figure 5.2 shows the time functions that result from zero and real complex frequencies. Thus, a complex frequency that is real includes the response that we studied as a transient solution in Chap. 3, except that we now include growing as well as decaying exponentials.

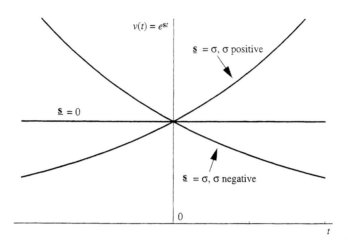

Figure 5.2 Time-domain responses for complex frequencies that are real.

When s is pure imaginary. When the complex frequency is pure imaginary, Eq. (5.1) takes the form of a sinusoidal function, such as we studied in Chap. 4. In this case,

$$v(t) = \text{Re}\{\underline{V} e^{j\omega t}\} = V_p \cos(\omega t + \theta) \tag{5.7}$$

where $\underline{V} = V_p e^{j\theta}$, with V_p the peak value and θ the phase of the sinusoidal function. Imaginary complex frequencies are used in the analysis of ac circuits.

Sinusoid = two complex frequencies. From another point of view, we may represent a sinusoidal function by two complex frequencies. This is implied by the "real part" operation in Eq. (5.1), because an alternate way to express the real part is through the identity in Eq. (5.8)

$$\text{Re}\{\underline{z}\} = \tfrac{1}{2}(\underline{z} + \underline{z}^*) \tag{5.8}$$

where $\underline{z}^*$ is the complex conjugate of $\underline{z}$. Thus Eq. (5.7) can be expressed in the form

$$V_p \cos(\omega + \theta) = \text{Re}\{\underline{V} e^{j\omega t}\} = \frac{\underline{V}}{2} e^{j\omega t} + \frac{\underline{V}^*}{2} e^{-j\omega t} \tag{5.9}$$

When we compare Eq. (5.9) with Eq. (5.1), we see that two complex frequencies, $\underline{s} = j\omega$ and $\underline{s} = -j\omega$, express a sinusoidal function. In our consideration of complex frequency, we will often find this second point of view to be useful.

Frequency and complex frequency. We have a slight semantic problem in speaking of complex frequency. When the complex frequency $\underline{s}$ is imaginary, the frequency, ω, is said to be real. Thus a sinusoidal function has a real frequency, but is described by a complex frequency that is imaginary. As shown above, a complex frequency that is real corresponds to a growing or decaying exponential function.

General interpretation of complex frequency. We now consider the meaning of $e^{\underline{s}t}$, with $\underline{s} = \sigma + j\omega$. In general,

$$v(t) = \text{Re}\{\underline{V}e^{\underline{s}t}\} = \text{Re}\{\underline{V}e^{(\sigma+j\omega)t}\}$$
$$= e^{\sigma t}\text{Re}\{\underline{V}e^{j\omega t}\} = V_p e^{\sigma t}\cos(\omega t + \theta) \quad (5.10)$$

Equation (5.10) expresses a time function that combines sinusoidal behavior with the exponential behavior that we hitherto have associated with transients. Equation (5.10) can also be considered a sinusoidal function in which the peak value of the sinusoid changes exponentially with time. Figure 5.3 shows the character of the time function in Eq. (5.10). The decaying sinusoid pictured in Fig. 5.3(b) occurs in oscillating systems that have energy losses, such as a pendulum. Growing oscillations such as those pictured in Fig. 5.3(a) can occur in electronic and mechanical systems that contain an energy source. Such responses represent unstable behavior.

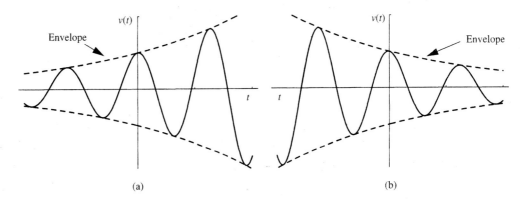

Figure 5.3 The function $\text{Re}\{e^{(\sigma+j\omega)}\}$ for (a) σ positive and (b) σ negative.

The $\underline{s}$-plane. A complex number, $\underline{s}$, can be represented by a point in the complex plane. In Fig. 5.4 we show such an $\underline{s}$-plane and identify the regions of complex frequency corresponding to associated time responses.

- The origin corresponds to a constant or dc function.
- Complex frequencies on the real axis correspond to growing and decaying exponential functions.
- Complex frequencies on the imaginary axis correspond to sinusoidal functions.
- The region to the right of the vertical axis, the right-half plane, corresponds to sinusoids that are growing exponentially.
- The left-half plane corresponds to sinusoids that are decreasing exponentially.

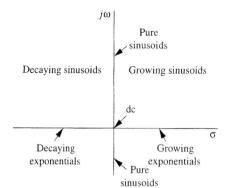

Figure 5.4 The s-plane and associated time functions.

EXAMPLE 5.2 Complex frequency

What complex frequency corresponds to the function

$$v(t) = 100e^{1000t} \sin(400\pi t + 75°) \text{ V} \tag{5.11}$$

SOLUTION:
There are two frequencies: $\sigma = 1000$ comes from the exponential part and $j\omega = \pm j400\pi$ comes from the sinusoidal part. Thus

$$\underline{s}_1 = 1000 + j400\pi \quad \text{and} \quad \underline{s}_2 = 1000 - j400\pi \tag{5.12}$$

WHAT IF? What if the function were cosine instead of sine?[2]

The Time Domain

The Frequency Domain

eigenfunction

Time domain and frequency domain. The introduction of complex frequency expands our concept of the frequency domain to cover a wider class of time-domain behavior. We now have a frequency-domain representation of exponentially growing or decreasing time functions that may oscillate. As hinted above, this expansion will allow frequency-domain techniques to be applied to transient problems as well as sinusoidal steady-state problems such as we studied in Chap. 4. Furthermore, these new functions allow the study of a wider class of transient problems.

Importance of e^{st}. The function e^{st} is a mathematical probe[3] we use to investigate the properties of electric circuits and other linear systems. At this point, we wish to look ahead and anticipate later results. We will find that all linear systems are characterized by certain complex frequencies. Once we determine these frequencies for a given system, for example, an RC circuit, we can determine from them the system response in the time domain or the frequency domain. We may thereby examine transient and steady-state responses. In the following section we show how to determine these

---
[2] The "sin/cos" and "75°" affect the phase but not the complex frequency.
[3] The mathematical term for e^{st} is the *eigenfunction* for a linear DE.

characteristic frequencies and how to derive from them the time-domain response of the system.

Summary. In this section we have generalized frequency to include growing or decreasing exponentials and exponentially growing or decreasing sinusoids. Such complex frequencies allow frequency-domain techniques to be applied to electric circuits and other linear systems. The characteristics of linear systems are often described in terms of complex frequency.

Check Your Understanding

1. What complex frequency or frequencies describe(s) an *RC* transient with a dc source if $R = 100 \, \Omega$ and $C = 10 \, \mu\text{F}$.
2. What would be the time between zero crossings for a function described by the complex frequency $\underline{s} = -2 + j10$?
3. A pure sinusoid may be described by a complex frequency that is (1) pure imaginary, (2) two complex frequencies that are pure imaginary, (3) either, or (4) neither. (Which?)
4. Complex frequencies near the origin in the $\underline{s}$-plane describe functions that vary slowly with time. (True or False?)

Answers. (1) 0 (for the source) and $-1000 \, \text{s}^{-1}$ (for the circuit); (2) 0.314 s; (3) either, depending on context; (4) true.

5.2 IMPEDANCE AND THE TRANSIENT BEHAVIOR OF LINEAR SYSTEMS

Generalized Impedance

generalized impedance

Impedance of *R*, *L*, and *C*. We may use complex frequency to generalize the concept of impedance. We will use the technique presented in Chap. 4, where only real frequency was considered. As before, we argue that $e^{\underline{s}t}$ is a function that is indestructible to linear operations such as addition, differentiation, and integration. To determine the impedance, therefore, we excite a circuit with a voltage $\text{Re}\{\underline{V}(\underline{s})e^{\underline{s}t}\}$ and observe a response, $\text{Re}\{\underline{I}(\underline{s})e^{\underline{s}t}\}$. The *generalized impedance* is defined as

$$\underline{Z}(\underline{s}) = \frac{\underline{V}(\underline{s}) e^{\underline{s}t}}{\underline{I}(\underline{s}) e^{\underline{s}t}} \, \Omega \tag{5.13}$$

We illustrate with an inductor, as shown in Fig. 5.5. The equations are

$$v(t) = L\frac{d}{dt}i(t) \Rightarrow \underline{V}(\underline{s})e^{\underline{s}t} = L\frac{d}{dt}\underline{I}(\underline{s})e^{\underline{s}t} = \underline{s}L\underline{I}(\underline{s})e^{\underline{s}t} \tag{5.14}$$

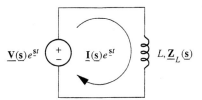

Figure 5.5 An inductor excited by $\underline{V}(\underline{s})e^{\underline{s}t}$.

where we have omitted the "real part of" for simplicity. The differentiation is performed only on the e^{st} function because this is the only function of time. Thus the generalized impedance of an inductor is

$$\underline{Z}_L(\underline{s}) = \frac{\underline{V}(\underline{s})e^{\underline{s}t}}{\underline{I}(\underline{s})e^{\underline{s}t}} = \underline{s}L \;\Omega \tag{5.15}$$

In like manner we can establish the generalized impedances of resistors and capacitors:

$$\underline{Z}_R(\underline{s}) = R \;\Omega, \quad \text{and} \quad \underline{Z}_C(\underline{s}) = \frac{1}{\underline{s}C} \;\Omega \tag{5.16}$$

These are familiar formulas for impedance, with $j\omega$ replaced by $\underline{s}$.

Application of generalized impedance. We will analyze the *RL* circuit shown in Fig. 5.6(a). In the frequency domain, Fig. 5.6(b), the impedances combine like resistors:

$$\underline{Z}(\underline{s}) = 2 \| (1 + \underline{s}/2) = \frac{1}{\dfrac{1}{2} + \dfrac{1}{1 + \underline{s}/2}} = \frac{2(\underline{s} + 2)}{(\underline{s} + 6)} \;\Omega \tag{5.17}$$

We will use the complex impedance function in Eq. (5.17) to explore the transient response of the circuit.

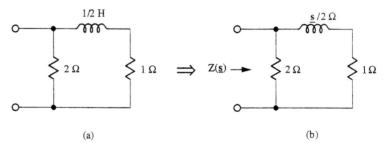

Figure 5.6 An *RL* circuit in (a) time domain; (b) frequency domain.

Transient Analysis

LEARNING OBJECTIVE 2.

To use generalized impedance to find the forced and natural response of a system

Forced and natural response. When we excite a circuit with a voltage or current source, we force a certain response from the circuit. For example, if we apply a dc source, we force a dc response. But, as we saw in Chap. 3, part of the transient response of the circuit takes a form that is natural to, and determined by, the circuit itself. Thus the response is always of the form, say, for a current:

$$i(t) = i_f(t) + i_n(t) \tag{5.18}$$

where $i_f(t)$ is the forced response and $i_n(t)$ is the natural response. We now explore the role of complex frequency and the impedance function in determining the forced and natural responses of a linear circuit.

Forced response from impedance. The forced response of the circuit may be determined directly from the impedance function when the forcing function, the input current or voltage, is of the form e^{st}. In this case we must evaluate the impedance function at the value of $\underline{s}$ of the forcing function. For example, if the voltage is a known function with a complex frequency $\underline{s} = \underline{s}_f$, the current must be of the same form and can be determined from Eq. (5.19)

$$\frac{\underline{V}_f e^{\underline{s}_f t}}{\underline{I}_f e^{\underline{s}_f t}} = \underline{Z}(\underline{s}_f) \Rightarrow \underline{I}_f e^{\underline{s}_f t} = \frac{\underline{V}_f}{\underline{Z}(\underline{s}_f)} \times e^{\underline{s}_f t} \qquad (5.19)$$

where $\underline{V}_f$ and $\underline{I}_f$ are phasors describing the source and response, respectively. The response in the time domain is thus

$$i_f(t) = \text{Re}\{\underline{I}_f e^{\underline{s}_f t}\} = \text{Re}\left\{\frac{\underline{V}_f}{\underline{Z}(\underline{s}_f)} e^{\underline{s}_f t}\right\} \qquad (5.20)$$

When the exciting function is not of the form e^{st}, the forced response must be determined by other methods, such as Laplace transform theory.

EXAMPLE 5.3 Forced response with voltage source

For the circuit in Fig. 5.6, a voltage source is connected to the input, $v_f(t) = 2e^{-t}$, for all time. Find the forced current response at the input.

SOLUTION:
The complex frequency of the input is $\underline{s}_f = -1 \text{ s}^{-1}$. The current must be of the form $I_f e^{-t}$, where I_f is a constant. From Eqs. (5.17) and (5.19)

$$\frac{2e^{-t}}{I_f e^{-t}} = \underline{Z}(-1) = \frac{2(-1+2)}{-1+6} = 0.4 \Rightarrow I_f = \frac{2}{0.4} = 5 \text{ A} \qquad (5.21)$$

Thus $i_f(t) = 5e^{-t}$ A. This is the forced response of the circuit to the exponential voltage source.

WHAT IF? What if $v(t) = 2e^{-2t}$?[4]

EXAMPLE 5.4 Forced response with a current source

Find the forced response of the circuit in Fig. 5.6 if excited by a current source with $i_f(t) = 1.5 \cos(2t + 42°)$.

SOLUTION:
The input can be represented by a phasor $i_f(t) = \text{Re}\{1.5 \angle 42° e^{j2t}\}$, so $\underline{s}_f = j2$. Thus the phasor representing the forced voltage response is

---
[4] Then $I_f = \infty$ by our method. More advanced methods must be used in that case.

$$\underline{V}_f = \underline{I}_f \times \underline{Z}(j2) = 1.5 \angle 42° \times \frac{2(j2+2)}{j2+6} = 1.34 \angle 68.6° \quad (5.22)$$

Thus the voltage produced in steady state is $v_f(t) = 1.34\cos(2t + 68.6°)$ V. All this should look familiar to you since this is merely ac circuit analysis such as we studied in Chap. 4.

WHAT IF? What if $\underline{s} = +j2$ and $\underline{s} = -j2$ are used?[5]

Finding the natural response. The impedance function can also assist us in finding the natural response of the circuit. Consider the case of a circuit excited by a voltage source and we wish to determine the current. The natural response of a linear circuit or system must be of the form

$$i_n(t) = \mathrm{Re}\{\underline{I}_n e^{\underline{s}_n t}\} \quad (5.23)$$

where $\underline{I}_n$ is an unknown phasor and $\underline{s}_n$ is the natural frequency of the system. We can determine the natural frequency from the impedance function. The natural frequency corresponds to the case where the exciting voltage is zero but the current is nonzero, being established solely by the circuit. The impedance function thus becomes

$$\underline{Z}(\underline{s}) = \frac{\underline{V}(\underline{s})e^{\underline{s}t}}{\underline{I}(\underline{s})e^{\underline{s}t}} = \frac{0}{\neq 0} \quad (5.24)$$

Equation (5.24) can be valid only if the natural frequency makes the impedance zero

$$\underline{Z}(\underline{s}_n) = 0 \quad (5.25)$$

Equation (5.25) allows us to determine the natural frequency or frequencies of the circuit for voltage excitation.

We illustrate with the circuit in Fig. 5.6 with the impedance function derived in Eq. (5.17). The circuit has only one natural frequency, which may be determined by setting the impedance to zero

$$\underline{Z}(\underline{s}_n) = \frac{2(\underline{s}_n + 2)}{(\underline{s}_n + 6)} = 0 \Rightarrow \underline{s}_n = -2 \quad (5.26)$$

The natural frequency in this first-order circuit corresponds to the negative of the reciprocal of the time constant with the input shorted. Since we are exciting the circuit with a voltage source, the time constant must be determined with the input shorted.

---

[5] It's a little more work, but you get the same answer.

EXAMPLE 5.5 Transient with voltage source

Find the current input to the circuit in Fig. 5.6 if the voltage source input of Fig. 5.7 is applied as an input.

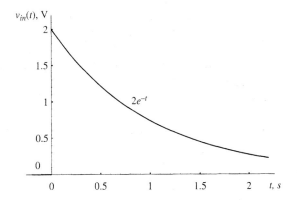

Figure 5.7 The input voltage is zero for negative time and a decreasing exponential for positive time.

SOLUTION:

The current will consist of a forced and a natural component, as in Eq. (5.18). The forced component is derived in Eq. (5.21) and the natural response will be of the form of Eq. (5.23) with $\underline{s}_n = -2$, as shown in Eq. (5.26). Thus the current produced by the voltage in Fig. 5.7 must be of the form

$$i(t) = 5e^{-t} + Ae^{-2t} \qquad (5.27)$$

where A is to be determined from the initial conditions. In this case, the initial current must be 1 A because the initial voltage is 2 V and the inductor acts as an open circuit.[6] Thus A comes from Eq. (5.27) at $t = 0$

$$1 = 5e^0 + Ae^0 \Rightarrow A = -4 \qquad (5.28)$$

The input current is therefore

$$i(t) = 5e^{-t} - 4e^{-2t} \text{ A} \qquad (5.29)$$

which is shown in Fig. 5.8.

---

[6] See the discussion on p. 137ff.

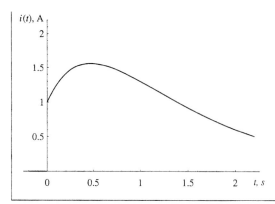

Figure 5.8 Current response of circuit in Fig. 5.6 to the voltage source in Fig. 5.7.

Open-circuit and short-circuit natural frequencies. We showed in Eq. (5.25) that the value of $\underline{s}$ that makes the impedance function go to zero corresponds to the natural frequency with a turned-OFF voltage source at the input.[7] In Fig. 5.9, for example, the inductor would see an equivalent resistance of 1 Ω if the 2-Ω resistor is shorted; thus, the short-circuit time constant of the circuit would be $L/R = 0.5$ s. This time corresponds to a natural frequency of $\underline{s}_n = -2$, Eq. (5.26).

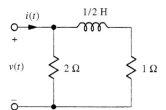

Figure 5.9 Circuit in Fig. 5.6(a) repeated.

Additional information can be derived from the value of $\underline{s}$ that makes the impedance function go to infinity. Infinite impedance would correspond to the input current $\mathbf{I}(\underline{s})$ going to zero, an open circuit, with the input voltage, $\mathbf{V}(\underline{s})$, remaining nonzero.

$$\mathbf{Z}(\underline{s}) = \frac{\mathbf{V}(\underline{s})e^{\underline{s}t}}{\mathbf{I}(\underline{s})e^{\underline{s}t}} = \frac{\neq 0}{0} \tag{5.30}$$

Thus the input impedance goes to infinite at the open-circuit natural frequency. For Eq. (5.17) this requires

$$\frac{2(\underline{s}+2)}{(\underline{s}+6)} = \frac{\neq 0}{0} \Rightarrow \underline{s}_n = -6 \tag{5.31}$$

This natural frequency corresponds to the negative of the reciprocal of the time constant of the circuit with the input open-circuited. The open circuit corresponds to exciting the circuit with a current source, which is turned OFF for the natural response.

---

[7] A turned-OFF voltage source is equivalent to a short circuit; see p. 59f.

EXAMPLE 5.6 Transient with a current source

Find the voltage produced in the circuit in Fig. 5.6 if the current in Fig. 5.10 were applied.

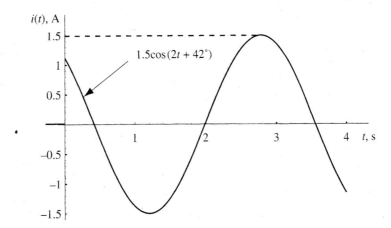

Figure 5.10 The current source is zero for negative time and a sinusoid for positive time.

SOLUTION:
The resulting voltage has a forced and a natural component. The forced response was calculated in Eq. (5.22). The natural response is of the form of Eq. (5.23) with $\mathbf{s}_n = -6$, as derived in Eq. (5.31), and voltage substituted for current. Combining the forced and natural responses, we find the input voltage to be

$$v(t) = 1.34 \cos(2t + 68.6°) + Be^{-6t} \text{ V} \tag{5.32}$$

where B is a constant to be determined from the initial conditions. The inductor acts initially as an open circuit, so the initial value of the voltage must be $1.5 \cos 42° \times 2\,\Omega = 2.23$ V. From Eq. (5.32) at $t = 0$,

$$2.23 = 1.34 \cos(0 + 68.6°) + Be^0 \Rightarrow B = 1.74 \text{ V} \tag{5.33}$$

Hence the voltage produced by the current in Fig. 5.10 in the circuit in Fig. 5.9 is

$$v(t) = 1.34 \cos(2t + 68.6°) + 1.74 e^{-6t} \text{ V} \tag{5.34}$$

which is shown in Fig. 5.11.

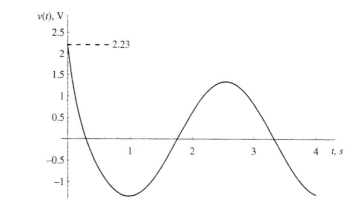

Figure 5.11 Voltage response of the circuit in Fig. 5.9 to the current source in Fig. 5.10.

pole zero

Summary. The value of $\underline{s}_n$ that sets the impedance function to zero is called a *zero* of the impedance function and corresponds to the short-circuit natural frequency of the circuit. The value of $\underline{s}_n$ that makes the impedance function go to infinity is a *pole* of the impedance function and corresponds to the open-circuit natural frequency of the circuit. These natural frequencies are normally called the zeros and poles of the impedance function.

Poles and zeros. The impedance function of a complicated circuit takes the form of Eq. (5.17), except that the numerator and denominator in general are polynomials in $\underline{s}$. Thus the function can always be factored into the form

$$\frac{\underline{V}(\underline{s})e^{\underline{s}t}}{\underline{I}(\underline{s})e^{\underline{s}t}} = \underline{Z}(\underline{s}) = K\frac{(\underline{s}-\underline{z}_1)(\underline{s}-\underline{z}_2)}{(\underline{s}-\underline{p}_1)(\underline{s}-\underline{p}_2)(\underline{s}-\underline{p}_3)} \tag{5.35}$$

where K is a constant and we have assumed a second-order polynomial in the numerator and a third-order polynomial in the denominator.[8] In Eq. (5.35), the constants $\underline{z}_1$ and $\underline{z}_2$ are the zeros of the function because these are the values of $\underline{s}$ at which the function goes to zero. Likewise, the constants $\underline{p}_1$, $\underline{p}_2$, and $\underline{p}_3$ are the poles of the function because these are the values of $\underline{s}$ at which the function goes to infinity.

Natural frequencies. As we have shown above, the zeros correspond to the short-circuit natural frequencies and the poles correspond to the open-circuit natural frequencies of the circuit. Except for a multiplicative constant, the poles and zeros of the impedance function fully establish the behavior of the circuit. They give the natural frequencies, as we have illustrated above, and they also give directly the forced response to excitations of the form $e^{\underline{s}t}$, such as dc ($\underline{s}=0$) or ac ($\underline{s}=j\omega$).

---

[8] This indicates a circuit with three independent energy-storage elements.

EXAMPLE 5.7 Building the impedance function

A circuit has no zeros, one pole at $\underline{s} = -10 \text{ s}^{-1}$, and an input impedance of $1000\,\Omega$ at dc. Find the impedance of the circuit as a function of complex frequency, $\underline{s}$.

SOLUTION:
From Eq. (5.35), the form of the impedance must be

$$\underline{Z}(\underline{s}) = K\frac{1}{\underline{s} - \underline{p}} = K\frac{1}{\underline{s} - (-10)} \tag{5.36}$$

Since $\underline{Z}(0) = 1000$, $K = 10{,}000$ and the result is

$$\underline{Z}(\underline{s}) = \frac{10^4}{\underline{s} + 10} \tag{5.37}$$

WHAT IF? What if the circuit is a parallel RC combination? What are R and C?[9]

transfer function

Second-order circuit. The circuit in Fig. 5.12 is identical to that in Fig. 5.6 except that a 2-H inductor is added at the input and the voltage across the 1-Ω resistor is taken as an output.[10] The ratio of the output and the input is the *transfer function*, $\underline{T}(\underline{s})$

$$\frac{\underline{V}_{out}(\underline{s})\,e^{\underline{s}t}}{\underline{V}_{in}(\underline{s})\,e^{\underline{s}t}} = \underline{T}(\underline{s}) \tag{5.38}$$

The transfer function is like an impedance function except it relates circuit variables at different places in a system.

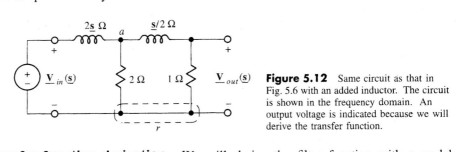

Figure 5.12 Same circuit as that in Fig. 5.6 with an added inductor. The circuit is shown in the frequency domain. An output voltage is indicated because we will derive the transfer function.

Transfer function derivation. We will derive the filter function with a nodal analysis. Since the output is open circuited, we will first determine the voltage at a, and then find $\underline{V}_{out}(\underline{s})$ with a voltage divider. Kirchhoff's current law for node a in the frequency domain becomes

Conservation of Charge

$$\frac{\underline{V}_a - \underline{V}_{in}}{2\underline{s}} + \frac{\underline{V}_a - (0)}{2} + \frac{\underline{V}_a - (0)}{\underline{s}/2 + 1} = 0 \tag{5.39}$$

---
[9] $R = 1000\,\Omega$ and $C = 100\,\mu\text{F}$.
[10] And we have labeled the circuit for nodal analysis.

which, after a bit of algebra, becomes

$$\underline{V}_a = \frac{\underline{s} + 2}{\underline{s}^2 + 7\underline{s} + 2} \times \underline{V}_{in} \qquad (5.40)$$

Using a voltage-divider relationship, we find the transfer function to be

$$\underline{T}(\underline{s}) = \frac{\underline{V}_a}{\underline{V}_{in}} \times \frac{\underline{V}_{out}}{\underline{V}_a} = \frac{\underline{s} + 2}{\underline{s}^2 + 7\underline{s} + 2} \times \frac{1}{\underline{s}/2 + 1} = \frac{2}{\underline{s}^2 + 7\underline{s} + 2} \qquad (5.41)$$

The methods we have developed for transient analysis work equally well for transfer functions, as shown by the following example.

EXAMPLE 5.8 Transient analysis of transfer function

The circuit in Fig. 5.12 has a switched-battery input, as shown in Fig. 5.13. Find the output voltage.

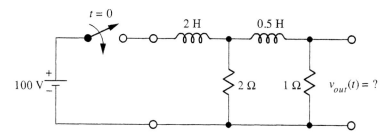

Figure 5.13 The circuit in Fig. 5.12 with a switched battery input. We will determine the output voltage.

SOLUTION:
The output voltage consists of a forced and a natural component. The forced component has a frequency of $\underline{s}_f = 0$, since the input is dc. The output voltage in the frequency domain will therefore be from Eqs. (5.38) and (5.41) at $\underline{s} = 0$

$$\frac{\underline{V}_f e^{0t}}{100 e^{0t}} = \underline{T}(0) = \left. \frac{2}{\underline{s}^2 + 7\underline{s} + 2} \right|_{\underline{s}=0} = 1 \qquad (5.42)$$

Thus the forced component of the output voltage is $\underline{V}_f = 100$ V. This occurs because the inductors become short circuits at dc.

The natural response corresponds to having nonzero output with zero input

$$\frac{\underline{V}_{out}}{\underline{V}_{in}} = \underline{T}(\underline{s}) = \frac{\neq 0}{0} \qquad (5.43)$$

which can be valid only if the denominator of Eq. (5.41) is set to zero

$$\underline{s}^2 + 7\underline{s} + 2 = 0 \Rightarrow \underline{s}_n = -0.298, -6.70 \text{ s}^{-1} \qquad (5.44)$$

Thus we have two natural frequencies that are real and negative. The output is therefore of the form

$$v_{out}(t) = 100 + Ae^{-0.298t} + Be^{-6.70t} \text{ V} \tag{5.45}$$

where A and B are constants. The two inductors block the initial voltage from the output; hence the initial conditions on $v_{out}(t)$ are

$$v_{out}(0^+) = 0 \text{ and } \left.\frac{dv_{out}(t)}{dt}\right|_{t=0^+} = 0 \tag{5.46}$$

We may substitute Eq. (5.45) into Eqs. (5.46) and solve simultaneously for A and B, with the result

$$v_{out}(t) = 100 - 104.66e^{-0.298t} + 4.66e^{-6.70t} \text{ V} \tag{5.47}$$

which is plotted in Fig. 5.14.

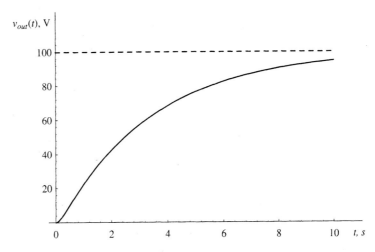

Figure 5.14 Response of circuit in Fig. 5.13.

The time constants derived from Eq. (5.47) do not correspond to any L/R combination in the circuit but are influenced by the interactions of all components. When the energy storage elements are all inductors or all capacitors, the natural frequencies are always real and negative.

Check your Understanding

1. What is the impedance of a 100-μF capacitor at a complex frequency of $\underline{s} = -100$ s^{-1}?

2. An impedance consists of $R = 100 \, \Omega$ and $C = 10 \, \mu\text{F}$ in series. What impedance does this present to an input voltage $v(t) = 10e^{-500t}$?

3. A 10-Ω resistor is connected in series with a 100-μF capacitor. At what complex frequency does this combination look like a short circuit?

4. An impedance with a zero at $\underline{s}=0$ will pass a dc current. (True or False?)

5. The open-circuit natural frequencies of a circuit correspond to the zeros of the impedance function. (True or False?)

6. A circuit with two inductors must have natural frequencies that are real. (True or False?)

Answers. (1) $-100\ \Omega$; (2) $-100\ \Omega$; (3) $-1000\ \text{s}^{-1}$; (4) true; (5) false; (6) true.

5.3 TRANSIENT RESPONSE OF RLC CIRCUITS

Types of Natural Responses in RLC Circuits

Transfer function of an RLC circuit. The circuit in Fig. 5.15 can exhibit an oscillatory response because the circuit contains an inductor and capacitor. The transfer function of this circuit can be determined from a voltage divider

$$\underline{T}(\underline{s}) = \frac{\underline{V}_{out}(\underline{s})\,e^{\underline{s}t}}{\underline{V}_{in}(\underline{s})\,e^{\underline{s}t}} = \frac{R\|(1/\underline{s}C)}{\underline{s}L + R\|(1/\underline{s}C)} = \frac{1}{LC}\frac{1}{\underline{s}^2 + (\underline{s}/RC) + (1/LC)} \tag{5.48}$$

where $\underline{T}(\underline{s})$ is the transfer function.

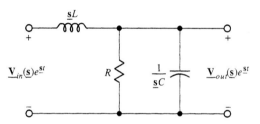

Figure 5.15 An *RLC* circuit in the frequency domain. The transfer function is the ratio of output to input voltage.

LEARNING OBJECTIVE 3.

To recognize conditions for undamped, underdamped, critically damped, and overdamped responses in second-order systems

Natural frequencies of the circuit. The poles of the transfer function correspond to an output voltage with no input voltage, and hence are the natural frequencies of the circuit with the input short-circuited. These poles will be investigated in detail because they reveal the various types of behavior that can result with an *RLC* circuit. We may determine the poles with the quadratic formula

$$\underline{s}^2 + \frac{\underline{s}}{RC} + \frac{1}{LC} = 0 \Rightarrow \underline{s}_n = -\frac{1}{2RC} \pm \sqrt{\left(\frac{1}{2RC}\right)^2 - \frac{1}{LC}}\ \text{s}^{-1} \tag{5.49}$$

There are four possibilities, depending on the relative values of *R*, *L*, and *C*. The possibilities are undamped, underdamped, critically damped, and overdamped responses.

undamped response

Undamped behavior. First we consider the case where the resistance is infinite, which is equivalent to removing the resistor from the circuit. Equation (5.49) gives imaginary roots:

$$\underline{s}_1 = +j\omega_0 \quad \text{and} \quad \underline{s}_2 = -j\omega_0\ \text{s}^{-1} \tag{5.50}$$

where $\omega_0 = 1/\sqrt{LC}$ rad/s. Thus, with no resistance, the short-circuit natural frequencies are imaginary, meaning that the natural response of the circuit is a sinusoid of constant amplitude. This may be understood as a lossless resonance, with ω_0 the resonant frequency. With no resistor, any energy imparted to the circuit alternates between the inductor and capacitor and produces a sinusoidal output. This is called an *undamped response* because the oscillations do not diminish with time.

damped oscillation, damping constant

Underdamped behavior. For large but finite resistance, the square root remains imaginary, and the roots of the quadratic are complex:

$$\underline{s}_{1,2} = -\alpha \pm \sqrt{\alpha^2 - \omega_0^2} \quad \text{or} \quad \underline{s}_{1,2} = -\alpha \pm j\omega \text{ s}^{-1} \tag{5.51}$$

where $\alpha = 1/2RC \text{ s}^{-1}$ and $\omega = \sqrt{\omega_0^2 - \alpha^2} \text{ s}^{-1}$. The two roots are complex conjugates and correspond to an exponentially decreasing sinusoid, a *damped oscillation*. Thus the natural response of the circuit is of the form

$$v_n(t) = Ae^{-\alpha t}\cos(\omega t + \theta) \tag{5.52}$$

where A and θ are constants. The constant α is called the *damping constant* and is the reciprocal of the time constant of the dying oscillation. The frequency of the oscillation, ω, is less than the resonant frequency, ω_0, because the exchanges of energy between inductor and capacitor are slowed down by loss in the resistor.

underdamped response

Condition for underdamped response. The exponentially decreasing oscillation is called an *underdamped response*. The condition for an underdamped response is

$$\alpha < \omega_0 \quad \text{or} \quad R > \frac{1}{2}\sqrt{\frac{L}{C}} \tag{5.53}$$

Thus for large values of resistance, the natural response of the circuit is dominated by the resonance of the inductor and capacitor, but the oscillations die out in time due to the loss in the resistor.

critical damping

Critical damping. *Critical damping* occurs when $\alpha = \omega_0$, and the roots of Eq. (5.49) become real and equal. This is interesting mathematically but unimportant in practice because it exists only for one exact value of resistance, given by Eq. (5.53) with an equality sign. Even if we wished to produce this type of response in a physical circuit, we would be able to achieve the required resistance only with great care or good fortune. Critical damping is important for us as the boundary between underdamped oscillations and overdamped behavior.

overdamped response

Overdamped behavior. When $\alpha > \omega_0$, the roots of Eq. (5.49) are real, negative, and unequal.

$$\underline{s}_1 = -\alpha + \sqrt{\alpha^2 - \omega_0^2} \quad \text{and} \quad \underline{s}_2 = -\alpha - \sqrt{\alpha^2 - \omega_0^2} \tag{5.54}$$

Thus the *overdamped response* has the form

$$v_n(t) = Ae^{\underline{s}_1 t} + Be^{\underline{s}_2 t} \tag{5.55}$$

where $\underline{s}_1$ and $\underline{s}_2$ are real, negative numbers. This response consists of two time constants, which are the negatives of the reciprocals of $\underline{s}_1$ and $\underline{s}_2$. Thus, for small resistance, increased loss eliminates the resonance between the inductor and capacitor.

Summary. The natural frequencies of the *RLC* circuit are derived from the roots of a quadratic equation. The roots may be imaginary, complex, real and equal, or real and unequal. Imaginary roots indicate an undamped oscillation. Complex roots indicate a damped oscillation. Real, unequal roots indicate two time constants and no oscillation. Real, equal roots represent the mathematical boundary between oscillatory and nonoscillatory behavior. These time-domain responses are derived through the application of frequency-domain techniques. The next section explores the various forms of circuit behavior.

RLC Circuit Transient Behavior

Forced response. We now investigate the time-domain response of the circuit shown in Fig. 5.16, in which a dc voltage is applied suddenly to the *RLC* circuit in the previous section. The output voltage consists of two components, a natural response and a forced response due to the dc input. We may determine the forced response by the methods presented in Chap. 3, by treating the capacitor as an open circuit and the inductor as a short circuit. Alternately, we may consider the input as $10e^{\underline{s}_f t}$, with $\underline{s}_f = 0$, and derive the forced response from the transfer function given in Eq. (5.48) with $\underline{s} = 0$. Using the latter method, we find the forced component of the output voltage to be

$$\frac{V_{out} e^{0t}}{V_{in} e^{0t}} = \underline{T}(0) \implies V_{out} = \underline{T}(0) \times 10 e^{0t} = 10 \text{ V} \quad (5.56)$$

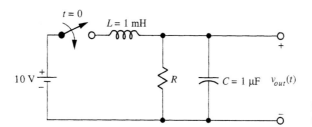

Figure 5.16 An *RLC* circuit with a source in the time domain.

since $\underline{T}(0) = 1$. Equation (5.56) indicates a dc output of 10 V due to the battery, a result that is independent of the resistance. The total response will therefore be

$$v_{out}(t) = 10 + v_n(t) \text{ V} \quad (5.57)$$

where $v_n(t)$ is the natural response. Several forms are possible for the natural response, depending on the resistance.

Natural response: undamped case. We consider first the response for $R = \infty$

$$v_{out}(t) = 10 + A \cos(\omega_0 t + \theta) \quad (5.58)$$

where $\omega_0 = 1/\sqrt{LC} = 3.16 \times 10^4$ and A and θ must be determined from the initial conditions.

Initial conditions. We may establish the initial conditions of the output voltage through the techniques presented in Chap. 3. The capacitor acts initially as a short circuit; hence the output voltage must be zero at $t = 0^+$

$$0 = 10 + A\cos(\theta) \tag{5.59}$$

The inductor acts initially like an open circuit. Consequently, the initial current through the capacitor is also zero, and the derivative of the output voltage must also be zero at $t = 0^+$

$$\left.\frac{dv_{out}(t)}{dt}\right|_{t=0^+} = 0 \;\;\Rightarrow\;\; (0) = -\omega_0 A \sin(\theta) \tag{5.60}$$

so $\theta = 0$ and Eq. (5.59) yields $A = -10$ V.[11] Hence the response is

$$v_{out}(t) = 10 - 10\cos(\omega_0 t) \text{ V} \tag{5.61}$$

This response is shown in Fig. 5.17.

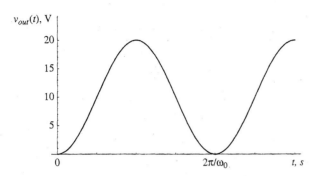

Figure 5.17 Response of the circuit in Fig. 5.16 with $R = \infty$. The peak voltage is twice the input voltage, a property of the circuit that finds many applications.

resonance charging

Resonance charging. The voltage on the capacitor reaches twice that of the input because the inductor gives a momentum to the current to continue charging past the equilibrium point. Since the stored energy in a capacitor is proportional to the square of its voltage, the maximum stored energy is four times that obtained by charging the capacitor through a resistor, and no losses occur. This technique is known as *resonance charging* of the capacitor, and applications are found in radar and power electronics.

Underdamped response. For the inductance and capacitance values given in Fig. 5.16, the value of the resistance for critical damping is given by Eq. (5.53) with an equality sign

$$R_c = \frac{1}{2}\sqrt{\frac{L}{C}} = \frac{1}{2}\sqrt{\frac{10^{-3}}{10^{-6}}} = 15.8 \;\Omega \tag{5.62}$$

where R_c is the resistance for critical damping. We consider first the underdamped response, which requires a resistance larger than the critical value. We assume $R = 5R_c = 79.1 \;\Omega$. The resonant frequency and damping coefficient are from Eq. (5.51)

$$\omega_0 = \frac{1}{\sqrt{LC}} = 31{,}600 \text{ s}^{-1} \quad \text{and} \quad \alpha = \frac{1}{2RC} = 6320 \text{ s}^{-1} \tag{5.63}$$

---

[11] An identical answer results from $\theta = \pi$ and $A = +10$.

and thus the natural frequencies, determined from Eq. (5.51), are

$$\underline{s}_1 = -6320 + j31{,}000 \quad \text{and} \quad \underline{s}_2 = -6320 - j31{,}000 \tag{5.64}$$

The natural response is given by Eq. (5.52) and the forced response is the same as before; here the total response is

$$v_{out}(t) = 10 + Ae^{-6320t} \cos(31{,}000t + \theta) \text{ V} \tag{5.65}$$

The initial conditions are the same as for the undamped response. Hence at $t = 0^+$

$$0 = 10 + A\cos(\theta) \tag{5.66}$$

and

$$\left.\frac{dv_{out}(t)}{dt}\right|_{t=0+} = 0 \Rightarrow 0 = A[-6320\cos(\theta) - 31{,}000\sin(\theta)] \tag{5.67}$$

Equations (5.66) and (5.67) yield $\theta = -11.5°$ and $A = -10.21$ V. The output voltage is therefore

$$v_{out}(t) = 10 - 10.21e^{-6324t} \cos(31{,}000t - 11.5°) \text{ V} \tag{5.68}$$

which is shown in Fig. 5.18.

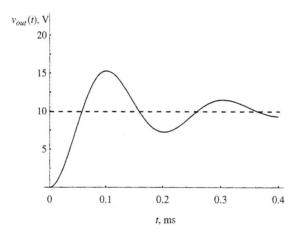

Figure 5.18 Underdamped response for the circuit in Fig. 5.16 with $R = 79.1 \, \Omega$.

Comparison of Figs. 5.17 and 5.18 shows two major differences. First, the losses in the resistor cause the oscillations to die out rapidly. Second, the output approaches a constant of 10 V when losses are present. This is the same as the average output voltage of the lossless circuit, except that, with no losses, the voltage never settles down.

Overdamped response. To obtain an overdamped response, we will use a resistance one-half the critical value given by Eq. (5.62): $R = R_c/2 = 7.91 \, \Omega$. With this resistance, $\alpha = 63{,}200$, and Eq. (5.54) gives the natural frequencies

$$\underline{s}_1 = -8470 \text{ s}^{-1} \quad \text{and} \quad \underline{s}_2 = -118{,}000 \text{ s}^{-1} \tag{5.69}$$

The natural response is given by Eq. (5.55) and the forced response is the same as before; thus the output voltage has the form

$$v_{out}(t) = 10 + Ae^{-8470t} + Be^{-118,000t} \text{ V} \tag{5.70}$$

where A and B are constants. The initial conditions are the same as before; hence we may solve for A and B from the equations

$$0 = 10 + A + B \quad \text{and} \quad 0 = -8470A - 118,000B \tag{5.71}$$

which yield $A = -10.77$ V and $B = 0.77$ V. Thus the output voltage is

$$v_{out}(t) = 10 - 10.77e^{-849t} + 0.77e^{-118,000t} \text{ V} \tag{5.72}$$

This response is shown in Fig. 5.19. The third term in Eq. (5.72) has a brief influence to give the zero derivative at the origin, but otherwise the response is that of a first-order transient of the RL part of the circuit.

Parallel and series RLC circuits. The circuit in Fig. 5.15 is a parallel RLC circuit when excited by a voltage source. With the source turned OFF, the resistor, inductor, and capacitor are connected in parallel. In the circuit shown in Fig. 5.20, the resistor, inductor, and capacitor are connected in series. The analysis of the series RLC circuit for the transfer function, natural frequencies, and transient response follows the same lines as for the parallel circuit, but the results differ in detail. We leave the analysis of the series RLC circuit for a homework problem.

Relationship with Laplace transform theory. The Laplace transform gives an alternative mathematical foundation to the techniques used in this chapter and is closely associated with linear system theory. Although the Laplace transform yields the total response of a circuit or system, including initial conditions, its primary importance lies in furnishing the tools for describing linear systems that we have developed in this chapter by simpler means.

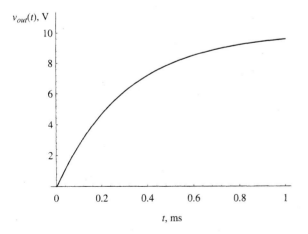

Figure 5.19 Overdamped response for the circuit in Fig. 5.16 with $R = 7.91 \, \Omega$.

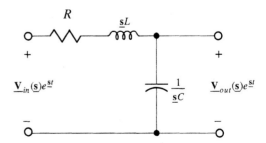

Figure 5.20 This is a series RLC circuit when driven by a voltage source.

The Frequency Domain

Frequency domain. The functions such as $\underline{V}(\underline{s})$ and $\underline{I}(\underline{s})$, which we have been treating as constants, can be considered *transforms* of the time-domain voltage and current. Similarly, generalized impedances and transfer functions are frequency-domain transforms of the linear differential equations that describe the circuit or system in the time domain. In the next section we show how these transforms are related to the spectra of the time-domain functions.

Summary. In this section we used complex frequency to investigate transient behavior of first- and second-order circuits. Natural responses are decreasing exponentials or exponentially decreasing sinusoids. The natural frequencies and the character of the natural response may be determined from the poles and zeros of the impedance or transfer functions. In the next section, we focus on filtering electrical signals in the frequency domain. Finally, we will illustrate how the transfer function describes the properties of a linear-system component.

Check Your Understanding

1. Describe the largest number and the character of the natural frequencies of the following circuits: (a) Two resistors and a capacitor; (b) One resistor and two capacitors; (c) Three resistors and two inductors; (d) Two resistors, a capacitor, and an inductor.
2. Determine the natural frequencies of the circuit in Fig. 5.16 if the resistor has the value for critical damping.

Answers. (1) (a) one natural frequency, real and negative; (b) one natural frequency, real and negative; (c) two natural frequencies, real and negative; (d) two natural frequencies, either complex with negative real part or both real and negative; (2) $\underline{s} = -31{,}623 \text{ s}^{-1}$ (repeated).

5.4 FILTERS AND BODE PLOTS

Spectra

Exploring the frequency domain. In the early parts of Chap. 4, we considered the steady-state response of circuits to the application of a single, fixed frequency. In Sec. 4.4, we let the frequency vary from dc to a high frequency and examined the effects on the amplitude and phase of the circuit response. In this chapter we have investigated the response of circuits to the complex frequencies associated with a wide variety of signals that occur in physical systems. In these ways we have begun to explore the concept of the frequency domain.

In the present section, we continue to explore the frequency domain by considering signals that contain more than one frequency. When we speak of the *spectrum* of a signal, we are thinking of the signal as a collection of many frequencies. Here we explain spectra briefly to give a context for electric filters.

spectrum, fundamental harmonics

Spectra. The simplest spectrum is a single sinusoid. Examples would be the power system, having a frequency of 50 or 60 Hz, or a single pure musical tone. The "emergency broadcast signal" used on radio stations consists of two nonharmonically related frequencies. Musical instruments playing a single note generate harmonic spectra. The frequency corresponding to the musical tone is the fundamental frequency. The excitation producing the fundamental produces also the higher "harmonics": the second harmonic has a frequency that is twice the fundamental; the third harmonic three times the fundamental, etc. Other signals having harmonic spectra are repetitive signals generated by electric motors, TV sets, and radar systems.

Nonrepetitive signals, such as those generated by lightning strikes and by Brownian motion of charges in semiconductors, also produce spectra. Such spectra contain many frequencies, with some frequencies much stronger than others.

Time and frequency domains. We are familiar with a time-domain description of a signal, for instance, a voltage like $v(t)$. The same signal is fully described in the frequency domain by its spectrum, $\underline{V}(\underline{s})$, which gives the amplitude and phase of all sinusoids in the signal. The response of a circuit or system to a signal may be described in either the time or frequency domain. In the previous section we used complex frequency to develop techniques appropriate to the time domain. In this section we use complex frequency to develop techniques appropriate to the frequency domain.

> **LEARNING OBJECTIVE 4.**
>
> To understand the concept of electric filters and the use of Bode plots to describe signals and filters

Filter Functions

Electric filters are used extensively in electronic circuits and occasionally in power circuits as well. To examine the filtering properties of circuits, we consider signals to be composed of real frequencies: $\underline{s} = j\omega$, and thus the voltage spectrum would be $\underline{V}(j\omega)$. We begin with the low-pass filter.

Low-pass filter circuit. Consider the circuit shown in Fig. 5.21.

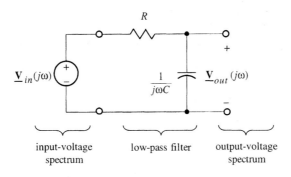

Figure 5.21 A low-pass filter with input- and open-circuit output voltage spectra.

As we will soon show, the RC circuit acts as a low-pass filter. The resistor and capacitor are represented by their impedances because the circuit and signals are represented in the frequency domain.

Since we are interested in the open-circuit output voltage spectrum, we may analyze the circuit as a voltage divider

$$\underline{V}_{out}(j\omega) = \frac{\frac{1}{j\omega C}}{R + \frac{1}{j\omega C}} \times \underline{V}_{in}(j\omega) = \frac{1}{1 + j\omega RC} \times \underline{V}_{in}(j\omega) \qquad (5.73)$$

The filter function, $\underline{F}(j\omega)$, is defined as the ratio of output to input voltage

$$\mathbf{F}(j\omega) = \frac{\mathbf{V}_{out}(j\omega)}{\mathbf{V}_{in}(j\omega)} = \frac{1}{1 + j\omega RC} \tag{5.74}$$

If we know the input spectrum and the filter function, the output spectrum may be obtained as shown in Fig. 5.22.

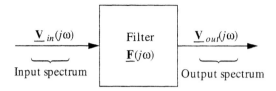

$$\mathbf{V}_{out}(j\omega) = \mathbf{V}_{in}(j\omega) \times \mathbf{F}(j\omega)$$

Figure 5.22 The output spectrum of the filter is the product of the input spectrum and the filter function, $\mathbf{F}(j\omega)$.

The amplitude response. For simplicity, think of the input spectrum as having one frequency, ω_1, and thus being a simple sinusoid with an amplitude and phase

$$\mathbf{V}_{in}(j\omega) = V_{in} \angle \theta_{in} \tag{5.75}$$

where V_{in} is a real number, the amplitude of the sinusoid, and θ_{in} is its phase. The filter function at $\omega = \omega_1$ also has a magnitude and angle

$$\mathbf{F}(j\omega_1) = |\mathbf{F}(j\omega_1)| \angle \theta_F \tag{5.76}$$

where for the low-pass filter function in Eq. (5.74) the magnitude is

$$|\mathbf{F}(j\omega_1)| = \frac{1}{\sqrt{1 + (\omega_1 RC)^2}} \tag{5.77}$$

and the phase shift in the filter is

$$\theta_F = -\tan^{-1} \omega_1 RC \tag{5.78}$$

as shown in Fig. 5.23.

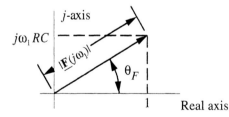

Figure 5.23 The conversion of the filter function from rectangular to polar form requires the hypotenuse and angle of a right triangle. Since the $1 + j\omega RC$ term is in the denominator in Eq. (5.74), the phase shift is negative in Eq. (5.78).

The output voltage, $\mathbf{V}_{out}(j\omega)$ has the same frequency, $\omega = \omega_1$, and the equation in Fig. 5.22 shows the output spectrum to be

$$\mathbf{V}_{out}(j\omega_1) = V_{out} \angle \theta_{out} = V_{in} \angle \theta_{in} \times |\mathbf{F}(j\omega_1)| \angle \theta_F \tag{5.79}$$

From the rules for multiplying complex numbers in polar form, we find the amplitude of the output voltage to be the product of the amplitude of the input voltage and the magnitude of the filter function

$$V_{out} = V_{in} \times |\mathbf{F}(j\omega_1)|$$
$$= V_{in} \times \frac{1}{\sqrt{1 + (\omega_1 RC)^2}} \quad (5.80)$$

and the phase of the output is the phase of the input plus the phase shift due to the filter

$$\theta_{out} = \theta_{in} + \theta_F = \theta_{in} - \tan^{-1} \omega_1 RC \quad (5.81)$$

Since these relationships are true at a single frequency, they must be true for all frequencies; thus it must be generally true that the magnitudes of the input and output power spectra are related as

$$|\underline{\mathbf{V}}_{out}(j\omega)| = |\underline{\mathbf{V}}_{in}(j\omega)| \times |\mathbf{F}(j\omega)|$$
$$= |\underline{\mathbf{V}}_{in}(j\omega)| \times \frac{1}{\sqrt{1 + (\omega RC)^2}} \quad (5.82)$$

How the filter works. The characteristics of the low-pass filter function, $\mathbf{F}(j\omega)$, can be understood from the effect of frequency on the impedance of the capacitor. Recall that the impedance of a capacitor has a magnitude of $1/\omega C$. At dc the capacitor will act as an open circuit; hence all the input voltage will appear at the output, independent of R. At very low ac frequencies, the impedance of the capacitor will still be very high compared to R. Specifically, as long as $1/\omega C \gg R$, the output voltage will be approximately equal to the input voltage, and thus the filter gain will be near unity. Thus $\mathbf{F}(j\omega) \to 1$ as $\omega \to 0$, as shown mathematically by Eq. (5.77).

low-pass filter

At very high frequencies, the impedance of the capacitor will approach that of a short circuit and hence little voltage will appear across the capacitor. As long as $R \gg 1/\omega C$, the current will be approximately $\underline{\mathbf{V}}_{in}/R$ and the magnitude of the voltage across the capacitor, which is the output voltage, will be $|\underline{\mathbf{V}}_{in}|/\omega RC$ and hence will approach very small values as ω increases. Thus $\mathbf{F}(j\omega) \to 0$ as $\omega \to \infty$. Consequently, the gain of the low-pass filter becomes very small at high frequencies. This means that a high-frequency signal would be greatly reduced, whereas low-frequency components would not be reduced by the filter; that is, it will "pass" only components of the signal at low frequencies. The filter amplitude characteristic, $\mathbf{F}(j\omega)$, is shown in Fig. 5.24, plotted on a log scale to show a large range of frequencies.

cutoff frequency, half-power frequency

Cutoff frequency. The region of transition between these two types of behavior is centered on the frequency where the impedance of the capacitor is equal to that of the resistance

$$R = \frac{1}{\omega_c C} \Rightarrow \omega_c = \frac{1}{RC} \quad \text{and} \quad f_c = \frac{1}{2\pi RC} \quad (5.83)$$

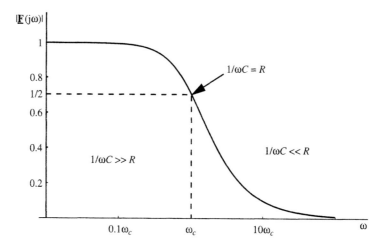

Figure 5.24 Low-pass filter amplitude characteristic.

where ω_c is called the *cutoff frequency* in rad/s and f_c is the cutoff frequency in Hz. The filter function can be expressed in terms of the cutoff frequency

$$\mathbf{F}(j\omega) = \frac{1}{1 + j\omega RC} = \frac{1}{1 + j(\omega/\omega_c)} = \frac{1}{1 + j(f/f_c)} \quad (5.84)$$

At $\omega = \omega_c$, or $f = f_c$, the filter function has the value $1/(1 + j1)$, which has a magnitude of $1/\sqrt{2} = 0.707$.[12] The significance of a voltage gain of 0.707 lies in the relationship between input and output power. Power is proportional to the square of the voltage; hence at ω_c the output power is reduced by a factor of 2 from what it is at low frequencies. For this reason, the cutoff frequency, expressed in either rad/s or hertz, is often called the *half-power frequency*.

Phase effects. Figure 5.24 shows only the magnitude of the filter function, $|\mathbf{F}(j\omega)|$. The filter function is a complex function with real and imaginary parts. Thus the filter affects both the magnitude and the phase of sinusoids passing through it. For example, at the cutoff frequency the filter function has the value $1/(1 + j1) = 0.707 \angle -45°$. This phase shift is an unavoidable by-product of the filtering process that can cause problems in some applications. In audio systems few problems occur due to phase shift because the ear is largely insensitive to phase, but in video systems such phase shifts can degrade performance.

Bode Plots

dB, Bode plot

Decibels of gain. The gain of amplifiers and the loss of filters are frequently specified in decibels, dB. This unit refers to a logarithmic measure of the ratio of output power to input power, as defined in

$$\text{gain in dB} = 10 \log\left(\frac{P_{out}}{P_{in}}\right) \text{ dB} \quad (5.85)$$

---

[12] This has nothing to do with the rms value of a sinusoid.

Because the power gain is proportional to the square of the voltage gain, the voltage gain in dB is defined to be

$$\text{gain in dB} = 10 \log \left|\frac{\mathbf{V}_{out}}{\mathbf{V}_{in}}\right|^2 = 20 \log \left|\frac{\mathbf{V}_{out}}{\mathbf{V}_{in}}\right| \text{ dB} \tag{5.86}$$

Loss in dB. A loss can also be described by Eq. (5.86) but the "gain in dB" is negative: for example, the gain of a low-pass filter, Eq. (5.84), at $f = f_c$ is $1/\sqrt{2}$; Eq. (5.86) yields

$$G_{dB} = 20 \log \left|\frac{\mathbf{V}_{out}}{\mathbf{V}_{in}}\right|_{\omega=\omega_c} = 20 \log |\mathbf{F}(j\omega_c)| = 20 \log \frac{1}{\sqrt{2}} = -3.0 \text{ dB} \tag{5.87}$$

cutoff frequency, half-power frequency

Thus, we could say that the gain of the filter at its half-power frequency is negative 3.0 dB or, equivalently, that its loss is 3.0 dB. For this reason, the critical frequency of a filter is often called its *3-dB frequency* or *half-power frequency*. We note in comparing Eqs. (5.85) and (5.86) that power ratios require a factor of 10 in the dB calculation and voltage ratios a factor of 20.

EXAMPLE 5.9

dB of gain

A low-pass filter gives an output voltage of 500 μV with an input of 10 mV at 1000 Hz. Find the dB gain of the amplifier.

SOLUTION:
Using Eq. (5.86), we find

$$G_{dB} = 20 \log \left(\frac{500 \times 10^{-6}}{10 \times 10^{-2}}\right) = -26.0 \text{ dB} \tag{5.88}$$

WHAT IF? What if the frequency is 0 Hz (dc)?[13]

Signal power in dB. The dB scale is also useful for expressing the power in a signal. In this context, the power is expressed as a logarithmic ratio between the signal power and an assumed power level, usually 1 W or 1 mW. Thus dBw, dB relative to 1 W, is defined as

$$\text{dBw} = 10 \log \frac{P}{1 \text{ watt}} \tag{5.89}$$

where P is a power and dBw is that same power relative to 1 W, expressed in dB. Similarly, dBm normalizes signal power to a milliwatt, expressed in dB.

---
[13] Then the dB of voltage gain is 20 log 1 = 0 dB.

EXAMPLE 5.10 Satellite transmitter

A satellite transmitter has an output power of 1000 W. Express this power in dBw.

SOLUTION:
From Eq. (5.89) this power would be

$$\text{dBw} = 10 \log \frac{1000 \text{ W}}{1 \text{ W}} = 30 \text{ dBw} \quad (5.90)$$

WHAT IF? What if you want it in dBm?[14]

Why dB is useful. There are three reasons why electrical engineers use the dB scale for describing gains, losses, and signal levels. First, because of the compressive nature of the logarithmic function, the numbers involved in a dB calculation are moderate compared to a linear scale. This feature of dB measure is particularly useful when plotting quantities that vary greatly in magnitude.

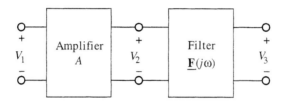

Figure 5.25 Cascaded systems. The output of the amplifier is the input to the filter.

cascaded systems

Cascaded systems. The second virtue of dB measure is the adding of dB gains when a signal is passed through cascaded systems. Figure 5.25 shows an amplifier *cascaded* with a filter, meaning that the output of the amplifier is the input to the filter. The combined voltage gain of the cascaded system is

$$\frac{|\mathbf{V}_3|}{|\mathbf{V}_1|} = \frac{|\mathbf{V}_2|}{|\mathbf{V}_1|} \times \frac{|\mathbf{V}_3|}{|\mathbf{V}_2|} = A \times |\mathbf{F}(j\omega)| \quad (5.91)$$

where A is the voltage gain of the amplifier. The dB gain of the cascaded system is

$$G_{\text{dB}} = 20 \log \frac{|\mathbf{V}_3|}{|\mathbf{V}_1|} = 20 \log A + 20 \log |\mathbf{F}(j\omega)| \quad (5.92)$$
$$= A_{\text{dB}} + F_{\text{dB}}$$

where A_{dB} is the gain of the amplifier in dB and F_{dB} is the gain of the filter in dB. Thus there is adding of dB gains for cascaded systems. In these days of computers this adding of dB gains seems no great benefit, but the custom of using dB became universal among electrical engineers in an earlier day and will no doubt continue in the future.

---

[14] 60 dBm. In general dBm = dBw + 30 dB.

Bode Plots

LEARNING OBJECTIVE 5.
To be able to describe filters with Bode plots

A log-log plot. A third virtue of dB measure involves a special way of plotting the characteristic of a filter, amplifier, or spectrum. The Bode plot (named for H. W. Bode, 1905–1982) is a log power versus log frequency plot. For example, Fig. 5.26 shows the Bode plot of the function

$$\underline{\mathbf{A}}(jf) = \frac{100}{1 + j(f/10)} \qquad (5.93)$$

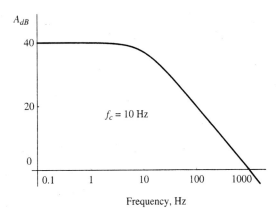

Figure 5.26 Bode plot.

which combines an amplifier (voltage gain of $100 = 40$ dB) with a low-pass filter ($f_c = 10$ Hz). In Fig. 5.26 the vertical axis is proportional to log power expressed in dB. The horizontal axis expresses frequency on a log scale, although normally the frequency, not the logarithm of the frequency, is marked on the graph. The log plot gives equal spacing on the graph to equal ratios of frequency—1:10, 10:100, etc.

decade

Advantages of Bode plots. The Bode plot is useful in two ways. As mentioned earlier, the log scales allow the representation of a wide range of both power and frequency. In this case, the vertical scale of 0 to 40 dB represents a range of power gain from 1 to 10^4, or a voltage gain from 1 to 100. Similarly, in the horizontal scale, we have represented four *decades*, factors of 10, in frequency. These wide ranges are possible because of the compressive nature of the logarithmic function.

asymptotic Bode plot

Asymptotic Bode plots. The other advantage of the Bode plot is that a filter characteristic can often be well represented by straight lines. The straight lines are evident in Fig. 5.26: the characteristic is quite flat for frequencies well below $f_c = 10$, and slopes downward with constant slope for frequencies well above this value. Only in the vicinity of f_c does the exact Bode plot depart significantly from these straight lines. In the following section we examine in more detail the relationship between the exact Bode plot and the straight-line approximation, which is called the asymptotic Bode plot.

Bode plot for a low-pass filter. As shown in Eq. (5.84), the low-pass filter has the characteristic

$$\underline{\mathbf{F}}(jf) = \frac{1}{1 + j(f/f_c)} \qquad (5.94)$$

The amplitude Bode plot represents the magnitude of the voltage gain and hence requires the absolute value

$$F_{dB}(jf) = 20 \log \left| \frac{1}{1+j(f/f_c)} \right| = 20 \log \left| \frac{1}{\sqrt{1+(f/f_c)^2}} \right|$$
$$= -10 \log \left[1 + \left(\frac{f}{f_c}\right)^2 \right]$$
(5.95)

We may generate the exact Bode plot using normalized frequency f/f_c for numerical calculation, Fig. 5.27(a). In Fig. 5.27(b) we show that the exact Bode plot approaches a straight line above and below the cutoff frequency, $f/f_c = 1$. At the cutoff frequency, the filter gain is -3.0 dB, as noted earlier.

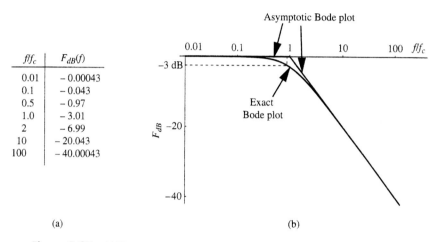

Figure 5.27 (a) Exact calculations; (b) exact and asymptotic Bode plot for the low-pass filter.

The asymptotes. We may derive the asymptotic Bode plot by taking the limiting forms of Eq. (5.95). For frequencies well below the cutoff frequency, $f \ll f_c$, the gain is approximately

$$F_{dB} \to -10 \log(1 + \text{small}) \approx -10 \log(1) \approx 0, \quad f \ll f_c \quad (5.96)$$

This accounts for the flat section. For frequencies well above the cutoff frequency, the gain approaches

$$F_{dB} \to -10 \log \left(\frac{f}{f_c}\right)^2 = -20 \log f + 20 \log f_c, \quad f \gg f_c \quad (5.97)$$

Recalling that our horizontal variable is $\log f$, we see that Eq. (5.97) has the form of a straight line, $y = mx + b$, where $m = -20$, x is $\log f$, and b is $20 \log f_c$. Thus the high-frequency asymptote will be a straight line when plotted against log frequency. When $\log f = \log f_c$, the y of the straight line has zero value and hence the line passes through the horizontal axis at $f = f_c$. The slope is -20; usually we say that the slope is negative

20 dB/decade because dB of gain and decades of frequency are the units for a Bode plot.

The low- and high-frequency asymptotes of the exact Bode plot combine to form the asymptotic Bode plot. As shown in Fig. 5.27(b), the exact Bode plot is well represented by this approximation, the largest error being −3.0 dB at the cutoff frequency. For most applications, the asymptotic Bode plot gives an adequate picture of the filter characteristics.

EXAMPLE 5.11 Low-pass filter

A data acquisition system has a low-pass filter with a cutoff frequency of 50 Hz. From the asymptotic Bode plot, find the frequency where the gain is −6 dB.

SOLUTION:
This frequency falls in the part of the characteristic described by Eq. (5.97)

$$-6 = -20 \log f + 20 \log 50$$
$$20 \log f = 39.98 \Rightarrow f = 10^{1.999} = 99.76 \text{ Hz} \tag{5.98}$$

WHAT IF? What if you use the exact Bode plot?[15]

Frequency Response

Poles, zeros, and frequency response. The poles and zeros of a transfer function define the frequency response of the circuit in the sinusoidal steady state. In this context, the system acts as a filter with $\underline{s} = j\omega$ and the poles and zeros correspond to the critical frequencies defining the Bode plot.

EXAMPLE 5.12 Bode plot of filter function

Give the filter function and Bode plot of the network in Fig. 5.12.

SOLUTION:
The filter function $\underline{F}(j\omega)$ is the transfer function given in Eq. (5.41) with $\underline{s} = j\omega$

$$\underline{F}(j\omega) = \frac{2}{(j\omega)^2 + 7(j\omega) + 2} = \frac{2}{(j\omega + 0.298)(j\omega + 6.70)} \tag{5.99}$$

where we have factored the denominator using the results of Eq. (5.44). Equation (5.99) can be placed in the form

$$\underline{F}(j\omega) = \frac{1}{(1 + j\omega/0.298)(1 + j\omega/6.70)} \tag{5.100}$$

---

[15] The exact −6 dB frequency is 86.33 Hz.

When we convert Eq. (5.100) to dB we obtain two terms

$$F_{dB} = 20 \log \left| \frac{1}{(1 + j\omega/0.298)(1 + j\omega/6.70)} \right|$$
$$= -10 \log \left[1 + \left(\frac{\omega}{0.298} \right)^2 \right] - 10 \log \left[1 + \left(\frac{\omega}{6.70} \right)^2 \right] \quad (5.101)$$

This is the sum of two low-pass filters, one having a critical frequency of 0.298 rad/s and the other with a critical frequency of 6.70 rad/s.

The asymptotic Bode plot of this filter function is shown in Fig. 5.28. Thus we see that the poles of the transfer function correspond to the critical frequencies of the filter response; indeed, these critical frequencies are often called the "poles" of the filter.

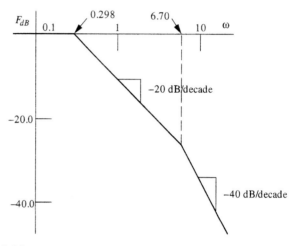

Figure 5.28 Bode plot of circuit in Fig. 5.12. The poles of the transfer function correspond to the critical frequencies of the filter, as indicated on the Bode plot.

High-Pass Filter

high-pass filter

High-pass filter circuit. Figure 5.29 shows a high-pass filter, which blocks low frequencies and passes high frequencies. The filter characteristic can be determined by analyzing the circuit as a voltage divider in the frequency domain

$$\mathbf{F}(j\omega) = \frac{\mathbf{V}_{out}(j\omega)}{\mathbf{V}_{in}(j\omega)} = \frac{R}{R + 1/j\omega C} = \frac{j\omega RC}{1 + j\omega RC} = \frac{j(f/f_c)}{1 + j(f/f_c)} \quad (5.102)$$

where $f_c = 1/2\pi RC$ is the cutoff frequency.

How the filter works. At low frequencies, $f \ll f_c$, the capacitor has a high impedance and allows little current; hence little voltage will develop across the resistor. Indeed, at zero frequency, dc, the capacitor acts as an open circuit and no voltage appears at the output. At high frequencies, $f \gg f_c$, the capacitor impedance will be small, and

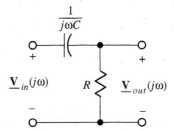

Figure 5.29 High-pass filter in the frequency domain.

essentially the input voltage appears across the resistor. Consequently, the filter passes high frequencies and blocks low frequencies. The transition between these two regimes occurs when the impedance of the capacitor is comparable to that of the resistor, and the cutoff frequency occurs where they are equal, as given again by Eq. (5.83).

Asymptotic Bode plot for high-pass filter. The asymptotic Bode plot for the filter function given in Eq. (5.102) may be derived by the same method as we used for the low-pass filter. The Bode plot is derived from the absolute value of the filter function,

$$F_{dB} = 20 \log \left| \frac{j(f/f_c)}{1 + j(f/f_c)} \right| = 20 \log \frac{f/f_c}{\sqrt{1 + (f/f_c)^2}} \qquad (5.103)$$

At frequencies well below the cutoff frequency, the frequency term in the denominator can be neglected and the asymptote becomes

$$F_{dB} \rightarrow 20 \log \frac{f}{f_c} = 20 \log f - 20 \log f_c \qquad (5.104)$$

Plotted against $\log f$, this is a straight line with a slope of $+20$ dB/decade, passing through the horizontal axis at $f = f_c$, as shown in Fig. 5.30. For high frequencies, the filter function approaches unity, or 0 dB, also shown in Fig. 5.30. The exact Bode plot for the filter combines these two asymptotes and makes a smooth transition near the cutoff frequency. At the cutoff frequency, the exact Bode plot lies 3 dB below the intersection of the low- and high-frequency asymptotes.

Combining filter characteristics. Figure 5.31 shows a combination of a low-pass filter, an amplifier, and a high-pass filter. As derived in Eq. (5.92) and shown in Fig. 5.31, the dB gain of cascaded circuits is the sum of the dB gain of the components

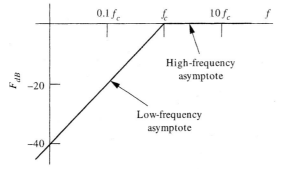

Figure 5.30 Asymptotic Bode plot for a high-pass filter.

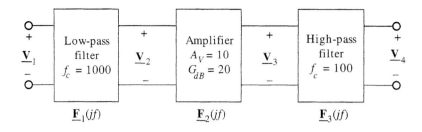

$$\mathbf{F}(jf) = \mathbf{F}_1(jf) \times \mathbf{F}_2(jf) \times \mathbf{F}_3(jf)$$

$$G_{dB} = F_{1(dB)} + G_{2(dB)} + F_{3(dB)}$$

Figure 5.31 Two filters and an amplifier in cascade.

$$G_{dB} = F_{1(dB)}(f) + G_{2(dB)}(f) + F_{3(dB)}(f) \qquad (5.105)$$

bandpass filter

where G_{dB} is the dB gain of the system. Thus the Bode plots will add to give the combined Bode plot of the cascaded system. Figure 5.32 shows the effect of adding up the individual Bode plots. The amplifier has a constant gain of 20 dB and raises the sum of the filter Bode plots by that amount. The high-pass filter blocks the low frequencies, and the low-pass filter blocks the high frequencies. The result is a *bandpass filter*; the "band" being passed in this case contains the frequencies between 100 and 1000 Hz. Thus we combine filter effects by adding the Bode plots.

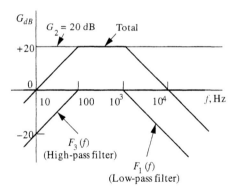

Figure 5.32 Adding Bode plots.

EXAMPLE 5.13 Bandpass filter

A bandpass filter for an audio equalizer is created by combining a low-pass filter with a cutoff frequency of 2 kHz with a high-pass filter with a cutoff of 1 kHz. What is the gain in dB at 1.5 kHz?

SOLUTION:
From the asymptotic Bode plots we must say 0 dB. But the exact value from Eqs. (5.103) and (5.95) is

$$G_{dB} = -20 \log \frac{1.5/1}{\sqrt{1 + (1.5/1)^2}} - 10 \log[1 + (1.5/2)^2] \quad (5.106)$$
$$= -1.60 - 1.94 = -3.54 \text{ dB}$$

Thus, this type of filter does not work well when the relative bandwidth is small.

A narrowband *RLC* filter. Example 5.13 presents a filter that passes the frequencies in a certain band by combining high-pass and low-pass *RC* filters. This approach works well only when the relative passband of frequencies is rather large. However, most communication systems, such as radios, require filters that pass a narrow band of frequencies. This is normally accomplished with *RLC* filters such as appears in Fig. 5.33. This particular *RLC* filter produces a parallel resonance between the inductor and capacitor at about 1125 kHz, which is near the middle of the AM radio band. The resistance shown in parallel with the inductor is not present in the physical circuit but is placed in the circuit model to represent the losses of the inductor. For frequencies far away from the resonant frequency, the impedance of either the inductor or the capacitor becomes small and the filter response drops. At the resonant frequency, the inductance and capacitance resonate and the output is maximum.

Figure 5.34 shows the bandpass characteristics of this filter in linear and dB scales. It might be noted that this filter has an inherent loss, even at its maximum output,

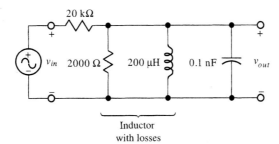

Figure 5.33 *RLC* bandpass filter.

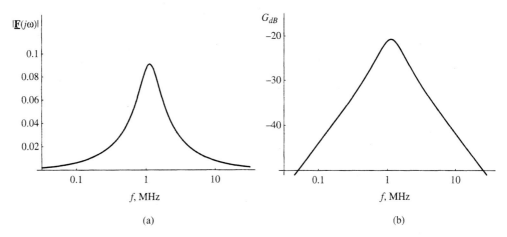

Figure 5.34 Narrowband *RLC* filter characteristics: (a) linear plot; (b) dB plot.

−20.8 dB in this case. Thus, such a filter must be used in conjunction with an amplifier to compensate for this loss. This is no disadvantage in most communication circuits, for the radio already must provide considerable amplification—this merely requires a bit more. We might mention in closing this section that the filter characteristic of the *RLC* filter in Fig. 4.41 is similar to that in Fig. 5.33. The parallel form of the filter is more easily accomplished with realistic inductors and capacitors, and hence this is the form of narrowband filter commonly used in radios.

Check Your Understanding

1. A filter will normally change the phase as well as the amplitude spectrum of the input signal. (True or False?)

2. An *RC* high-pass filter will pass a dc signal. (True or False?)

3. If one decade is a factor of 10, what ratio corresponds to one-fourth of a decade?

4. If the input to an amplifier is 5 mW and the output is 0.1 W, what is its gain in dB?

5. If an amplifier reduces the signal voltage by a factor of 4, what is its gain in dB?

Answers. (1) True; (2) false; (3) 1.778; (4) 13.0 dB; (5) −12.0 dB.

5.5 SYSTEMS

LEARNING OBJECTIVE 5.

To use system notation to describe the properties of composite systems

System Notation

In the introduction to this chapter, we described a system as having several components that together accomplish some purpose. System components have one or more inputs and outputs, and the outputs of some components are inputs to other components. This chapter deals with linear systems, in which the outputs are proportional to the inputs, including derivatives and integrals of the inputs.

Voltage divider. System concepts and notation will be illustrated by a simple example, the voltage divider shown in Fig. 5.35.

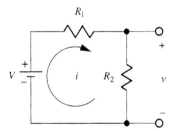

Figure 5.35 A voltage divider with current and output voltage defined.

Since the output has no load, the same current flows in the two resistors. The cause and effect relationships are as follows: The battery, V, causes a current, i, in the two resistors, R_1 and R_2, and the current i causes a voltage across R_2, which is the output

$$V \rightarrow i \rightarrow v \quad (5.107)$$

The specifics are given by the equations

$$i = \frac{V}{R_1 + R_2} \quad (5.108)$$

and

$$v = iR_2 \quad (5.109)$$

System notation. System notation replaces the equations by system components, as shown in Fig. 5.36. The convention is that the output variable is equal to the input variable, for example, V, times the term written in the box, $1/(R_1 + R_2)$, to give the output variable from that box, i.

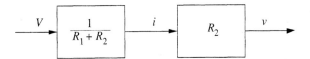

Figure 5.36 The convention in system notation is that the output of each component is the input times the value in the box.

It follows from this convention that intermediate variables, in this case i, can be eliminated by multiplying together the values in adjacent boxes, as in Fig. 5.37.

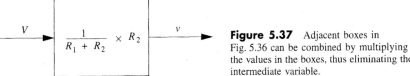

Figure 5.37 Adjacent boxes in Fig. 5.36 can be combined by multiplying the values in the boxes, thus eliminating the intermediate variable.

Summers. For the system notation to represent linear equations, we need also to represent sums and differences. For example, we can write the equations of the voltage divider in Fig. 5.35 as

$$i = \frac{V - v}{R_1} \quad (5.110)$$

plus Eq. (5.109). Equation (5.110) can be represented by two system components, a "summer," which in this case will use the minus sign to indicate subtraction,[16] followed by a box representing division by R_1, as shown in Fig. 5.38.

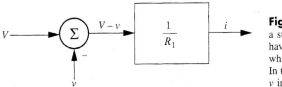

Figure 5.38 The circle with the "Σ" is a summer. In general, either input may have a $\pm$ associated with it, depending on what sum or difference is to be represented. In this case the V input requires a $+$ and the v input requires a $-$.

---

[16] If $V + v$ were to be represented by the summer, a "+" sign would have been used beside the v input to the summer in Fig. 5.38.

Thus Equations (5.109) and (5.110) may be represented by the system diagram in Fig. 5.39.

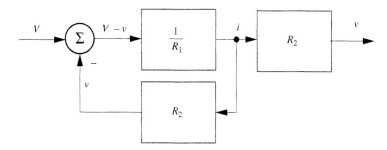

Figure 5.39 An alternative system for the voltage divider, using a loop. The black dot suggests that the variable i has influence through two effects.

Samplers. Figure 5.39 represents the equations of a voltage divider. One new feature is the "sampler," represented by the black dot between the two boxes, which indicates that variable i plays two roles in the equations of the system. Specifically, it produces the output voltage, and it also produces the voltage that subtracts from the input voltage, V, to produce the current in R_1, as shown in Fig. 5.39.

Loops. A second new feature in Fig. 5.39 is that we no longer have only a direct path from input, V, to output, v, but we also have a loop. The effect of i comes back through R_2 and the summer to interact with the input. Such loops are common in systems, because feedback loops are introduced for a variety of reasons. Here we merely show how to determine the input–output relationship when a loop is present.

Main amplifier

We will analyze the loop in Fig. 5.39 but will use notation that is commonly used in the analysis of feedback systems, shown in Fig. 5.40. The neutral symbol, x, is used for variables because systems often mix electrical with mechanical and other variables. The "main amplifier" is the gain from input to output without regard for the loop. For the loop in Fig 5.39, A is $1/R_1$.[17]

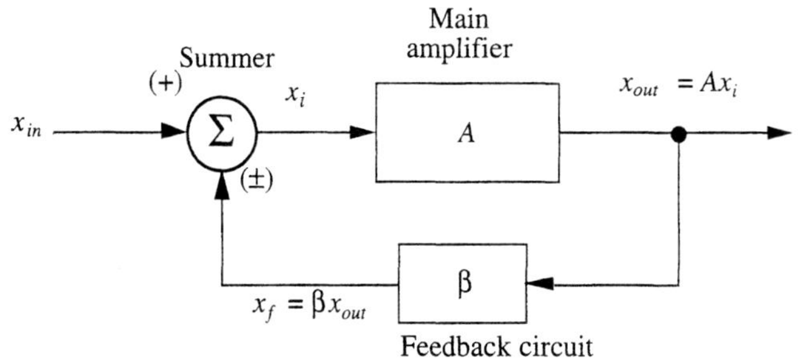

Figure 5.40 System with loop.

Analysis of the system. Our goal is to determine the gain of a system with a loop, $A_f = x_{out}/x_{in}$. The equations of the system are

---

[17] We could also let A be $1/R_1 \times R_2 = R_2/R_1$, in strict adherence to the definition. As you will see, both give the same result. Here we use $1/R_1$ to make the system configuration the same as in Fig. 5.40.

1. Main amplifier:

$$x_{out} = A x_i \tag{5.111}$$

2. Feedback circuit:

$$x_f = \beta x_{out} \tag{5.112}$$

3. Summer:

$$x_i = +x_{in} \pm x_f \tag{5.113}$$

In Eq. (5.113), we have used a gain of $+1$ for the input signal but allowed for either $+1$ or -1 for the feedback signal. We can eliminate two of the variables in Eqs. (5.111)–(5.113) and solve for the ratio of the output and input signals to obtain the gain with feedback, A_f, with the result

$$A_f = \frac{x_{out}}{x_{in}} = \frac{A}{1 - (\pm 1)(\beta)(A)} \tag{5.114}$$

Loop gain. Another form for Eq. (5.114) is

$$A_f = \frac{A}{1 - L} \tag{5.115}$$

loop gain

where

$$L = (\pm 1)(\beta)(A) \tag{5.116}$$

is called the loop gain. The *loop gain* is the product of the gains of all the system components around the loop, including signs. The basic idea of the loop gain is suggested in Fig. 5.41. To calculate or measure the loop gain, break the loop at some convenient point, insert a test signal, x_t, and calculate or measure the return signal, x_r. The loop gain is the ratio $L = x_r/x_t$. The loop may be broken, both in analysis and in the laboratory, only at a point where the function of the various system components is unimpaired. Care must be taken to terminate the break with the same impedance as the circuit saw before the break.

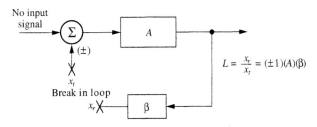

Figure 5.41 Determining the loop gain by breaking the loop at some point, inserting a test signal, and measuring the return signal.

EXAMPLE 5.14 Voltage divider

What is the gain of the voltage divider modeled by the system diagram in Fig. 5.39? Use the approach of Eqs. (5.115) and (5.116).

SOLUTION:
The gain of the summer is -1 so the loop gain is

$$L = (-1)(A)(\beta) = -\frac{1}{R_1} \times R_2 = -\frac{R_2}{R_1} \tag{5.117}$$

Using Eq. (5.115), we find the gain to the output of the part with the loop to be

$$\frac{i}{V} = \frac{A}{1-L} = \frac{1/R_1}{1-(-R_2/R_1)} = \frac{1}{R_1+R_2} \tag{5.118}$$

and therefore the gain to the output is

$$\frac{v}{V} = \frac{1}{R_1+R_2} \times R_2 = \frac{R_2}{R_1+R_2} \tag{5.119}$$

Clearly we have reduced the system diagram to that shown in Fig. 5.37, derived directly from circuit analysis.

WHAT IF? What if we let $A = R_2/R_1$, as suggested in footnote 17?[18]

EXAMPLE 5.15

An alternative system representation for Eqs. (5.109) and (5.110) is shown in Fig. 5.42. Show that this gives the same results as the system in Fig. 5.39.

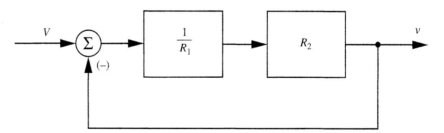

Figure 5.42 The equations of the voltage divider can also be presented by this system diagram.

SOLUTION:
The direct gain from input to output is

$$A = \frac{1}{R_1} \times R_2 \tag{5.120}$$

---

[18] The gain comes out the same, provided you use the same value for L.

The loop gain, as before, is

$$L = (-1)\frac{R_2}{R_1} \tag{5.121}$$

From Eq. (5.115)

$$A_f = \frac{v}{V} = \frac{A}{1-L} = \frac{R_2/R_1}{1 - \left(-\frac{R_2}{R_1}\right)} = \frac{R_2}{R_1 + R_2} \tag{5.122}$$

This is the same result as before.

Representing Differential Equations In System Notation

LEARNING OBJECTIVE 6.

To determine the transient response of linear systems to the sudden application of a dc input

The system representations and techniques we have introduced thus far apply to linear algebraic equations but cannot represent linear differential equations. They fail in the time domain for lack of appropriate multiplication boxes to represent derivatives. But in the frequency domain, differentiation becomes multiplication by $\underline{s}$ and hence differential equations can be represented by system models. For this reason linear systems most commonly are described by block diagrams of system components represented in the frequency domain.

A temperature transducer. To illustrate, we use a temperature sensor. We assume that the sensor output is linear with a sensitivity of K volts/degree C and a time constant τ. This tells us that the sensor response to a sudden change in temperature is as shown in Fig 5.43.

The response shown in Fig. 5.43 is described by a first-order differential equation[19]

$$\tau \frac{d}{dt} v(t) + v(t) = KT(t) \tag{5.123}$$

System model. We may transform Eq. (5.123) to the frequency domain by letting

$$\frac{d}{dt} \Rightarrow \underline{s} \quad v(t) \Rightarrow \underline{V}(\underline{s}) \quad T(t) \Rightarrow \underline{T}(\underline{s}) \tag{5.124}$$

with the result

$$\tau \underline{s}\underline{V}(\underline{s}) + \underline{V}(\underline{s}) = K\underline{T}(\underline{s}) \tag{5.125}$$

---

[19] Equation (5.123) suggests that $v = 0$ at $T = 0°C$, which may not be true. Actually the temperature and voltage in Eq. (5.123) represent incremental changes in temperature and voltage. Total temperature and voltage do not concern us here since we are modeling the dynamic behavior of the system.

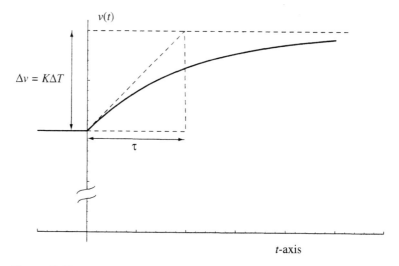

Figure 5.43 Response of temperature transducer to sudden jump of temperature, ΔT.

Equation (5.125) can be solved for the system function of the sensor

$$\mathbf{F}(\mathbf{s}) = \frac{\mathbf{V}(\mathbf{s})}{\mathbf{T}(\mathbf{s})} = \frac{K}{\tau \mathbf{s} + 1} = \frac{K/\tau}{\mathbf{s} + 1/\tau} \tag{5.126}$$

which is represented by the system diagram in Fig. 5.44.

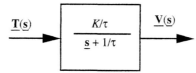

Figure 5.44 In the frequency domain, the sensor is represented by the box, which describes both the static and dynamic properties of the device. The input and output spectra represent incremental changes in input temperature and output voltage.

EXAMPLE 5.16 A temperature transducer with a sensitivity of 10 mV/°C and a time constant of 0.5 s is followed by an RC low-pass filter with a time constant of 0.87 s. Find the response of this system to a sudden temperature increase of 10°C.

SOLUTION:
Using the results in Fig. 5.44, we may represent the system as shown in Fig. 5.45.

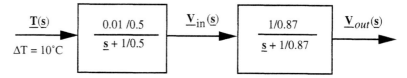

Figure 5.45 System diagram for temperature sensor followed by low-pass filter. We will determine the output voltage in the time domain as the result of a sudden 10°C increase in input temperature.

The combined system function is found by multiplying the values in the boxes

$$\mathbf{F(s)} = \frac{\mathbf{V}_{out}(\mathbf{s})}{\mathbf{T(s)}} = \frac{2.30 \times 10^{-2}}{(\mathbf{s}+2)(\mathbf{s}+1.15)} \tag{5.127}$$

From the poles of the system function we know that the time-domain response is of the form

$$v_{out}(t) = A + Be^{-2t} + Ce^{-1.15t} \tag{5.128}$$

where A, B, and C are constants to be determined by the initial conditions. The first constant, A, is the forced response to the steady input temperature after the temperature increase. If we assume the output voltage to be zero volts before the sudden increase, the voltage afterward will approach

$$A = K\Delta T = 10^{-2} \times 10 = 0.1 \text{ V} = 100 \text{ mV} \tag{5.129}$$

Initial conditions. To evaluate B and C we need $v_{out}(t)$ and $dv_{out}(t)/dt$ at $t = 0^+$. From physical considerations we reason that both these initial conditions must be zero. Both components act as low-pass filters: the first due to thermal inertia of the sensor and the second due to the capacitor. In the time domain, both give a delay, so the output of the sensor and the RC filter must be 0 at $t = 0^+$. Furthermore, the output of the RC filter will not begin to increase until a finite voltage builds up at its input; thus the derivative of v_{out} will also be zero at $t = 0^+$.

The initial conditions, combined with Eq. (5.128), yield

$$\begin{aligned} 0 &= 100 + B + C \\ 0 &= 0 + (-2)B + (-1.15)C \end{aligned} \tag{5.130}$$

which give the results $B = 135$ mV and $C = -235$ mV. The response to the $\Delta T = 10°C$ is therefore

$$v_{out}(t) = 100 + 135e^{-2t} - 235e^{-1.15t} \text{ mV} \tag{5.131}$$

which is shown in Fig. 5.46.

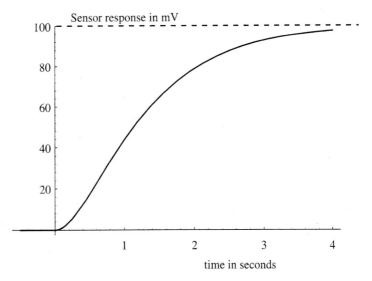

Figure 5.46 The sensor response to a sudden increase in temperature. Note that the total delay corresponds roughly to the sum of the two time constants.

> **WHAT IF?** What if the change in temperature were $-1\,°C$?[20]

CHAPTER SUMMARY

This chapter introduces the use of complex frequency in linear circuit and system analysis. We generalize the concept of frequency to include exponential growth and decay, plus damped or growing oscillations. We demonstrate the significance of the poles and zeros of the impedance and system functions in establishing the transient and steady-state response of a system. We explore the various transient responses of which second-order systems are capable. We introduce the concepts of electric filters and Bode plots to describe the response of circuits to sinusoids. Finally, we give a brief introduction to the notations, concerns, and techniques of system theory.

Objective 1. To recognize the class of functions that can be described by complex frequencies. Complex frequencies can represent constant signals, growing and decaying exponentials, sinusoidal functions, and growing and decaying sinusoidal functions. These functions have a broad practical significance but also are useful as a mathematical probe to explore the behavior of linear systems.

Objective 2. To use generalized impedance to find the forced and natural response of a system. We generalize the impedance expressions from Chap. 4 using complex frequency to include a broader class of behavior than simple sinusoids. The poles and zeros of an impedance or transfer function determine both the natural and forced responses when the circuit is excited at a complex frequency.

Objective 3. To recognize conditions for undamped, underdamped, critically damped, and overdamped responses in second-order systems. We study *RLC* circuits as an example of a second-order system. The four types of response are named and described.

Objective 4. To understand the concept of electric filters and the use of Bode plots to describe signals and filters. Electric filters use the frequency-dependent impedance of inductors and capacitors to eliminate or diminish signals in bands of frequencies. Bode plots are log-log plots of a filter response or perhaps a spectrum. Often the filter characteristic is well approximated by straight lines.

Objective 5. To use system notation to describe the properties of composite systems. Hybrid and nonelectrical linear systems can be described by complex-frequency techniques. The idea of system analysis is to focus on the dynamic interaction of system components without excessive attention to physical details.

Objective 6. To determine the transient response of linear systems to the sudden application of a dc input. The generalized impedance or transfer functions of linear systems reveal the natural and forced response to a wide class of inputs.

---

[20] $v_{out}(t) = -10 - 13.5e^{-2t} + 23.5e^{-1.15t}$ mV.

We combine these with the initial conditions to give a full transient response to the sudden applications of a dc input.

Throughout this chapter, frequency-domain concepts are emphasized. Input and output signals have been the frequency-domain transforms of time-domain differential equations representing dynamic behavior.

GLOSSARY

Complex frequency $\underline{s}$, p. 216, a time-domain variable has a complex frequency, $\underline{s} = \sigma + j\omega$, when it can be expressed in the form $v(t) = \text{Re}\{\underline{V}e^{\underline{s}t}\}$.

Critical damping, p. 233, when the natural frequencies of a circuit are real and equal. Critical damping is the boundary between oscillatory and nonoscillatory behavior.

Damped oscillation, p. 233, a response when circuit complex frequencies are complex conjugates and correspond to an exponentially decreasing sinusoid; also termed an underdamped response.

Damping constant, p. 233, the reciprocal of the time constant of a damped oscillation.

Generalized impedance, p. 221, the phasor voltage divided by phasor current when phasors are functions of complex frequency.

Overdamped response, p. 233, a response consisting of two or more time constants, but no oscillation. The natural frequencies of the system are real and negative.

Resonance charging, p. 235, charging a capacitor through an inductor.

Stable system, p. 216, a system in which oscillations die out.

System, p. 229, several components that together accomplish some purpose.

System function, p. 229, the ratio of the transforms of input and output as a function of complex frequency.

Undamped response, p. 232, a transient response in which oscillations do not diminish with time.

Underdamped response, p. 233, a response when circuit natural frequencies are complex conjugates and correspond to an exponentially decreasing sinusoid; also termed a **damped oscillation**.

PROBLEMS

Section 5.1: Complex Frequency

5.1. The half-life of the exponential decay of carbon 14 is 3730 years. What complex frequency in s^{-1} describes this process?

5.2. A decaying sinusoid has a complex frequency $\underline{s} = -2 + j1$ in s^{-1} and has its maximum value at $t = 0$. What is the angle of the phasor representing this signal in the frequency domain?

5.3. A decaying sinusoid crosses zero every 10 ms and each positive peak is 90% of the previous positive peak. What complex frequency describes this function?

5.4. A time-domain function is shown in Fig. P5.4. The function is of the form
$v(t) = A + Be^{\sigma t}\cos(\omega t)$
(a) From the graph, determine A, B, σ, and ω.
(b) Estimate the complex frequencies that can be used to describe this function?

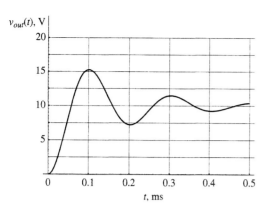

Figure P5.4

5.5. The world's population is currently increasing at a rate so as to double every 44 years. What complex frequency describes this process?

5.6. Evaluate $5e^{(1+j2)t} + \dfrac{d}{dt} e^{(1+j2)t}$ at $t = 2$.

5.7. Estimate the complex frequency associated with the waveform in Fig. P5.7.

5.8. Estimate the complex frequency associated with the waveform in Fig. P5.8.

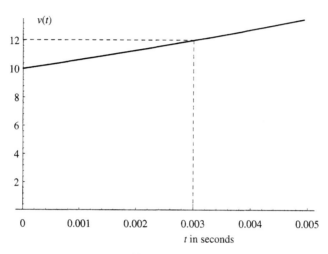

Figure P5.7

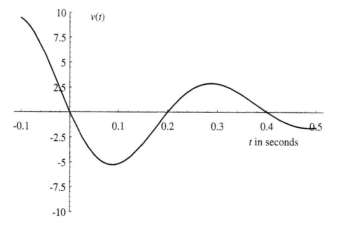

Figure P5.8

Section 5.2: Impedance and the Transient Behavior of Linear Systems

5.9. Consider a resistor, R, in series with a capacitor, C.
  (a) Determine the generalized input impedance as a function of complex frequency and, from that, the pole and zero of the circuit.
  (b) What is the interpretation of the pole?
  (c) What is the interpretation of the zero?

5.10. The circuit shown in Fig. P5.10 is excited by a voltage source that is zero for negative time and exponential for positive time, as indicated.
  (a) Determine the impedance of the circuit, $\mathbf{Z(s)}$.
  (b) What is the complex frequency of the source for $t > 0$?
  (c) What is the natural frequency of the circuit?
  (d) Determine the forced response of the current.
  (e) Determine the total response of the current for $t > 0$, including the initial condition.

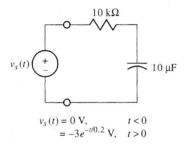

Figure P5.10

5.11. A circuit has the impedance function

$$\mathbf{Z(s)} = 10^4 \frac{\mathbf{s}^2 + 4\mathbf{s} + 1}{\mathbf{s}(\mathbf{s} + 2)} \; \Omega$$

  (a) What is the impedance at dc?
  (b) What are the natural frequencies of the circuit if the input is shorted?
  (c) What are the natural frequencies if the input is open-circuited?
  (d) In addition to resistors, this circuit has one inductor, one capacitor, two inductors, two capacitors, or one capacitor and one inductor? Which combinations are possible (may be more than one)? Explain your answer.
  (e) If the circuit is excited with a current source of value $-2e^{-t}$ A that is suddenly turned on at $t = 0$, what is the forced response of the voltage at the input to the circuit?
  (f) What is the total response for $t > 0$ for this input voltage if the initial value of the voltage and its rate of change are zero?

5.12. A system has a transfer function

$$\mathbf{T(s)} = \frac{K}{(\mathbf{s} - \mathbf{s}_1)(\mathbf{s} - \mathbf{s}_2)}.$$

The input voltage is zero for negative time and $+10$ V for positive time. With this input, the output is

$$v_{out}(t) = 5 + 10e^{-t} - 5e^{-2t} \text{ V}.$$

  (a) Find K, $\mathbf{s}_1$, and $\mathbf{s}_2$.
  (b) If the input voltage is $3e^{-t/2}$ V for $t > 0$, find the output voltage for $t > 0$, assuming the same initial conditions.

5.13. Determine the open-circuit and short-circuit natural frequencies for the circuits shown in Fig. P5.13 and interpret these in terms of the time constants of the circuits.

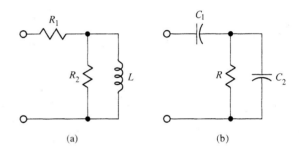

Figure P5.13

5.14. Figure P5.14(a) shows the input function to the low-pass filter shown in Fig. P5.14(b). Determine the equation of the output voltage in the time domain.

5.15. The circuit shown in Fig. P5.15 is excited by a voltage source that is zero for negative time, but $v(t) = 100e^{-10,000t}$ V for positive time. Determine the current for positive time, assuming zero current at $t = 0$.

5.16. The impedance of the circuit in Fig. P5.16 has one zero at $\mathbf{s} = 0$ and one pole at $\mathbf{s} = -1$ s^{-1}. At $\mathbf{s} = +1$ s^{-1}, the impedance into the circuit has a value of 15 Ω. The circuit is excited by a current source that is zero for negative time and has a value of $i(t) = 0.5e^{-2t}$ A for positive time. Determine the input voltage, as shown, assuming $v(0+) = +2$ V.

5.17. An RC network has the impedance function

$$\mathbf{Z(s)} = 100 \frac{\mathbf{s}^2 + 25\mathbf{s} + 100}{\mathbf{s}(\mathbf{s} + 8)} \; \Omega$$

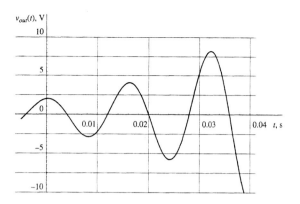

(a)

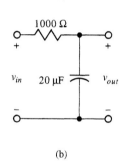

(b)

Figure P5.14

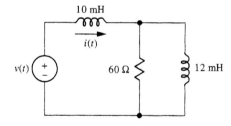

Figure P5.15

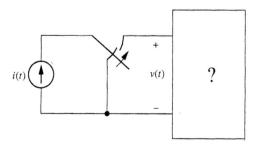

Figure P5.16

(a) Does the circuit allow a dc current?
(b) Find the impedance of the circuit at a frequency of 5 Hz.
(c) If the circuit is excited with a current source, what is (are) the natural frequency (frequencies) of the circuit?
(d) If the circuit source is $3e^{-10t}$ A, what input voltage results?
(e) What are the short-circuit natural frequencies of the network?

5.18. The system diagram shown in Fig. P5.18 gives the relationship between input and output voltage in the frequency domain. No voltage is applied prior to $t = 0$.
(a) Does this system pass a dc signal? How do you know?
(b) If the input is $v_{in}(t) = 2e^{-10t}$ V for $t > 0$, and $v_{out}(0^+) = 0$, find the output for $t > 0$.
(c) If the input is $v_{in}(t) = 2\cos(2t)$ V for $t > 0$, and $v_{out}(0^+) = 0$, find the output for $t > 0$.

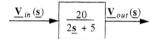

Figure P5.18

5.19. For the circuit in Fig. P5.19,
(a) Find the input impedance of the circuit.
(b) What is the zero of the impedance? What is the significance or interpretation of this zero.
(c) What is the pole of the impedance? What is the significance, or interpretation of this pole.

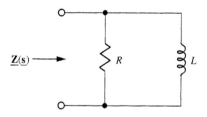

Figure P5.19

5.20. The circuit shown in Fig. P5.19 has $R = 10\,\Omega$, $L = 0.1$ H and has no stored energy for $t < 0$.
(a) At $t = 0$, a voltage source $v_s(t) = 5e^{-50t}$ V is applied. Find the total current response into the circuit.
(b) At $t = 0$, a current source $i_s(t) = 1.5e^{-50t}$ A is applied. Find the total voltage response of the circuit.

Section 5.3: Transient Response of RLC Circuits

5.21. For the *RLC* circuit of Fig. 5.16, what would be the natural frequencies of the network if *R* and *L* were exchanged? What would be the initial value and initial derivative of the output response to the sudden application of an input voltage?

5.22. For the series *RLC* circuit of Fig. 5.20, with $L = 1$ mH and $C = 1$ μF, find the following:
 (a) Determine the transfer function, $\mathbf{T(s)}$.
 (b) From the transfer function, determine the natural frequencies of the circuit if excited by a voltage source.
 (c) What is the value of resistance for critical damping of the circuit? Is this different from the critical value given in Eq. (5.62) for the parallel *RLC* circuit?
 (d) For resistances one-half and twice the critical value established in part (c), determine the natural frequencies of the circuit and write the corresponding time responses with unknown constants.
 (e) Consider that a 10-V source is suddenly applied to the input in the manner shown in Fig. 5.16. What are the initial value and initial derivative of the output voltage? Work out the complete response for the overdamped case calculated in part (d).

5.23. For the circuit shown in Fig. P5.23, find the following:
 (a) Find the input impedance, $\mathbf{Z(s)}$.
 (b) Determine the open-circuit natural frequencies.
 (c) What value (range) of resistance corresponds to behavior that is (1) undamped, (2) underdamped, (3) critically damped, and (4) overdamped?

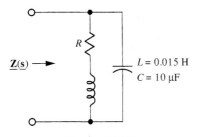

Figure P5.23

5.24. A circuit impedance has the pole–zero pattern shown in Fig. P5.24 and has an impedance magnitude of 30 Ω at $\mathbf{s} = +1 \text{ s}^{-1}$. The circuit is

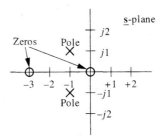

Figure P5.24

excited by a voltage source that is zero for negative time and has a value of $v(t) = 5\cos(2t)$ V for positive time. Determine the current into the circuit for positive time, assuming zero for the initial value and initial derivative of the current.

5.25. Figure P5.25 shows a circuit with a disconnected source.
 (a) Find the generalized impedance of the circuit, $\mathbf{Z(s)}$.
 (b) What is the impedance at $\mathbf{s} = 0$?
 (c) What is the impedance at $\mathbf{s} = \infty$?
 (d) If the switch is closed, what are the natural frequencies of the circuit?
 (e) What is the character of the response (undamped, underdamped, critically damped, overdamped)?
 (f) What is the initial value of the current in the resistor?
 (g) What is the final value of the current in the resistor?

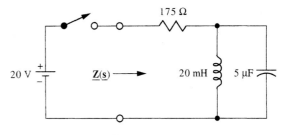

Figure P5.25

5.26. Figure P5.26 shows an undamped *LC* circuit that can resonant charge the capacitor to twice the power-supply voltage. Determine the inductance and capacitance to store 2 J of energy in the capacitor in 1 ms from the time of switch closure.

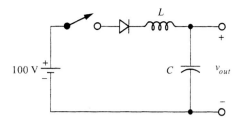

Figure P5.26

5.27. A circuit has the impedance function

$$\mathbf{Z}(\mathbf{s}) = 12 \frac{\mathbf{s}}{\mathbf{s}^2 + 9\mathbf{s} + 6} \; \Omega$$

(a) What is the impedance at dc?

Section 5.4: Filters

5.28. Make asymptotic Bode plots of the magnitude of the following filter functions:

(a) $\mathbf{F}(\omega) = \dfrac{25}{1 + j(\omega/250)}$

(b) $\mathbf{F}(\omega) = 50 \times \dfrac{j\omega}{100 + j4\omega}$

(c) $\mathbf{F}(f) = 10^{-2} \times \dfrac{1 + j(f/10)}{1 + j(f/1000)}$

5.29. Show that the loss of the low-pass filter in Fig. 5.21 is approximately 1 dB at a frequency one octave (factor of 2) below the cutoff.

5.30. The gain in dB of a filter is shown in the Bode plot of Fig. P5.30.

(a) Is this a high-pass, low-pass, or band-pass filter?

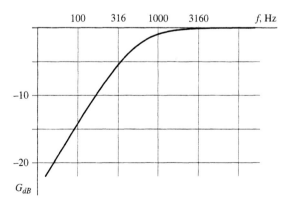

Figure P5.30

(b) What are the natural frequencies of the circuit if the input is shorted?

(c) What are the natural frequencies if the input is open-circuited?

(d) In addition to resistors, this circuit has one inductor, or one capacitor, or two inductors, or two capacitors, or one capacitor and one inductor? Which combinations are possible (may be more than one)? Explain your answer.

(e) If the circuit is excited with a current source of value $-2e^{-2t}$ A that is suddenly turned on at $t = 0$, what is forced response of the voltage at the input to the circuit?

(f) What is the total response for the input voltage if the initial value of the voltage and its rate of change are zero?

(b) What is the gain in dB at 500 Hz?

(c) Estimate the cutoff frequency of the filter.

(d) If the input voltage to the filter were $v_{in}(t) = 5 \cos(2000t)$ V, what would be the output in the form $v_{out}(t) = A \cos(2000t + \theta)$? In other words, find A and θ.

5.31. The input to a filter is the time function $v_{in}(t) = 10 + 5 \sin(8000t)$ V. The asymptotic Bode plot of the filter characteristic is shown in Fig. P5.31.

(a) Is the filter high-pass or low-pass?

(b) What is the frequency in hertz of the fundamental of the input?

(c) What is the input "power" in volt^2?

(d) What is the output "power" in volt^2?

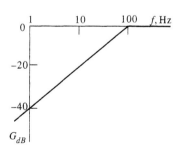

Figure P5.31

5.32. The input signal to an RC low-pass filter is $v(t) = 5 + 5 \sin(1000t)$. Draw the circuit diagram of a filter such that the power in the output is 90% dc power. The filter uses a 10-μF capacitor.

Section 5.5: System Analysis

5.33. A feedback system is shown in Fig. P5.33.
 (a) What is the system function with feedback?
 (b) Find A for critical damping, that is, to put the system on the boundary between under- and overdamped behavior.
 (c) For a value of A 10 times that computed in part (b), what is the form of the natural response of the system?

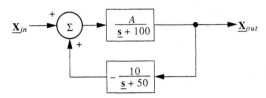

Figure P5.33

5.34. A linear system is described by the differential equation

$$\frac{d^2 x_{out}}{dt^2} + 5 \frac{dx_{out}}{dt} + 16 x_{out} = 5 \frac{dx_{in}}{dt},$$

where x_{out} is the output signal and x_{in} is the input signal. A fraction of the output, call it β, is fed back to the input and subtracted from the input in a negative feedback system.
 (a) Draw a block diagram of the system in the frequency domain.
 (b) What range of β gives transient behavior that is overdamped?

5.35. The block diagram for a linear process is shown in Fig. P5.35.
 (a) Determine the maximum value of B for an overdamped response.
 (b) Find the transfer function for $B = 0.1$.
 (c) What value(s) of $\underline{s}$ allows output with no input?

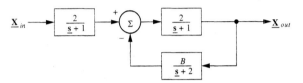

Figure P5.35

5.36. The circuit shown in Fig. P5.36 has a main amplifier and a feedback network. The main amplifier has a gain of 500, infinite input impedance, zero output impedance, and a -3-dB point at 5 kHz (one pole only). Find C such that the poles are real and negative and differ by a factor of 2.

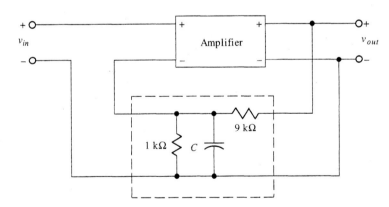

Figure P5.36

Answers to Odd-Numbered Problems

5.1. 5.89×10^{-12} s^{-1}.

5.3. $-5.27 + j314$.

5.5. $+0.0158$.

5.7. $+60.8$.

5.9. (a) $\underline{Z}_{eq} = \dfrac{\underline{s}RC + 1}{\underline{s}C}$; (b) At dc we can have voltage but no current, short circuit natural frequency; (c) At $\underline{s} = -1/RC$ we can have current but no voltage, open circuit natural frequency.

5.11. (a) ∞; (b) -0.268, -3.73 s^{-1}; (c) 0, -2 s^{-1};
(d) Must have two energy storage elements and block dc: 2 Cs or LC is possible; (e) $-4 \times 10^4 e^{-t}$ V;
(f) $v(t) = 2 \times 10^4 (1 + e^{-2t} - 2e^{-t})$ V.

5.13. (a) short circuit: $-\dfrac{R_1 \| R_2}{L}$, open circuit: $-\dfrac{R_2}{L}$;
(b) short circuit: $-\dfrac{1}{R(C_1 + C_2)}$, open circuit: 0, $-\dfrac{1}{RC_2}$.

5.15. $i(t) = 5(e^{-10^4 t} + e^{-1.1 \times 10^4 t})$ A.

5.17. (a) No dc current; (b) $116 \angle -27.2°$; (c) 0, -8 s^{-1};
(d) $-750 e^{-2t}$ V; (e) $-5, -20$ s^{-1}.

5.19. (a) $\underline{Z}_{L+R}(\underline{s}) = \dfrac{\underline{s}RL}{\underline{s}R + L}$; (b) $\underline{s} = 0$. Says that the circuit will allow input current without input voltage at zero frequency (dc); (c) $\underline{s} = -R/L$. Says the circuit will allow voltage at the input with no input current at the given natural frequency, which corresponds to the time constant for the R-L circuit.

5.21. Same as before, 0 initial value but V/RC for initial derivative.

5.23. (a) $\dfrac{R + \underline{s}L}{\underline{s}^2 LC + \underline{s}RC + 1}$;

(b) $\underline{s} = -\dfrac{R}{2L} \pm \sqrt{\left(\dfrac{R}{2L}\right)^2 - \dfrac{1}{LC}}$;
(c) 77.5 Ω is critical damped, undamped is 0, underdamped up to 77.5 Ω, overdamped over 77.5 Ω.

5.25. (a) $\dfrac{175(\underline{s}^2 + 10^7) + 2 \times 10^5 \underline{s}}{\underline{s}^2 + 10^7}$ Ω;
(b) 175 Ω, by inspection of the circuit; (c) 175 Ω, by inspection of the circuit; (d) $\underline{s} = -571 \pm j3110$ s^{-1}; (e) underdamped; (f) 20/175 A; (g) 20/175 A.

5.27. (a) 0 Ω; (b) 0 s^{-1}; (c) -0.725, -8.27 s^{-1}; (d) two inductors since it's second order and it approaches a short at dc, or could be LC; (e) $-6 e^{-2t}$ V;
(f) $v(t) = -6 e^{-2t} + 4.99 e^{-0.725 t} + 1.01 e^{-8.27 t}$ V.

5.29. -0.969 dB.

5.31. (a) high pass; (b) 1270 Hz; (c) 112.5 V^2; (d) 12.5 V^2.

5.33. (a) $\dfrac{A(\underline{s} + 50)}{(\underline{s} + 100)(\underline{s} + 50) + 10A}$; (b) 62.5;
(c) $-75.0 \pm j75.0$ s^{-1}.

5.35. (a) $B < 0.125$;
(b) $\underline{F}(\underline{s}) = \dfrac{4(\underline{s} + 2)}{(\underline{s} + 1)[(\underline{s} + 1)(\underline{s} + 2) + 0.2]}$;
(c) $\underline{s} = -1.28, -1.78$.

CHAPTER 6

Power in AC Circuits

AC Power and Energy Storage: The Time-Domain Picture

Power and Energy in the Frequency Domain

Transformers

Chapter Summary

Glossary

Problems

objectives

1. To understand how to calculate the time-average and effective values of periodic waveforms
2. To understand the energy and power requirements of resistors, inductors, and capacitors in the sinusoidal steady state
3. To understand how to calculate apparent, real, and reactive power flow in an electric network based on time-domain analysis
4. To understand how to calculate apparent, real, and reactive power in the frequency domain
5. To understand how to correct the power factor of a load
6. To understand how use a Thévenin equivalent circuit to establish the load to withdraw maximum power from an ac source
7. To understand the voltage, current, and impedance transforming properties of a transformer
8. To understand the role of transformers in the transmission of electric power
9. To know safe practice around electrical equipment

AC circuits are used universally for the generation, distribution, and consumption of electric power. This chapter examines energy processes in ac circuits. Additionally, we look into some matters of eminent practical importance, such as what meters indicate, what transformers do, and how to be safe around electrical equipment.

6.1 AC POWER AND ENERGY STORAGE: THE TIME-DOMAIN PICTURE

Importance of Power and Energy

Power and energy are important in ac problems for several reasons. Energy concerns you as a user of electricity because the local electric utility makes you pay for the energy you use. More important, energy plays a vital role in describing the behavior of physical systems. The roller coaster problem of freshman physics—in which you compute the speed at some point on the track from a difference in height—exemplifies how energy considerations often sweep away many details of a problem and lead directly to a useful result. Indeed, the more experience you gain in analyzing physical systems, the more you should become impressed with the importance of energy in revealing the true workings of a system. Some feel, as does your author, that any analysis is incomplete until energy relationships are explored and understood. Finally, we should point out that modern civilization is characterized by the ready availability and varied uses of electric power. Lighting, electric motors, communication systems, and laptop computers, just to name a few useful applications, ultimately get their energy from the electric power system.

In this chapter, we investigate energy and power relationships in electric circuits. We begin by looking at averages because time-average power is frequently our focus. Then we examine energy and power relationships in the time domain. The frequency domain follows, and we succeed in expressing energy and power relationships with phasors. Finally, we introduce two topics of eminent practical value, transformers and electrical safety.

Average Values of Electrical Signals

LEARNING OBJECTIVE 1.
To understand how to calculate the time-average and effective values of periodic waveforms

What is an average? Everybody knows how to calculate a numerical average. We compute the average of 12, 9, and 15 by adding the numbers (36), dividing by the number of values we are averaging (3), and getting the average (12). In general, the average of n numbers, x_i, $i = 1, 2, ..., n$, is

$$X_{avg} = \frac{\sum_{i=1}^{n} x_i}{n} \quad \text{or} \quad nX_{avg} = \sum_{x=1}^{n} x_i \tag{6.1}$$

In the second form of Eq. (6.1), we see that the average value multiplied by the number of samples is equal to the sum of the numbers.

average

That is how to compute an arithmetic average, but what is the definition of an average? An *average* is a number[1] that characterizes in some aspect a body of information. The arithmetic average, for example, gives us some idea of the size of the numbers, such as the average price of gasoline in Kansas. There are many kinds of averages. The grade-point average provides an example near to the heart of most college students. This average characterizes the academic performance of a student even though many important factors, such as course load and difficulty, are ignored. Thus, an average

---
[1] More precisely, a *statistic*.

is a number that characterizes a body of information in one particular way, omitting all the rest of the information.

time average

Computing time averages. The time average of a periodic function can be generalized from the arithmetic average. The periodic voltage in Fig. 6.1 provides an example. Because the function is periodic, the average over all time will be the same as the average over one period; hence we can limit our attention to the time from $t = 0$ to $t = T$, as shown in Fig. 6.2. We can define the time average by analogy with the second form of Eq. (6.1): n becomes T, the period; X_{avg} becomes V_{avg}; and the summation of the numbers becomes the summation of all the heights of the voltage, which is the integral over the time period from 0 to T. Thus, we have the definition of *time-average* voltage:

$$TV_{avg} = \int_0^T v(t)dt \Rightarrow V_{avg} = \frac{1}{T}\int_0^T v(t)dt \tag{6.2}$$

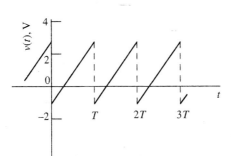

Figure 6.1 Find the time-average value.   **Figure 6.2** One period of the waveform.

The first form of Eq. (6.2) can be interpreted in terms of area, as shown in Fig. 6.3. Because V_{avg} is a constant, the product on the left side represents the area on the $v(t)$ graph of a rectangle having base T and height V_{avg}. The right side is the area under the $v(t)$ curve, counting the area above the x-axis as positive and the area below the x-axis as negative. This geometric interpretation is shown in Fig. 6.3. Equation (6.2) requires the two areas to be equal. The second form of Eq. (6.2) serves for computing averages.

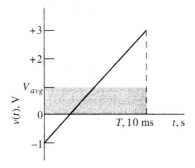

Figure 6.3 The average multiplied by the period has the same area as the original waveform, counting the area below the x-axis as negative.

EXAMPLE 6.1 Calculating the time average

Find the time-average value of the waveform in Fig. 6.1 for $T = 10$ ms.

SOLUTION:

We apply Eq. (6.2). The present example can be handled as the areas of triangles, but we will use calculus to illustrate the more general method. First, we derive the equation of $v(t)$ in the slope-intercept form: The intercept is -1; the slope is $3 - (-1)$ divided by 10 ms, or 400 V/s. Thus, the equation is

$$v(t) = -1 + 400t, \quad 0 < t < 10 \text{ ms} \tag{6.3}$$

We now substitute into Eq. (6.2), with the result

$$V_{avg} = \frac{1}{10 \text{ ms}} \int_0^{10 \text{ ms}} (-1 + 400t) dt = 10^2 \left(-t + \frac{400t^2}{2} \right)_0^{10^{-2}} = +1 \tag{6.4}$$

This is the value indicated on Fig. 6.3 and its validity is apparent.

WHAT IF?

What if the time scale were seconds rather than milliseconds? Would that change the average?[2]

dc value, dc component

Some special averages. Several results follow from the definition of time average. The time average of a dc (constant) voltage or current is the dc value. For this reason, the time-average value of a signal, such as that shown in Fig. 6.3, is often called the *dc value* or *dc component* of the signal. Another result is that the average value of a sinusoidal waveform is zero. The sinusoidal function has equal areas above and below the time axis and thus has zero time-average (or dc) value.

If we have the sum of two signals, say, two voltage sources connected in series, the average of the sum is the sum of the averages of the component signals. This follows from Eq. (6.2) because the integral distributes to the two functions.

$$\begin{aligned}(v_1 + v_2)_{avg} &= \frac{1}{T} \int_0^T [v_1(t) + v_2(t)] dt \\ &= \frac{1}{T} \left[\int_0^T v_1(t) dt + \int_0^T v_2(t) dt \right] = V_{1\,avg} + V_{2\,avg}\end{aligned} \tag{6.5}$$

One application of Eq. (6.5) would be the sum of a dc and a sinusoidal signal; the average would be the dc value because the sinusoidal part averages to zero.

---

[2] No. Both areas are scaled by the same amount.

Effective or Root-Mean-Square (RMS) Value

Time-average power. We will now consider the time-average power in a dc circuit, Fig. 6.4, where the resistor is "hot" because the electrical energy into the resistor appears as heat. From Chap. 2, we know that the power into the resistor is V_{dc}^2/R, but we will derive this result here from more general considerations, which we then apply to the heating of a resistor with an ac source. The instantaneous power (energy/time) into a circuit element is given by Eq. (1.30) as

$$p(t) = v(t)i(t) \tag{6.6}$$

For our dc circuit in Fig. 6.4, both voltage and current are constant, so the instantaneous power into the resistor is $v \times i = V_{dc} \times V_{dc}/R$, a constant. The time-average power, P, into R is thus

$$P = \frac{1}{T}\int_0^T p(t)dt = \frac{1}{T}\int_0^T \frac{V_{dc}^2}{R}dt = \frac{V_{dc}^2}{TR}\int_0^T dt = \frac{V_{dc}^2}{R} \tag{6.7}$$

The time-average power determines how hot the resistor will become. The movement of charge through the resistor imparts thermal energy to the material. The input electrical power appears as a heat source internal to the resistor; the temperature of the resistor depends on this input and on its thermal coupling to the environment. The more power into the resistor, the hotter it will become. Physical resistors are rated for $\frac{1}{4}$, $\frac{1}{2}$, 1, 2 watts, and so on. The power rating indicates how much power the resistor can handle without burning out or changing its resistance value significantly.

Power in ac circuits. Figure 6.5 shows the same circuit with an ac source. The instantaneous power is

$$p(t) = v(t)i(t) = V_p \cos(\omega t) \times \left[\frac{V_p}{R}\cos(\omega t)\right] = \frac{V_p^2}{R}\cos^2(\omega t)$$
$$= \frac{V_p^2}{2R}[1 + \cos(2\omega t)], \qquad \omega = \frac{2\pi}{T} \tag{6.8}$$

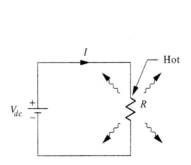

Figure 6.4 The heat is a measure of the average power.

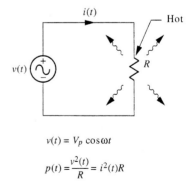

Figure 6.5 The heat is a measure of the time-average power.

where we have used the trigonometric identity

$$\cos^2 A = \tfrac{1}{2}[1 + \cos(2A)] \tag{6.9}$$

with $A = \omega t$.

Figure 6.6 shows a plot of the instantaneous power, Eq. (6.8). Although the charges move back and forth in the resistor, the power is always nonnegative, as shown mathematically from the squaring of the voltage in Eq. (6.8). The time-average power to the resistor, P, is the time average of the power curve:

$$P = \frac{1}{T}\int_0^T p(t)dt = \frac{V_p^2}{2R} = \frac{I_p^2 R}{2} \tag{6.10}$$

where $I_p = V_p/R$. The average value shown in Fig. 6.6 is half the peak, which is the average between the peak and valley.

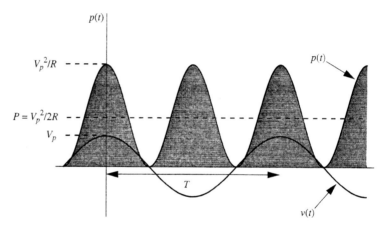

Figure 6.6 The energy flows into the resistor in spurts. The power is nonnegative at all times.

Energy flow to the resistor. The nonnegative power results physically because the moving electrons heat the resistance material regardless of their direction of flow. The energy does not flow smoothly into the resistor, but flows in spurts, twice each cycle of the ac waveform, as shown in Fig. 6.6. Thus, the resistor is heated cyclically. In most resistors, this time variation is unimportant because thermal inertia smoothes out the heating variations and the temperature remains essentially constant. However, these variations can be a problem in incandescent lighting, where the thermal inertia of the filament is small. The power frequencies of 50 or 60 Hz were established high enough to make the flickering of the light (at 100 or 120 Hz) barely noticeable.

EXAMPLE 6.2 Power rating of a resistor

A 2-watt, 500-Ω resistor is operated at its maximum power with ac. Find the peak voltage across the resistor.

SOLUTION:
From Eq. (6.10), the requirement is

$$2 = \frac{V_p^2}{2 \times 500} \Rightarrow V_p = \sqrt{2000} = 44.7 \text{ V} \tag{6.11}$$

WHAT IF? What if you want the maximum ac current in a 5-W, 20-Ω resistor?[3]

Effective or root-mean-square (rms) value of a sinusoid. The analytic computation of Eq (6.10) appears in Eq. (6.12). The integral uses the trigonometric identity given in Eq. (6.9):

$$P = \frac{1}{T}\int_0^T \frac{V_p^2}{R}\cos^2(\omega t)dt = \frac{V_p^2}{RT}\int_0^T\left[\frac{1}{2} + \frac{1}{2}\cos(2\omega t)\right]dt$$

$$= \frac{V_p^2}{RT}\left[\frac{t}{2} + \frac{\sin(2\omega t)}{4\omega}\right]_0^T = \frac{V_p^2}{2R} \tag{6.12}$$

effective value

Equation (6.12) leads to the effective value of the ac voltage. When we speak of effective, the "effect" to which we refer is the heating effect, or more generally the time-average energy conversion from electrical to nonelectrical form. The *effective value* of an ac waveform is the equivalent dc value that would heat the resistor as hot as the ac waveform heats it. For a sinusoid, this is

$$P = \frac{V_e^2}{R} = \frac{V_p^2}{2R} \Rightarrow V_e = \frac{V_p}{\sqrt{2}} \tag{6.13}$$

where V_e is the effective voltage. Equation (6.13) equates the time-average power from an equivalent dc source with magnitude V_e, the effective value, to the time average of power due to the ac source. Equation (6.13) defines the effective value of an ac source: The effective value of the sinusoidal voltage (or current, if we were dealing with current) is $1/\sqrt{2}$ or 0.707 times the peak value. For example, the ac voltage with a peak value of 44.7 V in the previous example would be equivalent in heating effect to a $44.7/\sqrt{2} = 31.6$-V battery, so this would be the effective value of the ac voltage. But we warn you that Eq. (6.13) applies only for the sinusoidal waveform, as we will illustrate shortly.

root-mean-square (rms) value

Effective value in general. The effective value is often referred to as the root-mean-square (rms) value. The general definition of effective or rms value of a periodic function is

---

[3] $I = 0.5$ A, rms.

$$P = \frac{V_e^2}{R} = \frac{1}{T}\int_0^T \frac{v^2(t)}{R}\,dt \Rightarrow V_e = \sqrt{\frac{1}{T}\int_0^T v^2(t)\,dt} \qquad (6.14)$$

where T is the period. Here we have illustrated the definition of rms for voltage, $v(t)$, but a similar expression would apply for the effective value of a current. The effective, or rms, value is the square *root* of the *mean* (that is, the average, as in "mean sea level") of the *square* of the function.[4] The practical importance of rms values is suggested by the fact that ac voltmeters and ammeters are calibrated to indicate rms for a sinusoid.

EXAMPLE 6.3 Effective value of nonsinusoidal voltage

Compute the rms value of the waveform in Fig. 6.1.

SOLUTION:
We derived the equation of the voltage during the first period to be

$$v(t) = -1 + 400t, \quad 0 < t < 10 \text{ ms} \qquad (6.15)$$

Substituting into Eq. (6.14) and integrating, we obtain

$$V_e = \sqrt{\frac{1}{10 \text{ ms}} \int_0^{10 \text{ ms}} (-1 + 400t)^2 \, dt} = 1.53 \text{ V} \qquad (6.16)$$

WHAT IF? What if we added 1 to the voltage, making $v(t) = 400t$?[5]

Measuring ac voltage. Many ac voltmeters would not indicate the true rms (1.53 V) of the waveform in Fig. 6.1. A meter designed to square and average the instantaneous waveform could be complicated and expensive. Thus, for simple voltmeters, some other property of the waveform is measured and the rms is inferred from that, assuming a sinusoidal shape. For example, a common type of meter actually responds to the peak-to-peak value of the waveform, but the meter scale is marked to indicate the peak-to-peak value divided by $2\sqrt{2}$, which would be the rms for a sinusoid. Such a meter would indicate $4/2\sqrt{2} = 1.41$ V for the waveform in Fig. 6.1, not the true rms value of 1.53 V. Thus, meter readings require careful interpretation when measuring nonsinusoidals.

Summary. We can represent the power-producing capability of a waveform with an average value called the effective (or rms) value of the waveform. This is defined as the value that, when squared, is equal to the time average of the square of the waveform. For a dc waveform, the effective value is equal to the dc value. For a sinusoidal wave-

---
[4] The name *rms* gives the formula.
[5] $V_e = 2.31$ V.

form, the effective value is $1/\sqrt{2}$ times the peak value. For other waveforms, the effective value may be calculated by squaring and averaging. Electrical meters are designed to indicate the effective value of a sinusoidal waveform, but may not indicate the effective value for other wave shapes.

Power and Energy Relations for R, L, and C

LEARNING OBJECTIVE 2.

To understand the energy and power requirements of resistors, inductors, and capacitors in the sinusoidal steady state

Resistance. We have discussed the power relationship for resistance in deriving the effective value of a sinusoid. As shown in Fig. 6.6, the energy flows unilaterally into the resistor, not smoothly, but in lumps.

$$p(t) = P_R[1 + \cos(2\omega t)] \quad (6.17)$$

where the time-average power into the resistor is

$$P_R = \frac{V_p^2}{2R} = \frac{V_e^2}{R} = I_e^2 R \quad (6.18)$$

where V_e and I_e are the effective voltage and current, respectively.

EXAMPLE 6.4 Resistance of a light bulb

Find the resistance of a 120-V, 100-watt light bulb.

SOLUTION:

The 120-V value represents the effective value of the standard voltage for lighting. According to Eq. (6.18), this indicates a resistance value of

$$R = \frac{V^2}{P_R} = \frac{(120)^2}{100} = 144 \, \Omega \quad (6.19)$$

The energy consumed by the bulb, operated for 24 hours, would be 24 hours × 0.100 kW, or 2.4 kWh (kilowatt-hours). This represents the total energy consumed, and at 6 cents/kWh, the bulb would operate for about 15 cents per day.

WHAT IF? What if the light bulb is a 130-V, 60-watt long-life bulb? Find its resistance.[6]

Inductance. We calculate the instantaneous and time-average magnetic energy stored by an inductor. The current and voltage for an inductor are

$$\begin{aligned} i_L(t) &= I_p \sin(\omega t) \\ v_L(t) &= L\frac{di}{dt} = +\omega L I_p \cos(\omega t) \end{aligned} \quad (6.20)$$

---

[6] 282 Ω.

The instantaneous power into the inductor is the product of voltage and current:

$$p_L(t) = v_L(t)i_L(t) = +\omega L I_p \cos(\omega t) I_p \sin(\omega t)$$
$$= +\frac{\omega L I_p^2}{2} \sin(2\omega t) \tag{6.21}$$

where we have used the trigonometric identity $[\sin(\omega t)][\cos(\omega t)] = [\sin 2(\omega t)]/2$. The time-average power is zero, for the ideal inductor has no loss. Thus, an inductor gives back on the average as much energy as it receives.[7] For a 60-Hz source, the energy would pulsate in and out of the inductor 120 times per second. For this reason, heavy electrical equipment, such as transformers and motors, often hums audibly at 120 Hz.

Stored energy in an inductor. The magnetic energy stored in an inductor is given in Eq. (3.5) as

$$w_m(t) = \tfrac{1}{2} L i_L^2(t) \tag{6.22}$$

For the sinusoidal source, we find the instantaneous stored energy to be

$$w_m(t) = \tfrac{1}{2} L [I_p \sin(\omega t)]^2 = \frac{L I_p^2}{4}[1 - \cos(2\omega t)] \tag{6.23}$$

Figure 6.7 shows the power and stored energy in an inductor with a sinusoidal current. The stored energy is nonnegative and pulsates at twice the ac source frequency. While the stored energy is increasing, the power into the inductor is positive. During this period of time, the ac source supplies energy and the inductor acts as a load. While the stored energy is decreasing, the power into the inductor is negative, indicating that the inductor now acts as a source, returning energy to the ac source.

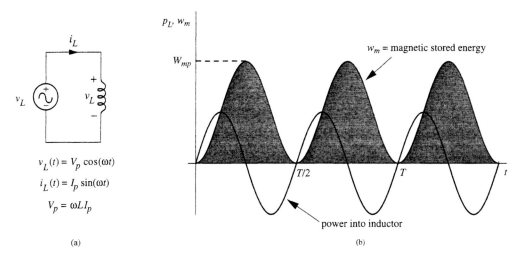

Figure 6.7 Stored energy and power into an inductor.

---

[7] This is true only for an ideal inductor; a real inductor would have resistive losses.

The time-average stored energy is one-half of $\frac{1}{2}LI_p^2$, the peak stored energy shown by Eq. (6.23). This can be written

$$W_m = \tfrac{1}{2}LI_e^2 \qquad (6.24)$$

where W_m is the time-average stored magnetic energy, and I_e is the effective value of the current. The time-average energy is important for two reasons. Although the time-average power into the inductor is zero, the time-average stored energy must be supplied to the inductor when the ac source is originally connected to the inductor. Additionally, the time-average stored energy indicates the magnitude of the energy pulsations in the inductor. Specifically, we know that the ac source must lend twice this amount of energy to the inductor twice each cycle.

EXAMPLE 6.5 **Stored energy in an inductor**

An ideal 0.1-H inductor has 30-Hz ac voltage applied to it, and the time-average stored energy is 5 J. What is the maximum power into the inductor?

SOLUTION:
From Eq. (6.24), we determine the effective value of the inductor current to be

$$I_e = \sqrt{\frac{2W_m}{L}} = \sqrt{\frac{2 \times 5}{0.1}} = 10 \text{ A} \qquad (6.25)$$

Equation (6.21) gives the instantaneous power into the inductor, and the maximum power is simply

$$p_{max} = \frac{\omega L I_p^2}{2} = \omega L I_e^2 = 2\pi \times 30 \times 0.1 \times 10^2 = 1885 \text{ W} \qquad (6.26)$$

WHAT IF? What if the frequency is changed to 60 Hz but the inductance and stored energy remain the same?[8]

Capacitance. We will calculate the instantaneous and time-average electric energy stored by a capacitor. The voltage and current for a capacitor are

$$v_C(t) = V_p \cos(\omega t)$$
$$i_C(t) = C\frac{dv_C}{dt} = -\omega C V_p \sin(\omega t) \qquad (6.27)$$

The instantaneous power into the capacitor is the product of voltage and current:

---
[8] 3770 W.

$$p_C(t) = v_C(t)i_C(t) = [V_p \cos(\omega t)][-\omega C V_p \sin(\omega t)]$$
$$= -\left(\frac{\omega C V_p^2}{2}\right)\sin(2\omega t) \tag{6.28}$$

where we have again used the trigonometric identity $[\sin(\omega t)][\cos(\omega t)] = [\sin 2(\omega t)]/2$. The average power is zero, which indicates that the capacitor is lossless.[9]

Stored energy in a capacitor. The pulsation of the power at twice the ac frequency corresponds to the shuttling of electric energy between source and capacitor. The stored electric energy in a capacitor is given in Eq. (3.23) as

$$w_e(t) = \tfrac{1}{2} C v_C^2(t) \tag{6.29}$$

For sinusoidal steady state, we find the instantaneous stored energy to be

$$w_e(t) = \tfrac{1}{2} C[V_p \cos(\omega t)]^2 = \frac{CV_p^2}{4}[1 + \cos(2\omega t)] \tag{6.30}$$

The stored energy is nonnegative and pulsates at twice the ac source frequency. While the stored energy is increasing, the power into the capacitor is positive. During this period, the ac source supplies energy and the capacitor acts as a load. While the stored energy is decreasing, the power into the capacitor goes negative, indicating that the capacitor is acting as a source, returning energy to the ac voltage source.

The time-average stored energy is one-half of $\tfrac{1}{2} CV_p^2$, the peak stored energy. This can be written

$$W_e = \tfrac{1}{2} CV_e^2 \tag{6.31}$$

where W_e is the time-average stored electric energy, and V_e is the effective value of the voltage.

EXAMPLE 6.6 Stored energy in a capacitor

Find the maximum current in a 120-μF capacitor that stores a time-average energy of 14 joules if the ac frequency is 60 Hz.

SOLUTION:
From Eq. (6.31), the effective value of the voltage is

$$\tfrac{1}{2} CV_e^2 = 14 \text{ J} \Rightarrow V_e = \sqrt{\frac{28}{120 \times 10^{-6}}} = 483 \text{ V} \tag{6.32}$$

So the effective current would be

$$I_e = \frac{V_e}{X_C} = \omega C V_e = 120\pi \times 120 \times 10^{-6} \times 483 = 21.9 \text{ A} \tag{6.33}$$

---

[9] For an ideal capacitor; a real capacitor would have some loss, though not as much as a real inductor.

Hence, the maximum current would be $21.9\sqrt{2} = 30.9$ A.

> **WHAT IF?** What if the effective current were 20.0 A? Find the peak stored energy.[10]

General Case for Power in an AC Circuit

LEARNING OBJECTIVE 3.

To understand how to calculate apparent, real, and reactive power flow in an electric network based on time-domain analysis

We have dealt with resistance, inductance, and capacitance separately, and shown the role of each in power and energy relationships. We now consider the general case of circuits containing combinations of resistors, inductors, and capacitors. We think in terms of the circuit shown in Fig. 6.8, an *RLC* circuit, although our results and interpretations will apply to all ac circuits.

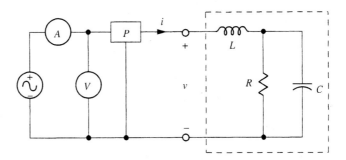

Figure 6.8 General *RLC* load with ammeter (*A*), voltmeter (*V*), and wattmeter (*P*).

This circuit will be analyzed in the frequency domain beginning on page 281. Here we assume it is described by an impedance, $\mathbf{Z} = |\mathbf{Z}| \angle \theta$. We are dealing here with power and energy in the time domain, so we express the relationship between voltage and current generally:

$$v(t) = V_p \cos(\omega t) \quad \text{and} \quad i(t) = I_p \cos(\omega t - \theta) \tag{6.34}$$

where $V_p = |\mathbf{Z}| I_p$. Note that θ is the angle of the impedance, which makes θ in this section have the opposite sign from Eq. (4.1). Although impedance is a frequency-domain concept and is computed in the frequency domain, the magnitude and angle of the impedance are real numbers that can be used in the time domain.

We investigate the time-average power into the *RLC* load through the use of two trigonometric identities. The instantaneous power is

$$p(t) = v(t)i(t) = V_p \cos(\omega t) I_p \cos(\omega t - \theta)$$
$$= \frac{V_p I_p}{2}[\cos\theta + \cos(2\omega t - \theta)] \tag{6.35}$$

The second form of Eq. (6.35) follows the application of the trigonometric identity:

---

[10] 11.7 J.

$$(\cos A)(\cos B) = \frac{\cos(A - B) + \cos(A + B)}{2} \tag{6.36}$$

with A identified with ωt and B with $\omega t - \theta$. We may expand the second cosine in Eq. (6.36) further with the identity

$$\cos(C - D) = \cos C \cos D + \sin C \sin D \tag{6.37}$$

where $C = 2\omega t$ and $D = \theta$. Thus,

$$p(t) = \frac{V_p I_p}{2}[\cos\theta + \cos\theta\cos(2\omega t) + \sin\theta\sin(2\omega t)] \tag{6.38}$$

$$= \underbrace{P[1 + \cos(2\omega t)]}_{\text{one-way flow}} + \underbrace{Q \sin(2\omega t)}_{\text{two-way flow}}$$

where

$$P = \frac{V_p I_p}{2}\cos\theta = V_e I_e \cos\theta \tag{6.39}$$

and

$$Q = \frac{V_p I_p}{2}\sin\theta = V_e I_e \sin\theta \tag{6.40}$$

The second forms of Eqs. (6.39) and (6.40) are preferred because meters indicate effective values and because we usually speak in terms of effective values when discussing power relationships.

real power, power factor

Interpretation of P. The second form of Eq. (6.38) reveals the nature of power flow in sinusoidal steady state. The time average of the instantaneous power given by Eq. (6.38) is called the *real power*, P, given by Eq. (6.39). This power is the time-average rate of energy conversion from electrical to nonelectrical form, and this is the power that costs you money on your electric bill. Equation (6.39) reminds us of the power in the dc case, with effective instead of dc values for voltage and current, except that we now have the $\cos\theta$ factor. This added term is the *power factor*, PF, defined by

$$PF = \cos\theta = \frac{P}{V_e I_e} = \frac{P}{S} \tag{6.41}$$

where $S = V_e I_e$.

apparent power

In words, the power factor is the real power divided by the product of the effective voltage and the effective current, which is called the *apparent power*, S. The power factor is also the cosine of the phase angle between the voltage and current.

When voltage and current have the same phase, $\theta = 0$, the power factor is unity, and the power into the load is $V_e I_e$. When the voltage has a different phase from the current, the power factor is less than unity and the time-average power is decreased proportionally. This occurs symmetrically, whether the circuit is inductive (positive θ) or capacitive (negative θ). When the phase is $\pm 90°$, no time-average power is transferred to the load.

reactive power

Interpretation of $P[1 + \cos(2\omega t)]$. Comparison of Eq. (6.38) with Eq. (6.8) suggests that the $P[1 + \cos(2\omega t)]$ term represents the lumpy flow of energy from the source to the load, as shown in Fig. 6.6. This term in Eq. (6.38) represents a one-way flow of energy, from source to load, in contrast with the Q term, which represents a two-way flow of energy.

Interpretation of $Q \sin(2\omega t)$. The second term in Eq. (6.38) represents *reactive power*, energy lent periodically to the reactive elements in the load. Equation (6.40) gives the magnitude of the reactive power, Q. This term averages zero and represents energy shuttled between source and load. The cases of pure resistance, inductance, and capacitance, the power and energy relationships are summarized in Table 6.1:

TABLE 6.1 Power and Energy Relationships for Pure Resistance, Inductance, and Capacitance

Element	θ	P	Q		Comment
resistance	$0°$	$I_e^2 R$	0		real power only
inductance	$+90°$	0	$V_e I_e \sin(+90°)$ $= +\omega L I_e^2$ $= +\omega W_{mp}$	(6.42)	positive reactive power
capacitance	$-90°$	0	$V_e I_e \sin(-90°)$ $= -\omega C V_e^2$ $= -\omega W_{ep}$	(6.43)	negative reactive power

In deriving the results in Table 6.1, we have used the following: $V_e = X_L I_e = \omega L I_e$, Eq. (4.95), and $W_{mp} = \frac{1}{2} L I_p^2 = L I_e^2$ as the peak magnetic stored energy in the inductance, Eq. (6.23); $I_e = V_e/|X_C| = \omega C V_e$, Eq. (4.96), and $W_{ep} = \frac{1}{2} C V_p^2 = C V_e^2$ as the peak electric stored energy, Eq. (6.30).

Conservation of Energy

Reactive power. From Eqs. (6.42) and (6.43), we see that the reactive power is a measure of the energy that the source has to lend to a reactive element in sinusoidal steady state. For an inductance, the reactive power is positive, and for a capacitance, the reactive power is negative.[11] The reactive power for a circuit containing several inductances and capacitances is the sum of reactive powers to each; hence, the reactive powers of inductances and capacitances tend to cancel. Adding Eqs. (6.42) and (6.43), we obtain

$$Q = Q_L + Q_C = \omega(W_{mp} - W_{ep}) \qquad (6.44)$$

This subtraction is the basis of power-factor correction, which we will illustrate presently.

---

[11] The inductance is said to consume reactive power, and the capacitor is said to supply reactive power.

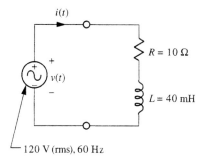

Figure 6.9 Series *RL* circuit.

The analysis of the *RLC* load pictured in Fig. 6.8 will be resumed in the next section. Here, for simplicity, we work out power and energy calculations for the series *RL* circuit shown in Fig. 6.9.

EXAMPLE 6.7 Power and energy of the series *RL* circuit

Find the power factor, the time-average power, the reactive power, and the peak stored energy for the series *RL* load shown in Fig. 6.9.

SOLUTION:

We are working in the time domain for our power calculations, but frequency-domain techniques are appropriate for finding the voltage and current that we need in the power calculations. We use the voltage source as our phase reference, so it would be represented by the phasor $\underline{V} = 120\sqrt{2} \angle 0°$. To calculate the current, we need the impedance,

$$\underline{Z} = 10 + j\,120\pi \times 0.04 = 10 + j\,15.1 = 18.1 \angle 56.4°\ \Omega \tag{6.45}$$

so $|\underline{Z}| = 18.1\ \Omega$ and $\theta = 56.4°$. The magnitude of the current is thus $120\sqrt{2}/18.1 = 6.63\sqrt{2}$ A and the phase angle of the current is $-56.4°$. Thus, the time-domain voltage and current are

$$v(t) = 120\sqrt{2}\ \cos(120\,\pi t)\ \text{V}$$
$$i(t) = 6.63\sqrt{2}\ \cos(120\,\pi t - 56.4°)\ \text{A} \tag{6.46}$$

The power factor is $PF = \cos 56.4° = 0.553$. From Eq. (6.39), we compute the time-average power delivered to the *RL* load by the ac voltage source to be

$$P = 120 \times 6.63 \times 0.553 = 440\ \text{W} \tag{6.47}$$

Because the inductor receives no time-average power, this power must represent electrical energy converted to thermal energy in the resistor. This interpretation is confirmed by direct calculation of the power into the resistor, Eq. (6.18).

$$P_R = I_e^2 R = (6.63)^2 (10) = 440\ \text{W} \tag{6.48}$$

The reactive power from Eq. (6.40) is

$$Q = V_e I_e \sin\theta = 120 \times 6.63\ \sin(+56.4°) = +663\ \text{VAR} \tag{6.49}$$

where VAR stands for "volt-ampere reactive," a unit intended to distinguish the reactive power from the real. The magnetic energy storage represented by the inductance has a peak value of

$$W_{mp} = LI_e^2 = 0.04 \times (6.63)^2 = 1.76 \text{ J} \tag{6.50}$$

You may confirm that Eq. (6.42) is satisfied. The instantaneous magnetic energy storage fluctuates between zero and the peak energy calculated in Eq. (6.50). This energy must be lent twice each cycle to the load by the source by means of reactive power flow.

WHAT IF? What if the inductor were replaced by a 200-μF capacitor?[12]

You may have noticed that we used rms values in every power and energy calculation in the foregoing example. However, we were careful to use peak values for time functions. This required inserting and taking out some $\sqrt{2}$'s that never entered into the calculations. Many texts and most practitioners drop them and use rms values for everything, but we favor the more explicit approach and continue to use effective or peak values where each is more natural.

Summary. We have defined real and reactive power as

$$\underbrace{P = V_e I_e \cos\theta}_{\text{real power}} \quad \text{and} \quad \underbrace{Q = V_e I_e \sin\theta}_{\text{reactive power}} \tag{6.51}$$

where V_e and I_e are the effective values of the voltage and current, respectively. The definitions refer to circuits generally, but when applied directly to individual resistances, inductances, and capacitances, we have

$$P_R = I_e^2 R \quad \text{and} \quad Q_L = I_e^2 X_L \quad \text{and} \quad Q_C = I_e^2 X_C \tag{6.52}$$

where X_C is numerically negative. Thus a resistance requires only real power; inductors and capacitors require reactive power of opposite signs.

We could push our investigation of power and energy in the time domain a little further, but we would rather move on to the frequency domain. The important question we ask is: Can such power calculations be made in the frequency domain without explicitly considering the time functions? The next section shows the answer to be yes; indeed, the frequency-domain viewpoint yields efficiency in calculation and suggests new insights.

Check Your Understanding

1. What is the peak value of a sinusoidal current if a standard ac ammeter indicates 5 A?
2. A 120-V electric iron (for ironing clothes) converts approximately 1200 W of electrical power to heat. Estimate the rms current to the iron.
3. For a resistor, the time-average power is one-half the peak instantaneous power. (True or False?)

---

[12] $P = 522$ W, $Q = -692$ VAR, $W_{ep} = 1.84$ J.

4. In an ac circuit, the peak stored energy in a capacitor is 10 µJ. What is the time-average stored energy?

5. What is the time-average value of $v(t) = 10 + 5\cos(100t)$ V?

6. If the power factor is 0.75, lagging, by what angle does the current lag the voltage?

7. Is the reactive power into a capacitor positive or negative?

8. For a dc voltage, a dc voltmeter measures 10 V. What is the time-average voltage? What is the rms value of the voltage? For an ac voltage, an ac voltmeter measures 10 V. What is the time-average value of the voltage? What is the rms value of the voltage?

Answers. (1) 7.07 A; (2) 10 A; (3) true; (4) 5 µJ; (5) 10 V; (6) 41.4°; (7) negative; (8) 10 V, 10 V, 0 V, 10 V.

6.2 POWER AND ENERGY IN THE FREQUENCY DOMAIN

Introduction. The time-average (real) power in an ac circuit, given in Eq. (6.39), involves the peak (or rms) values of the voltage and current and the power factor, which is the cosine of the phase angle between the voltage and current. All these quantities can be expressed in the frequency domain; indeed, the frequency-domain concept of impedance provides the most efficient way to determine the magnitude and the phase angle of the current. Thus, we can calculate time-average power without transforming to the time domain. In this section, we show how to determine real, reactive, and apparent power in the frequency domain.

Real and Reactive Power from Phasors

Real power. The time-average power into the load is given by Eq. (6.39) as

LEARNING OBJECTIVE 4.

To understand how to calculate apparent, real, and reactive power in the frequency domain

$$P = \frac{V_p I_p}{2} \cos\theta = \frac{1}{2}|\underline{V}| \times \underbrace{|\underline{I}|\cos\theta}_{\text{in-phase current}} \quad \text{watts} \quad (6.53)$$

Figure 6.10(b) and Eq. (6.39) suggest an interpretation of the power factor. If we associate the $\cos\theta$ with the magnitude of $\underline{I}$, the term $|\underline{I}|\cos\theta$ is the projection of the current phasor onto the voltage phasor. This part of the current is *in phase* with the voltage, and we conclude that the time-average power is given by the product of the voltage and the current in phase with the voltage. Thus, in the circuit of Fig. 6.9, repeated in Fig. 6.10(a),

in-phase current

$$P = \tfrac{1}{2} 120\sqrt{2} \times 3.67\sqrt{2} = 440 \text{ W} \quad (6.54)$$

which agrees with Eq. (6.47).

out-of-phase current

Reactive power. Equation (6.40) for the reactive power can be interpreted as the product of the voltage with the *out-of-phase* component of the current, as shown in Fig. 6.10(b):

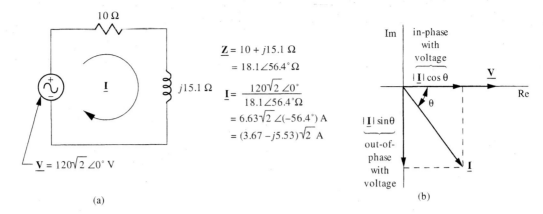

Figure 6.10 (a) Solution in the frequency domain. (b) The phasor diagram shows the "in-phase" and "out-of-phase" components of the current.

$$Q = \tfrac{1}{2}|\underline{V}| \times \underbrace{|\underline{I}|\sin\theta}_{\text{out-of-phase current}} \quad \text{VAR} \tag{6.55}$$

where we consider lagging current to produce a positive out-of-phase component.[13] In the circuit in Fig. 6.10(a),

$$Q = \tfrac{1}{2} \, 120\sqrt{2} \times 5.53\sqrt{2} = +663 \text{ VAR} \tag{6.56}$$

which agrees with Eq. (6.49). We showed in Eq. (6.42) the relation between reactive power and the stored-energy requirement of the inductor. We explore further the meaning of the reactive power in the following example.

EXAMPLE 6.8 **Real and reactive power in the *RLC* circuit**

Find the real and reactive power into the load in Fig. 6.11 with $R = 10 \ \Omega$, $L = 40$ mH, and $C = 120 \ \mu$F, and interpret these powers as lost or stored energy. The voltage is 120 V (rms) at 60 Hz.

SOLUTION:
Figure 6.11 shows the circuit in the frequency domain. The impedance as seen by the voltage source is

$$\underline{Z} = j15.1 + 10 \parallel (-j22.1) = 14.0 \angle 53.8° \ \Omega \tag{6.57}$$

so the current is

---

[13] Because θ is negative for lagging current, Q would be negative for lagging current if strict mathematics were followed. In a power system, however, lagging current is "normal" and reactive power is defined to be positive for lagging, or inductive, current.

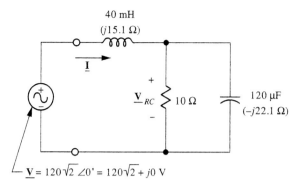

Figure 6.11 General *RLC* load.

$$|\underline{I}| = \frac{120\sqrt{2} \angle 0°}{14.0 \angle 53.8°} = 8.55\sqrt{2} \angle -53.8°$$
$$= (5.05 - j6.89)\sqrt{2} \text{ A} \tag{6.58}$$

The real power to the load is the product of the rms voltage with the in-phase component of the rms current, $P = 120 \times 5.05 = 606$ watts, and the reactive power to the load is the product with the out-of-phase rms current, $Q = 120 \times 6.89 = 827$ VAR. Lagging current gives positive reactive power.

To interpret these powers in terms of lost or stored energy, we must complete the phasor analysis of the circuit. We find the voltage across the $R\|C$ by multiplying the current by the parallel impedance:

$$\underline{V}_{RC} = (8.55\sqrt{2} \angle -53.8°) \times 10\|(-j22.1) = 77.9\sqrt{2} \angle -78.1° \text{ V} \tag{6.59}$$

The energy lost to the circuit will be in the time-average power to the resistor, which is

$$P_R = \frac{V_R^2}{R} = \frac{(77.9)^2}{10} = 606 \text{ watts} \tag{6.60}$$

where V_R is the rms voltage across the resistor. The peak electric energy stored in the capacitor is

$$W_{ep} = \tfrac{1}{2} CV_C^2 = \tfrac{1}{2}(120 \times 10^{-6}) \times (77.9\sqrt{2})^2 = 0.728 \text{ J} \tag{6.61}$$

where V_C is the peak voltage across the capacitor. The peak magnetic energy stored in the inductor is

$$W_{mp} = \tfrac{1}{2} LI_L^2 = \tfrac{1}{2}(0.040) \times (8.55\sqrt{2})^2 = 2.92 \text{ J} \tag{6.62}$$

where I_L is the peak current through the inductor. These stored energies relate to the reactive power through Eq. (6.44):

$$Q = \omega(W_{mp} - W_{ep}) = 120\pi \times (2.92 - 0.728) = 827 \text{ VAR} \tag{6.63}$$

which agrees with our earlier result.

> **WHAT IF?** What if the inductor and capacitor were exchanged? Find the real and reactive power for this case.[14]

Summary. The real (time-average) and reactive powers can be determined wholly in the frequency domain. The real power is the product of the voltage with the current in phase with the voltage, and the reactive power is the product of the voltage with the current out of phase with the voltage, with lagging current considered positive. The real power is the sum of the powers dissipated in the resistances of the load, and the reactive power is the product of the angular frequency with the difference between the magnetic and electric peak stored energies within the load.

Complex Power

complex power

Definition of complex power. The complex sum of the real and reactive power in Eqs. (6.39) and (6.40) is called the *complex power* and can be written

$$\underline{S} = \tfrac{1}{2}\underline{V}\underline{I}^* = P + jQ \quad \text{VA} \tag{6.64}$$

where $\underline{S}$ is the complex power in volt-amperes. The complex conjugate of the phasor current changes the sign of the imaginary part and thus introduces mathematically the customary change of sign in the out-of-phase current. The expression for the complex power in Eq. (6.64) also permits us to relax the requirement that the phasor voltage, $\underline{V}$, be a real quantity. Note that if

$$\underline{V} = |\underline{V}| \angle \theta_V \quad \text{and} \quad \underline{Z} = |\underline{Z}| \angle \theta \tag{6.65}$$

where θ_V is the phase of the voltage, no longer assumed zero, then

$$\underline{I} = \frac{\underline{V}}{\underline{Z}} = |\underline{I}| \angle (\theta_V - \theta) \Rightarrow \underline{I}^* = |\underline{I}| \angle (-\theta_V + \theta) \tag{6.66}$$

because the complex conjugate changes the sign of the current phase angle. The complex power, as defined in Eq. (6.64), would be

$$\begin{aligned}\underline{S} &= \tfrac{1}{2}\underline{V}\underline{I}^* = \tfrac{1}{2}|\underline{V}| \angle \theta_V \times |\underline{I}| \angle (-\theta_V + \theta) \\ &= \tfrac{1}{2}|\underline{V}||\underline{I}| \angle \theta = \tfrac{1}{2}|\underline{V}||\underline{I}| \,(\cos\theta + j\sin\theta)\end{aligned} \tag{6.67}$$

Thus, the phase reference drops out for the complex power defined in Eq. (6.64) and only the phase difference between voltage and current remains.

---

[14] $P = 282$ watts and $Q = -711$ VARs.

EXAMPLE 6.9 Complex power

Find the complex power into the circuit in Fig. 6.10(a).

SOLUTION:
The phasor voltage is $\underline{V} = 120\sqrt{2} \angle 0°$ V and the current we found in Fig. 6.10 to be $\underline{I} = 6.63\sqrt{2} \angle -56.4°$ A in polar form or $\underline{I} = (3.67 - j5.53)\sqrt{2}$ A in rectangular form. The complex power is

$$\underline{S} = \tfrac{1}{2}\underline{V}\underline{I}^* = \tfrac{1}{2}(120\sqrt{2} + j0)(3.67\sqrt{2} + j5.53\sqrt{2})$$

$$= 120 \times 3.67 + j120 \times 5.53 = \underbrace{440}_{P} + \underbrace{j663}_{Q} \text{ VA}$$

(6.68)

Thus, the real power is 440 W and the reactive power is +663 VAR. The positive sign indicates predominant magnetic energy storage. The $\sqrt{2}$'s all dropped out, as they always do in power calculations.

WHAT IF? What if the resistor were increased to 12 Ω? What would be the complex power in that case?[15]

Apparent power. The complex power is a complex number yielding information about the flow of time-average power and the shuttling of loaned energy between source and load in an ac circuit. The magnitude of the complex power, $S = |\underline{S}|$ is the *apparent power*. The apparent power results when you measure the voltage and current with meters and multiply the measured values without regard for phase.[16]

$$S = |\underline{S}| = \tfrac{1}{2}|\underline{V}||\underline{I}| = V_e I_e \text{ VA} \tag{6.69}$$

Apparent power is important as a measure of the operating limits in electrical equipment such as transformers, motors, and generators. Losses in the wires are proportional to the square of the current in the machine regardless of the phase, whereas losses in the magnetic materials are roughly proportional to the square of the operating voltage. Machine limits are established by losses. Because electrical machinery is operated with the voltage more or less constant, apparent power limits imply current limits. In the previous example the apparent power is $120 \times 6.63 = 796$ VA.

Summary. The complex power gives concise information about power and energy flow in an ac circuit. The real part of the complex power is the time-average power in watts. The real power heats resistors, turns motors, and makes the electric meter revolve. The imaginary part of the complex power is the reactive power in VARs, which is proportional to the electrical energy lent to the load by the ac source twice each cycle. The reactive power is considered positive when the load is inductive and negative

---

[15] $\underline{S} = 465 + j584$ VA.
[16] In Eq. (6.69), we need not use $|\underline{I}^*|$ because the magnitude is the same for $\underline{I}$ and $\underline{I}^*$.

when the load is capacitive, although the latter is rare in power systems. The magnitude of the complex power is the apparent power in volt-amperes and indicates the operating level of a power system.

power triangle

Power triangles. The apparent power, real power, and reactive power form a right triangle. Figure 6.12 shows this *power triangle* for the previous example. The angle at the origin is θ, the angle of the impedance, $+56.4°$ in this case. When the current lags the voltage, the power triangle is drawn above the x-axis.

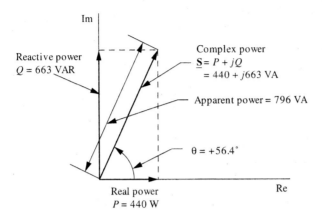

Figure 6.12 The power triangle pictures complex, apparent, real, and reactive power.

Because the various kinds of powers form a right triangle, and because the power factor is the cosine of an angle of that triangle, there follow a host of formulas relating power factor with the various types of powers. We now list several of these, which you can verify from the definitions and common trigonometric identities.

$$S = \sqrt{P^2 + Q^2} \tag{6.70}$$

$$PF = \frac{P}{S} = \frac{P}{\sqrt{P^2 + Q^2}} \tag{6.71}$$

$$Q = \pm S\sqrt{1 - (PF)^2} \quad (+ \text{ for lagging } PF) \tag{6.72}$$

Based on these formulas, and others that can easily be derived, a variety of problems in ac power systems can be analyzed, of which the following is typical.

Determining apparent power, real power, and reactive power from measurements. Figure 6.8 shows an ammeter (A), voltmeter (V), and wattmeter (P) metering an *RLC* load. We assume these meters are ideal and indicate measured values of I_m, V_m, and P_m. Because the meters indicate effective values, the measured apparent power would be the product of the meter indications:

$$S_m = V_m I_m \tag{6.73}$$

and Eq. (6.71) would give the measured power factor magnitude, $(PF)_m = P_m/S_m$.[17] The measured reactive power follows from Eq. (6.70).

---
[17] Whether leading or lagging cannot be determined from these measurements.

$$Q_m = \pm \sqrt{S_m^2 - P_m^2} = \pm \sqrt{(V_m I_m)^2 - P_m^2} \tag{6.74}$$

EXAMPLE 6.10 Motors

A 230-V motor has a mechanical output power of 3 horsepower (hp). The input current, voltage, and power are measured to be 226 V, 15.6 A, and 2920 W, respectively. Calculate the efficiency, the power factor, and reactive power. Draw a phasor diagram, assuming lagging current.

SOLUTION:
The measured apparent power is

$$S_m = V_m I_m = 226 \times 15.6 = 3530 \text{ VA} \tag{6.75}$$

The measured power factor is

$$PF_m = \frac{P_m}{S_m} = \frac{2920}{3530} = 0.828 \tag{6.76}$$

which corresponds to a phase angle of $\cos^{-1}(0.828) = 34.1°$, Eq. (6.41), assumed lagging current. The reactive power follows from Eq. (6.74):

$$Q = +\sqrt{S_m^2 - P_m^2} = \sqrt{(3530)^2 - (2920)^2} = +1980 \text{ VAR} \tag{6.77}$$

For the phasor diagram in Fig. 6.13, we have used the voltage for the phase reference, and shown the current lagging. The phase angle of the current follows from the power factor, $\theta = \cos^{-1}(0.828) = 34.1°$. The power triangle showing the real, reactive, and apparent powers would have the same angle but would be drawn above the real axis, similar to Fig. 6.12.

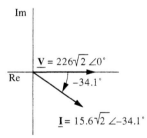

Figure 6.13 Phasors representing motor voltage and current.

WHAT IF? What if the current were leading? What would change?[18]

The vocabulary and units of power in ac circuits. We now summarize the four types of power in ac circuits:

---
[18] Only the signs of the phase angle and the reactive power.

P = real power = time-average power in watts
Q = reactive power in VARs,[19] volt-amperes reactive
S = apparent power in VAs, volt-amperes
$\underline{S}$ = complex power in VAs

The three units for ac power—watts, volt-amperes reactive, and volt-amperes—all have the scientific units for power, J/s. The related units of kW, kVAR, and kVA ($k = 10^3$) and MW, MVAR, and MVA ($M = 10^6$) are also common in the power industry. We use different units to clarify communication when speaking of the various kinds of "power" in ac circuits.

Reactive Power in Power Systems

Importance of reactive power. Power companies have to be careful about the reactive power load on their systems. As stated before, the limits of larger power equipment such as generators and transformers are described by the apparent power, the Pythagorean sum of the real and reactive powers. Thus, if the reactive power becomes large, a piece of equipment may become overloaded even though the real power is moderate. Also, the reactive energy must be transported from generator to user, often over great distances. Reactive power increases line current, and hence increases line losses. The power company might surcharge industrial customers whose requirements for reactive power are great to pay for line losses.

LEARNING OBJECTIVE 5.
To understand how to correct the power factor of a load

Power-factor correction. Often, the industrial consumer can save money by placing a bank of capacitors in parallel with an inductive load to store energy locally. In effect, they receive the stored energy from the power company only once and then keep it "in house" with the capacitors. The customer thus *corrects* his power factor by creating a resonance between the electric energy stored by the added capacitance and the magnetic energy used by motors or other heavy equipment.

power-factor correction

Power-factor correction: current method. The effect of adding parallel capacitance to correct the power factor can be understood from the phasor diagram of the currents. Figure 6.14(b) shows the current phasors with the voltage as the phase reference. The load current, I_{load}, lags the voltage due to the magnetic equipment in the load.

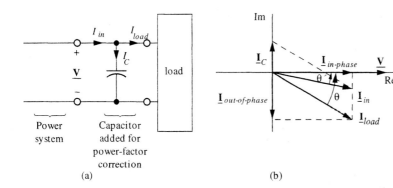

Figure 6.14 (a) The capacitor corrects the lagging power factor of the load; (b) the leading current of the capacitor cancels part or all of the lagging out-of-phase current of the load. The input current is thus reduced.

---
[19] Rhymes with "jars."

The in-phase current does useful work, and the out-of-phase current supplies the required magnetic energy. If the power factor is too low, capacitors are added in parallel with the load to draw leading current, which has the effect to improve the power factor from $PF = \cos\theta$ to $PF' = \cos\theta'$. The capacitive current is $I_C = \omega C V_e$ and the details amount to a simple problem in trigonometry, as shown by the following example.

EXAMPLE 6.11 Power factor correction (current method)

A 460-V, 60-Hz load uses 12 kW at a lagging power factor of 0.75. What capacitance should be placed in parallel with the load to correct the power factor to 0.9 lagging?

SOLUTION:
The rms load current is

$$I_{load} = \frac{P}{V \times PF} = \frac{12,000}{460 \times 0.75} = 34.8 \text{ A} \qquad (6.78)$$

at a lagging phase angle of

$$\theta = \cos^{-1}(0.75) = 41.4° \qquad (6.79)$$

With the voltage as the phase reference, the current is

$$\mathbf{I}_{load} = 34.8\sqrt{2} \angle -41.4° = (26.1 - j23.0)\sqrt{2} \text{ A} \qquad (6.80)$$

The parallel capacitor will draw leading current but will not change the in-phase component. For a power factor of 0.9 lagging ($\theta = -25.8°$), the resultant out-of-phase component must be

$$23.0 - I_C = 26.1 \tan 25.8° = 12.6$$
$$I_C = 10.4 \text{ A} \qquad (6.81)$$

where a lagging out-of-phase component is considered positive. The required capacitance is, therefore, 59.8 µF.

WHAT IF? What if the corrected power factor must be 0.95 lagging?[20]

Power-factor correction: power method. Power capacitors manufactured for power-factor correction are rated by voltage and reactive power.[21] Thus, the real and reactive power can be calculated directly, as shown by reworking the previous example on this basis.

---

[20] 83.2 µF.
[21] The frequency is assumed to be 60 Hz in the United States.

EXAMPLE 6.12 Power factor correction (power method)

ALTERNATE SOLUTION:
A calculation of the complex power follows from the real power and power factor:

$$\underline{S} = \frac{P}{PF} \angle (\pm \cos^{-1} PF) = \frac{12,000}{0.75} \angle +\cos^{-1}(0.75) \quad (6.82)$$
$$= 16,000 \angle (+41.4°) = 12,000 + j10,600$$

where + is used for lagging PF. The desired complex power is

$$\underline{S}' = \frac{12,000}{0.9} \angle \cos^{-1}(0.9) \quad (6.83)$$
$$= 12,000 + j5810$$

Because $\underline{S}' = \underline{S} + jQ_C$, the capacitor must contribute

$$Q_C = -4770 \text{ VAR} \quad (6.84)$$

as shown in Fig. 6.15. Thus we would use one 5-kVAR or five 1-kVAR, 460-V capacitors.

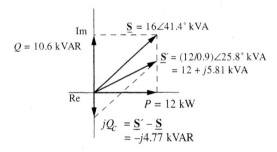

Figure 6.15 The power triangles before and after power-factor correction.

WHAT IF? What if you wanted to correct to 0.95 PF leading?[22]

Reactive Power in Electronics

LEARNING OBJECTIVE 6.
To understand how to use an ac Thévenin equivalent circuit to establish the load to withdraw maximum power from an ac source

Reactive power also plays an important role in electronics. Here we normally deal with small amounts of power, and must make full use of the power that is available. To see the role of reactive power, let us examine the conditions for maximum power transfer in an ac circuit.

Thévenin and Norton equivalent circuits for ac. In Chap. 2, we derived the Thévenin equivalent circuit using linearity and superposition. In Chap. 4, we reduced ac

---
[22] −14.5 kVAR (182 μF).

problems to equivalent dc problems through the use of phasors, through which sources and circuits are represented by complex numbers. All the techniques we developed for dc circuits remain valid for solving ac circuits, including Thévenin and Norton equivalent circuits. In Fig. 6.16, we show a Thévenin equivalent circuit with a phasor voltage source, $\underline{\mathbf{V}}_T$, and an output impedance, $\underline{\mathbf{Z}}_{eq}$, which we have expressed as a resistive and reactive part for benefit of the derivation that follows. Recall that the circuit replaced by the Thévenin equivalent circuit can be arbitrarily complicated.

Radio design. Let us consider a typical situation that might arise in electronics, that of getting maximum power out of a radio antenna. Specifically, consider the telescoping AM radio antenna on an automobile, as suggested in Fig. 6.17. Radio waves are radiated by a commercial station, perhaps at considerable distance, and these waves interact with the antenna to give a small voltage, typically 10 mV, rms, between the fender and the base of the antenna.

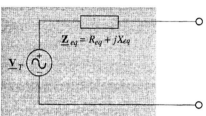

Figure 6.16 For an ac circuit, the Thévenin voltage is a phasor and the output resistance becomes an impedance.

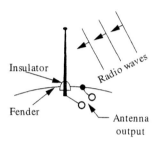

Figure 6.17 The antenna and the radio waves can be represented by the Thévenin equivalent circuit shown in Fig. 6.16.

Antenna impedance. The derivation of the output impedance of such an antenna is complicated, but we can say a little based on our understanding of electric and magnetic energy. As a circuit element, the antenna represents a wire that leads nowhere, that is, an open circuit. The radio waves tend to make current flow on the wire, but a small current produces a buildup of charge on the wire and the current stops. Thus, we anticipate that an antenna of this type would build up charge but carry little current. This suggests that electric energy storage would dominate magnetic energy storage and that the output impedance would be capacitive. We have shown typical values in Fig. 6.18.

Power transfer to the radio. Our task is to specify the input impedance of the radio, $\underline{\mathbf{Z}}_L = R_L + jX_L$, to receive maximum power out of the antenna. The average power into the load would be $|\underline{\mathbf{I}}|^2 R_L/2$, where $\underline{\mathbf{I}}$ is the current in the load. This current is the antenna voltage divided by the total impedance in the circuit, the sum of $\underline{\mathbf{Z}}_{eq}$ and $\underline{\mathbf{Z}}_L$.

$$\underline{\mathbf{I}} = \frac{\underline{\mathbf{V}}_T}{\underline{\mathbf{Z}}_{eq} + \underline{\mathbf{Z}}_L} = \frac{10\sqrt{2} \times 10^{-3} \angle 0°}{(10 + R_L) + j(X_L - 150)} \quad \text{A} \tag{6.85}$$

Because the phase does not matter in this power calculation, we will consider only the magnitude:

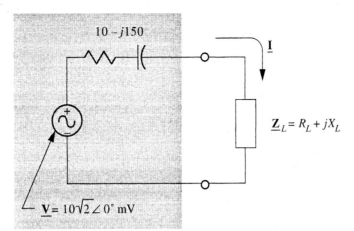

Figure 6.18 Find R_L and X_L to maximize power in R

$$|\mathbf{I}| = \frac{10 \times 10^{-3}\sqrt{2}}{\sqrt{(10 + R_L)^2 + (X_L - 150)^2}} \text{ A} \qquad (6.86)$$

Consequently, the time-average power delivered to the radio by the antenna would be

$$P(R_L, X_L) = \frac{|\mathbf{I}|^2 R_L}{2} = \frac{2 \times (10^{-2})^2 (R_L/2)}{(10 + R_L)^2 + (X_L - 150)^2} \text{ W} \qquad (6.87)$$

Maximizing the power. Equation (6.87) is the function to be maximized, so we might make an assault using the methods of differential calculus. Before taking derivatives, however, we should note the effect of the load reactance, X_L. Being in the denominator and being squared, the reactance term, $X_L - 150$, can only decrease the power. Clearly, the best we can do is set X_L to $+150 \, \Omega$. Once we do that, the power is a function of R_L only:

$$P(R_L + 150) = \frac{10^{-4} R_L}{(10 + R_L)^2} \Rightarrow R_L = 10 \, \Omega \text{ for maximum power} \qquad (6.88)$$

We have written by inspection the value of R_L that gives maximum power because the problem has been reduced to that solved back in Chap. 2, Eq. (2.19). Once the load reactance (X_L) is adjusted to balance the reactance in the source, the maximum power follows from equating the resistance of the load to the resistive part of the output impedance of the source. In summary,

$$\mathbf{Z}_L = \mathbf{Z}_{eq}^* = R_{eq} - jX_{eq} \qquad (6.89)$$

impedance matching

gives maximum power to R_L. This is called *matching* the load impedance to the source.

Thus, the maximum power transfer will occur when the load has the same resistance as the source but the opposite reactance. We therefore create a local resonance from the viewpoint of the load, in effect balancing the equivalent stored energies. In the case solved before, you can confirm that the power delivered to the radio by the antenna is 2.5×10^{-6} W. Not much power, but we know that it must be adequate because AM radios work.

EXAMPLE 6.13 Matching output cable

A 1-MHz voltage source with an output impedance of 50 Ω must be attached to a load with a cable that has a capacitance of 300 pF. What should be the load to draw maximum power from the source?

SOLUTION:
Figure 6.19 shows the equivalent circuit. We have expressed the unknown load impedance also as admittance because we have a choice of two loads, depending upon whether we use a series or parallel form.

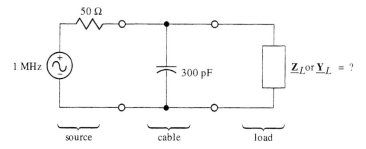

Figure 6.19 The load impedance has to match the output impedance of the source, including the capacitance of the cable.

The output impedance to the load is

$$\underline{Z}_{out} = 50\Omega \| 300 \text{ pF} @ 1\text{MHz} = 50 \| -j531 = 49.6 - j4.7 \Omega \quad (6.90)$$

Maximum power transfer requires $\underline{Z}_L = \underline{Z}_{out}^*$, Eq. (6.89):

$$\underline{Z}_L = 49.6 + j4.7 = 49.6 \text{ } \Omega \text{ in series with } 0.74 \text{ µH} \quad (6.91)$$

The small value of inductance might cause a problem, so we should consider a parallel realization for the load. For a parallel load, the output admittance would be appropriate for

$$\underline{Y}_{out} = \frac{1}{R} + j\omega C = 0.02 + j1.88 \times 10^{-3} \text{ mho} \quad (6.92)$$

The reciprocal of Eq. (6.89) is equally valid, so the load admittance for maximum power will be

$$\underline{Y}_L = 0.02 - j1.88 \times 10^{-3} \text{ mho} \quad (6.93)$$

which is 50 Ω in parallel with 84.5 µH.

WHAT IF? What if the frequency were 10 MHz?[23]

---

[23] The optimum load impedance would be 26.5 Ω in series with 0.4 µH or else 50 Ω in parallel with 1.7 µH.

A power system is designed to provide uniform voltage, independent of load. For this reason, the source impedance is always small relative to load impedance. Maximum power transfer is irrelevant to a power system.

Check Your Understanding

1. If the current leads the voltage in an ac circuit, the reactive power is positive or negative?
2. If the real power is 600 W and the reactive power is -300 VAR, what is the apparent power?
3. The complex power depends on the relative phase between voltage and current, not the absolute phase. True or False?
4. An electronic circuit has an open-circuit voltage of 100 mV (rms) and an output impedance of $20 + j10\ \Omega$. What load will draw maximum power from this source? What will be the power in this load?

Answers. (1) Negative; (2) 671 VA; (3) true; (4) $20 - j10\ \Omega$, 1.25×10^{-4} W.

6.3 TRANSFORMERS

Transformer Principles

LEARNING OBJECTIVE 7.
To understand the voltage, current, and impedance transforming properties of the transformer

Voltage transformation. A transformer is a highly efficient device for changing ac voltage from one value to another, for example, from 120 V to 6 V. Transformers come in all sizes, from the enormous transformers used in power substations to the small transformers used for doorbells.[24]

The transformer gives ac a feature lacking in dc power systems. Using a transformer, we can efficiently change ac voltage from small amplitudes to large amplitudes, or vice versa. Such changes are not simply accomplished with dc voltage.

Current and impedance transformation. As we will soon prove, a transformer also transforms current, and, as a consequence of transforming voltage and current, also transforms impedance. Indeed, the transformation of impedance is perhaps the most important property of the transformer.

Common transformer applications. Transformers are indispensable to electrical systems; here are a few common applications:

Impedance Level

- **Power transformers.** Transformers are vital to high-voltage power distribution systems. This is an example of voltage transformation, but the main purpose is impedance transformation. Power transformers are used to make the resistance and inductance of the transmission system look small relative to the load impedance; hence, the relative losses of the distribution system are reduced.

---

[24] Sometimes the word "transformer" is used for a device that employs a transformer but includes other controls or devices, such as the "transformers" used to power model railroad trains.

Impedance Level

- **Battery chargers.** Transformers reduce the power voltage from 120 V ac to some smaller value for charging of batteries through rectification.

- **Arc welders.** Transformers are used to match the ac system to the impedance of the molten metal, which is close to a short circuit, so that power can be delivered to it.

- **Doorbell and thermostat circuits.** Here, the transformer reduces the voltage to a small value for safety.

- **Electronic power supplies.** This application is similar to a battery charger. The voltage is changed to the level required by the electronic circuits.

Impedance Level

- **Stereo output transformers.** Transformers are used to match the electronic output circuit to the impedance of the speakers, as well as to allow optimum use of different speaker systems with the same amplifier.

- **Ignition coils.** The "coil" is a transformer that raises the voltage of the automotive electrical system to the high voltage required to create sparks in the cylinders.

- **Etcetera.** We could multiply this list manyfold. The point is that transformers are used everywhere, not just in power circuits. In this section, we present ideal transformers and discuss some applications. In Chap. 7, we describe the use of transformers in power distribution systems in more detail.

Transformer construction. Figure 6.20 shows a simple transformer. Two coils are coupled by time-varying magnetic flux, which is channeled by an iron core. If we construct such a device, connect the primary coil to an ac voltage source, and connect the secondary coil to a resistive load, we would find that the resistive load becomes hot. This would demonstrate the flow of electrical energy from the ac source through the transformer and into the load. Furthermore, we would find that the transformer does not become very hot, suggesting that the transformer is an efficient device for coupling load and source.

transformer primary and secondary

Transformer primary and secondary. The coil connected to the ac source is called the *primary* and that connected to the load the *secondary*. There is nothing special about the two sides, for the transformer can convey power either way. In most applications, however, the power flows in only one direction and hence these names are useful.

The ideal transformer. Here we define the ideal transformer as a circuit element and explore its properties in voltage, current, and impedance transformation. Primary and secondary voltage and current variables are defined in Fig. 6.21, which shows the circuit symbol for an ideal transformer. The primary variables form a load set and the secondary variables form a source set. These definitions are customary and indicate that the primary acts as a *load* to the power system supplying power to the transformer and the secondary acts as a *source* to the loads connected to it. We also define n_p and n_s, the turns on the primary and secondary, respectively.

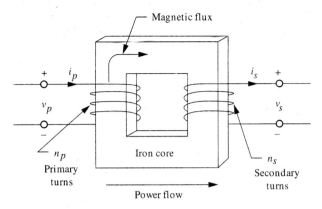

Figure 6.20 A simple electrical transformer.

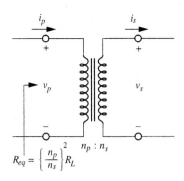

Figure 6.21 Circuit symbol for an ideal transformer. The primary voltage and current form a load set, and the secondary voltage and current form a source set.

ideal transformer

Voltage and current transformation in ideal transformers. The primary and secondary voltages in the *ideal transformer* are related by

$$\frac{v_p}{n_p} = \frac{v_s}{n_s} \tag{6.94}$$

Thus, the side of the transformer with the larger number of turns has the larger voltage; indeed, the voltage per turn is constant for a given transformer. The primary and secondary currents in the *ideal transformer* are related by

$$n_p i_p = n_s i_s \tag{6.95}$$

Thus, the side of the transformer with the larger number of turns has the *smaller* current. For example, a transformer to increase the voltage would have a primary with few turns of large wire (small voltage, large current) and the secondary would have many turns of small wire (large voltage, small current).

EXAMPLE 6.14 Ideal transformers

A transformer is required to produce 18 V (rms) and 650 mA from a 120-V rms line voltage. The primary uses 1000 turns of No. 30 wire. How many turns are required in the secondary and what would be the primary current?

SOLUTION:

Equation (6.94) is valid for instantaneous voltage in the time domain, so it transforms into the frequency domain as

$$\frac{\underline{V}_p}{n_p} = \frac{\underline{V}_s}{n_s} \Rightarrow \frac{120\sqrt{2}}{1000} = \frac{18\sqrt{2}}{n_s} \Rightarrow n_s = 150 \tag{6.96}$$

where $\underline{V}_p$ and $\underline{V}_s$ are the phasor primary and secondary voltages, respectively. The numerical version of Eq. (6.96) expresses only magnitudes because the phases must be the same. Using the same transformations in Eq. (6.95), we find the primary current to be

$$1000 I_p = 150 \times 650 \text{ mA} \Rightarrow I_p = 97.5 \text{ mA} \tag{6.97}$$

WHAT IF? What if the transformer primary uses only 500 turns?[25]

Equations (6.94) and (6.95) define the ideal transformer as a circuit element, with one restriction—no dc. The primary of the transformer is, like an inductor, a short circuit to dc and provides no coupling to the secondary unless voltage and current are changing.[26]

Conservation of Energy

Conservation of power. Multiplication of the left and right sides of Eqs. (6.94) and (6.95) shows that the instantaneous power into the primary, p_{in}, is equal to the instantaneous power out of the secondary, p_{out}

$$\frac{v_p}{n_p} \times n_p i_p = \frac{v_s}{n_s} \times n_s i_s \Rightarrow p_{in} = p_{out} \tag{6.98}$$

Thus, the *ideal* transformer has no losses and stores no energy.

Let us calculate the apparent power in the primary and secondary of the transformer in the previous example. The apparent power is the rms voltage times the rms current; hence,

$$S_p = 120 \times 97.5 \times 10^{-3} = 11.7 \text{ VA}$$

and

$$S_s = 18 \times 650 \times 10^{-3} = 11.7 \text{ VA} \tag{6.99}$$

Thus, apparent power is conserved by an ideal transformer. Because Eq. (6.98) shows that real power is conserved, it follows that reactive power is also conserved.

Impedance Level

Impedance transformer. Equations (6.94) and (6.95) reveal a very useful property of transformers—impedance transformation. Figure 6.22 shows an ideal transformer, a load resistor, R_L, connected to the secondary, and an equivalent resistance, R_{eq}, defined at the primary. The equations of the circuit are those of the ideal transformer in Eqs. (6.94) and (6.95) plus Ohm's law for R_L. If we divide Eq. (6.94) by Eq. (6.95), we obtain

$$\frac{1}{n_p^2} \frac{v_p}{i_p} = \frac{1}{n_s^2} \frac{v_s}{i_s} \tag{6.100}$$

---

[25] $n_s = 75$ turns, but $I_p = 97.5$ mA, the same.
[26] Practical transformers have a low frequency below which the coupling between primary and secondary begins to decrease, decreasing to zero at dc.

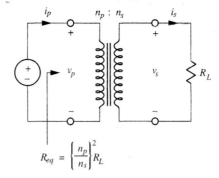

$$R_{eq} = \left(\frac{n_p}{n_s}\right)^2 R_L$$

Figure 6.22 The transformer will change R_L to an equivalent resistance R_{eq}.

But v_s/i_s is the load resistor, R_L, and v_p/i_p defines the equivalent resistance, R_{eq}, into the primary, so Eq. (6.100) leads to the value of the equivalent resistance

$$R_{eq} = \left(\frac{n_p}{n_s}\right)^2 R_L \qquad (6.101)$$

turns ratio

Thus, the impedance level is transformed by the square of the *turns ratio*, n_p/n_s.[27] By using a transformer, we can make a large impedance appear small, or we can make a small impedance appear large.

EXAMPLE 6.15 Using a transformer to match impedances

A source with an output resistance of 100 Ω must deliver power to a load of 500 Ω. Find the turns ratio, n_p/n_s, to maximize load power.

SOLUTION:
As shown in Eq. (6.89), we must use an equivalent load impedance equal to the complex conjugate of the output impedance of the source. As shown in Fig. 6.23, this requires an impedance of 100 Ω looking into the primary of the transformer; thus, we need a transformer having

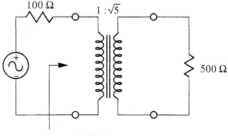

$R_{eq} = 100\ \Omega$ for maximum power

Figure 6.23 The transformer turns ratio maximizes the power to the 500-Ω load.

---

[27] The ratio of primary to secondary turns determines voltage, current, and impedance transformation properties. For this reason, the *turns ratio* is often stated as $1:n$ (or $n:1$), where n is not necessarily an integer and can be less than unity.

$\sqrt{100/500}$ for a turns ratio. Because we wish to make the 500-Ω resistor look smaller, we must connect it to the high-voltage side of the transformer, the side with the more turns, and look into the low-voltage side to see the smaller impedance. A routine calculation shows an increase of 80% of the power in the load compared to a straight connection.

WHAT IF? What if the required transformer were not available and a 1:2 transformer were used? What then would be the increase in load compared with a straight connection with no transformer?[28]

Impedance Level

Analysis of circuits containing transformers. The equations relating transformer voltages and currents, Eqs. (6.94) and (6.95), may be used to analyze circuits containing transformers. However, transforming impedances usually works better. Rather than writing equations for voltage and current, we deal with the ratio of voltage to current, that is, with the impedance level.

EXAMPLE 6.16 **Impedance transformation**

Solve for the load current, $\mathbf{I}_L$, in the ac circuit shown in Fig. 6.24(a).

SOLUTION:

We transform the load impedance into the primary and solve directly for the primary current. Figure 6.24(b) shows the transformed impedance, and the primary current is easily determined:

$$\mathbf{I}_p = \frac{30\sqrt{2} \angle 0°}{20 + j20 + 2^2(2 - j10)} = 0.872\sqrt{2} \angle 35.5° \text{ A} \qquad (6.102)$$

The secondary current, which is the load current, is greater by the turns ratio, Eq. (6.95), twice as great in this case. Thus the secondary current is $1.74\sqrt{2} \angle 35.5°$ A.

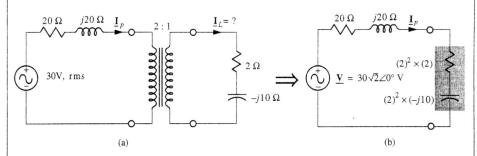

Figure 6.24 (a) Solve for $\mathbf{I}_L$; (b) equivalent circuit.

---

[28] Still 77.8%. Again this shows the importance of optimizing; small deviations do not matter.

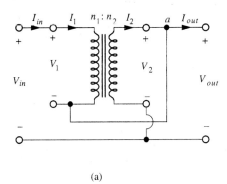
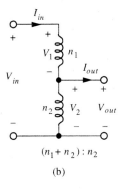

Figure 6.25 (a) The autotransformer connection; (b) the autotransformer connection drawn in the standard way.

Autotransformers. An autotransformer is a special transformer connection that is useful in power systems, motor starters, variable ac sources, and other applications. Figure 6.25(a) shows the autotransformer connection with the transformer primary and secondary drawn in the usual positions, and Fig. 6.25(b) shows the autotransformer drawn in a manner that clarifies the function of the transformer. In the step-down mode, which we have shown, the primary and secondary windings are connected in series for the new primary, and thus the input voltage is

$$V_{in} = V_1 + V_2 = \frac{n_1}{n_2} V_2 + V_2 = \frac{n_1 + n_2}{n_2} V_{out} \qquad (6.103)$$

because $V_2 = V_{out}$. Hence, the voltage turns ratio is $(n_1 + n_2):n_2$. The output current, from KCL at node a, is

$$I_{out} = I_1 + I_2 = I_1 + \frac{n_1}{n_2} I_1 = \frac{n_1 + n_2}{n_2} I_{in} \qquad (6.104)$$

because $I_1 = I_{in}$. Equation (6.104) gives the same turns ratio as Eq. (6.103), which is consistent with conservation of apparent power.

The principal virtue of the autotransformer connection is that the apparent power rating increases. In effect, some of the power bypasses the transformer; only part of the power is transformed, as illustrated by the following example.

EXAMPLE 6.17 Autotransformers

A 120/120-V, 12-kVA transformer is connected as an autotransformer to make a 240/120-V transformer. What is the apparent power rating of the autotransformer?

SOLUTION:
Figure 6.26 shows the transformer connection with rated voltage and current in the transformer. The current rating on both primary and secondary windings is 12 kVA/120 V = 100 A. In the autotransformer mode, the input apparent power is 240 × 100 = 24 kVA, and the output 120 × 200 = 24 kVA. Thus, the apparent power capacity of the 12-kVA transformer is doubled by the autotransformer connection. In effect, half the apparent power is transformed and half bypasses the transformer.

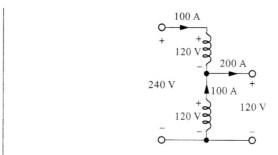

Figure 6.26 The autotransformer connection increases the kVA limit.

Transformer Applications in AC Power Systems

LEARNING OBJECTIVE 8.

To understand the role of transformers in the transmission of electric power

Impedance transformation. In this section, we explain why the impedance-transforming properties of transformers find their greatest application in electric power distribution. Figure 6.27 suggests the generation and delivery of electric power to a distant user over a distribution line. Although we have shown identical transformers at each end to simplify the analysis, in practice, the two transformers would not be identical because the power would be generated at a higher voltage than required by the consumer.

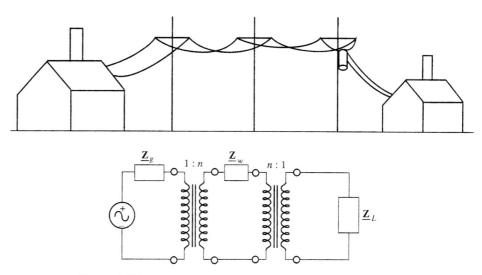

Figure 6.27 Simple power distribution system and its equivalent circuit.

Impedance Level

System characteristics. A power distribution system should offer constant voltage and high efficiency. The user wants constant voltage, independent of load, because her equipment is designed to operate at standard voltage. She requires, for example, that the voltage remain reasonably near 120 V whether she uses 1 kW or 10 kW of power. This requires, in turn, that the output impedance of the source, as seen from the user's point of view, be as low as possible. Of course, an ideal voltage source has zero output impedance, but the generator has an inherent output impedance, $\underline{Z}_g$, and the distribution line has an impedance, $\underline{Z}_w$, as shown in Fig. 6.27. The generator output impedance can be made quite small by good system design, but the resistance and reactance of the transmission line can only be reduced within limits due to the large distances.

The transmission system is required to be efficient to reduce costs and energy waste. The impedance-transforming properties of the transformers improve system characteristics, as shown in the following analysis.

Transmission system analysis. First, we transform $\underline{Z}_L$ with the transformer at the load. This raises the load impedance by a factor of n^2 and places it in series with $\underline{Z}_w$. We now can transform this series combination with the transformer at the generator, which lowers the impedances by n^2. The resulting equivalent circuit is shown in Fig. 6.28. Note that the load impedance appears with its true value in the final equivalent circuit, as does the generator impedance, but the impedance of the transmission line is reduced by the factor n^2. This will reduce its losses by the same factor.

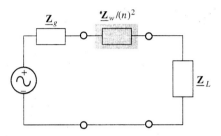

Figure 6.28 Equivalent circuit for the power distribution system.

This reduction in transmission line losses can be understood by considering the current in the transmission line. The transformer at the generator increases the voltage and decreases the current by its turns ratio, the total power being unchanged. The transmission-line losses will be $I^2 R_w$, where I is the rms current in the line and R_w is the wire resistance. Reduction of the line current by n therefore reduces these losses by n^2, and this reduction is reflected in the equivalent circuit in Fig. 6.28. For this reason, electrical power is distributed at extremely high voltages, up to 750 kV.

Impedance Level (5)

Figure 6.28 also suggests that the voltage regulation at the load (voltage relatively independent of load current) is improved by the use of a high-voltage transmission line. The load impedance appears with its true value, which shows that the same equivalent circuit would have resulted had we transformed all impedances to the load end of the circuit. Thus, the output impedance of the generator-transmission-line system is $\underline{Z}_g + \underline{Z}_w/n^2$. The effect of the transmission-line impedance is therefore reduced by the square of the turns ratio, to the improvement of load-voltage regulation.

EXAMPLE 6.18 Power distribution system

Power is generated at 24 kV, 100 miles from a town that uses 50 MW at 12 kV, as shown in Fig. 6.29. The transmission line has an impedance of $0.1 + j0.8 \, \Omega$/mile. What should be the transmission voltage for an efficiency of 98.5% for the transmission system?

SOLUTION:

The load current is $(50 \times 10^6)/(12 \times 10^3) = 4167$ A, and hence the transmission line current is

$$I = 4167 \times \frac{12}{V} \tag{6.105}$$

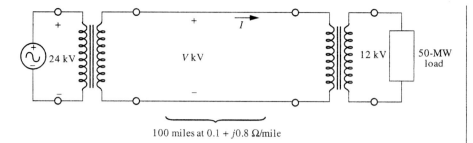

Figure 6.29 The transmission system is required to have 98.5% efficiency.

where V (in kV) is the line voltage to be determined. The line resistance is $0.1\,\Omega/\text{mile} \times 100$ miles $= 10\,\Omega$ and the allowed loss is 1.5% of the 50-MW load; hence,

$$\left(4167 \times \frac{12}{V}\right)^2 \times 10 = 0.015 \times 50 \times 10^6 \qquad (6.106)$$

Thus, the required voltage is 183 kV, and the transformer turns ratios are chosen accordingly.

WHAT IF? What if the line voltage is 138 kV? What is the efficiency?[29]

Multiple secondaries. Transformers may have multiple windings; for example, a transformer may have a 120-V primary and two secondaries, one with 12.6 V and one with 28 V. Such transformers are used in both power and electronic applications. Figure 6.30(a) shows a transformer with three windings, a primary with n_1 turns and secondaries with n_2 and n_3 turns, respectively, and appropriately defined voltage and current variables. The voltage/turn is constant for the transformer, so the voltages are related by a simple extension of Eq. (6.94)

$$\frac{v_1}{n_1} = \frac{v_2}{n_2} = \frac{v_3}{n_3} \qquad (6.107)$$

Thus, if we consider the primary voltage as given, we may calculate the secondary voltages based on the number of turns:

$$v_3 = \left(\frac{n_3}{n_1}\right) \times v_1 \quad \text{and} \quad v_2 = \left(\frac{n_2}{n_1}\right) \times v_1 \qquad (6.108)$$

However, Eq. (6.95) must be modified[30] to

$$n_1 i_1 = n_2 i_2 + n_3 i_3 \qquad (6.109)$$

---

[29] 97.4%.

[30] This equation is a form of Ampère's circuital law.

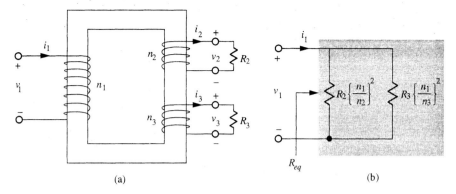

Figure 6.30 (a) The transformer has two secondaries. (b) Although the secondaries appear to be in series, the equivalent circuit shows them to be in parallel.

One application of Eqs. (6.108) and (6.109) is to determine the equivalent impedance looking into the primary, given that the two secondaries are loaded with resistors R_2 and R_3, respectively. If we divide Eq. (6.109) by the appropriate form of Eq. (6.108), we get the form

$$\frac{n_1 i_1}{v_1/n_1} = \frac{n_2 i_2}{v_2/n_2} + \frac{n_3 i_3}{v_3/n_3} \Rightarrow n_1^2 \times \frac{i_1}{v_1} = n_2^2 \times \frac{i_2}{v_2} + n_3^2 \times \frac{i_3}{v_3} \quad (6.110)$$

and because

$$R_{eq} = \frac{v_1}{i_1}, \qquad R_2 = \frac{v_2}{i_2}, \quad \text{and} \quad R_3 = \frac{v_3}{i_3} \quad (6.111)$$

Equation (6.110) becomes

$$\frac{1}{R_{eq}} = \left(\frac{n_2}{n_1}\right)^2 \times \frac{1}{R_2} + \left(\frac{n_3}{n_1}\right)^2 \times \frac{1}{R_3} \Rightarrow R_{eq} = \left(\frac{n_1}{n_2}\right)^2 \times R_2 \;\middle\|\; \left(\frac{n_1}{n_3}\right)^2 \times R_3 \quad (6.112)$$

In words, the primary sees the load impedances of both secondaries, transformed by the square of the turns ratios and connected in parallel. Figure 6.30(b) shows the equivalent circuit.

EXAMPLE 6.19 Two secondaries

A 16-Ω speaker and an 8-Ω speaker are driven off the secondaries of a stereo output transformer. Find the ratio $n_{16} : n_8$ such that equal power is given to both. Find the ratio $n_p : n_{16}$ to make the primary see 100 Ω.

SOLUTION:
To receive equal power,

$$\left(\frac{n_1}{n_{16}}\right)^2 \times 16 = \left(\frac{n_1}{n_8}\right)^2 \times 8 \Rightarrow \frac{n_{16}}{n_8} = \sqrt{\frac{16}{8}} = 1.414 \qquad (6.113)$$

For 100 Ω at the primary, each speaker must look like 200 Ω because they appear in parallel. Thus,

$$\left(\frac{n_1}{n_{16}}\right)^2 \times 16 = 200 \Rightarrow \frac{n_1}{n_{16}} = \sqrt{\frac{200}{16}} = 3.54 \qquad (6.114)$$

WHAT IF? What if the 8-Ω speaker were 4 Ω instead?[31]

Summary. Transformers are used for voltage, current, and impedance transformation. Applications abound in both electronics and power systems engineering. Realistic power systems, however, would use three-phase voltages for generation and distribution of power, as we discuss in the next chapter.

Residential AC Power

We do not recommend that you do it, but *if* you sought out the circuit breaker box[32] where you live, and *if* you removed the safety cover, you would discover three wires coming into the box from the transformer on the pole in the alley. One wire would be red, another black, and the third white.

Circuit for 120/240-V circuits. Figure 6.31(a) shows the usual arrangement at the transformer secondary. The white wire is the neutral and is grounded at the transformer

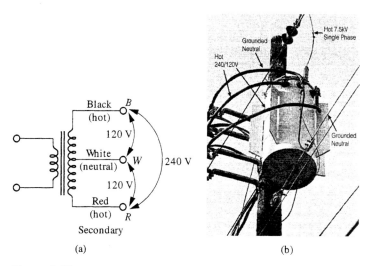

Figure 6.31 (a) A 120/240-V household power system; (b) pole transformer.

---

[31] $n_{16}/n_4 = 2$, $n_1/n_{16} =$ same.
[32] Or the fuse box in older structures.

by a wire that enters the moist earth and should also be grounded through the household plumbing. The black and red wires are "hot," each carrying 120 V relative to the neutral. The neutral voltage is at the midpoint between the two lines that carry 240 V, and hence 120 V is developed between each hot line and the neutral, which is grounded.

The 240-V power is used for heavy equipment such as air conditioners, electric stoves and clothes dryers, and certain power tools. Most appliances operate with 120 V, so the lighting and appliance circuits in the house are connected between a hot wire and neutral. The red color is not used in the household 120-V wiring: Black means hot and white neutral. Modern wiring codes require a third wire for a separate ground. The ground wires do not carry power like the neutral wire, but provide a separate ground connection independent of the power circuit.

EXAMPLE 6.20 Center-tapped transformer

A power transformer has a 7000-V primary and a center-tapped 240-V secondary. If the load on the 240-V circuit is 3000 W and the loads on each of the 120-V circuits are 1200 W, what is the current in the primary? Assume unity power factor for all loads, and consider the transformer ideal.

Conservation of Energy

SOLUTION:

Figure 6.32 shows the circuit. The "center-tap" on the secondary is a connection in the middle of the secondary winding, and is equivalent to Fig. 6.30(a) with $n_2 = n_3$ and the bottom of the top winding connected to the top of the bottom winding. The problem is most easily worked by conservation of energy. Because the total load on the secondary is 5400 W and the power factor is unity, the input current must be

$$7000 \times I = 5400 \Rightarrow I = 0.771 \text{ A} \tag{6.115}$$

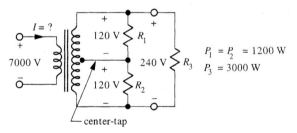

Figure 6.32 Center-tapped transformer circuit.

WHAT IF? What if the power factor of the 240-V load were 0.9, lagging, with the same power? What then would be the primary current?[33]

---

[33] Including now the reactive power of $Q = 1453$ VARs, the current is 0.799 A.

polarized

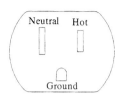

Figure 6.33 Grounded and polarized household outlet.

Grounding of appliances. Figure 6.33 shows a modern appliance outlet, which is both polarized and grounded with a three-wire system. The outlet and plug are said to be *polarized* because the hot and neutral connections differ in size. Some loads, such as table lamps, may work equally well, and be equally as safe, plugged in either way. Such a load would have an unpolarized plug, which would fit into the outlet either way. Other loads are equipped with polarized plugs that fit in only one way, thus controlling which wire connects to neutral. For safety, the neutral of the load (the male) cannot connect to the hot of the source because it is too large.

Home wiring. If you are wiring an appliance outlet (wall plug), pay close attention to polarity. Usually, the screw to which you should connect the hot (black) wire has a copper color and the screw to which you should connect the neutral (white) wire has a silver color. If you are installing a lighting fixture in a ceiling, the hot goes through the wall switch to the center of the receptacle and the neutral connects to the screw threads.

Electrical Safety

LEARNING OBJECTIVE 9.

To understand safe practice around electrical equipment

Although use of electrical power is vital to our modern way of life, the average citizen is poorly informed of the dangers of electrical power. One person might fear to touch the terminals of a 12-V auto battery, whereas another might think nothing of sticking a finger into a light socket to see if there is any power.[34] This section presents some basic information about electrical safety to enable you to recognize a dangerous situation and hopefully stay out of trouble.

The circuit theory of this subject is simple enough: Ohm's law is the key,

$$I = \frac{V}{R} \tag{6.116}$$

where R is you. Although we are accustomed to signs warning "DANGER, HIGH VOLTAGE," it is actually the current that affects our bodies. Many thousands of volts will do no more than startle, provided the current is small, as when we feel a small spark of static electricity after sliding across an auto seat. In the following, we speak first of the physiological effects of electric current on the human body. We then discuss body resistance and the resistance of typical surroundings. Finally, we offer some advice about electrical safety.

Physiological effects of current. Figure 6.34 shows the variety of effects that electrical current may have on the human body and the current levels at which they occur. Injury could be caused indirectly through being startled and losing muscular control or directly through burns. Death could result from suffocation, through loss of heart beat, or through severe burns. The large ranges in Fig. 6.34 represent variation between individuals in body size, condition, and tolerance to electrical shock. The region of the body through which the current passes is crucial; current passing through the lower part of the leg, for example, might be painful but would be unlikely to affect heart action.

---

[34] The battery is safe, but please keep your finger out of the light socket.

The surprising aspect of Fig. 6.34 is that small currents can have serious effects. This occurs because the communication system of the human body is electrical in nature and misbehaves under external electrical influence.

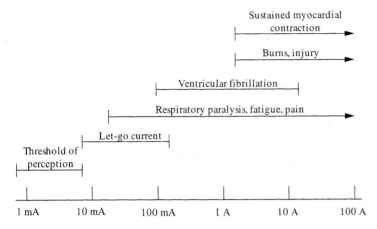

Figure 6.34 Physiological effects of electricity. (Adapted from John G. Webster, *Medical Instrumentation, Application and Design*, Copyright © 1978 Houghton Mifflin Company. Adapted with permission.)

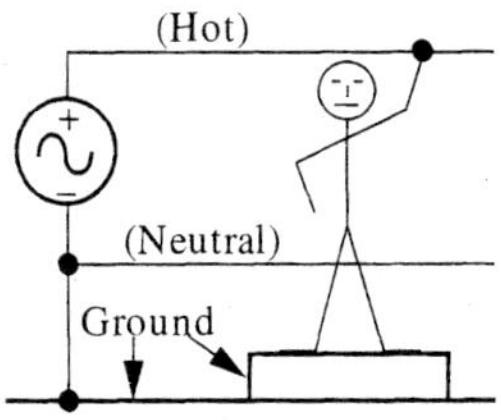

Figure 6.35 Most shocks occur between the hot wire and ground.

First aid for shock. When a person is experiencing electrical shock, time becomes an important factor, for the damaging effects are progressive. Hence, it is important to remove the source of electrical energy from a shock victim before EMS is called.[35] Particularly serious is the condition of ventricular fibrillation, where the heart loses its synchronized pumping action and circulation ceases. This condition is very dangerous because the heart may not resume normal action when the source of electrical power is removed; sophisticated medical equipment is required to restore coherent heart action.

Resistance. Earlier, we stated that "the resistance is you." This would be true if you came into simultaneous contact with both wires of an electrical circuit, say, by grabbing a wire with each hand, but most serious electrical shocks occur through the situation portrayed in Fig. 6.35. Here we have shown the victim in simultaneous contact with the hot wire and with "ground." Ground may be the moist earth, the plumbing of a house, or even a concrete floor that is in contact with the plumbing. In this case, the resistance in Eq. (6.116) includes not only the resistance of the body, but also the resistance of the shoes and the resistance between the shoes and earth ground.

Be aware of your surroundings. To assess the danger of a given situation, we must estimate the resistance of the "circuit" of which our body might become an unhappy part. Table 6.2 shows some basic information to allow such an estimate of the total resistance to ground.

---

[35] Cardiopulmonary resuscitation (CPR) should be administered only by trained personnel.

TABLE 6.2 Resistance for Safety Consideration

(a) For Various Skin-Contact Conditions

Condition (area to situ)	Resistance	
	Dry	Wet
Finger touch	40 kΩ–1 MΩ	4–15 kΩ
Hand-holding wire	15–50 kΩ	3–6 kΩ
Finger–thumb grasp	10–30 kΩ	2–5 kΩ
Hand-holding pliers	5–10 kΩ	1–3 kΩ
Palm touch	3–8 kΩ	1–2 kΩ
Hand around $1\frac{1}{2}$-in. pipe (or drill handle)	1–3 kΩ	0.1–1.5 kΩ
Two hands around $1\frac{1}{2}$-in. pipe	0.5–1.5 kΩ	250–750 Ω
Hand immersed		200–500 Ω
Foot immersed		100–300 Ω
Human body, internal, excluding skin = 200–1000 Ω		

(b) For Equal Areas (130 square cm) of Various Materials

Material	Resistance
Rubber gloves or soles	More than 20 MΩ
Dry concrete above grade	1–5 MΩ
Dry concrete on grade	0.2–1 MΩ
Leather sole, dry, including foot	0.1–0.5 MΩ
Leather sole, damp, including foot	5–20 kΩ
Wet concrete on grade	1–5 kΩ

Source: Adapted from Ralph Lee, "Electrical Safety in Industrial Plants," *IEEE Spectrum*, June 1971 © IEEE.

EXAMPLE 6.21

What would you experience if you were standing on moist ground with leather-soled shoes and you unwittingly grab hold of a 120-V wire?

SOLUTION:
Taking the lowest values in Table 6.2, we estimate the following resistances: 3 kΩ for the grasp, 200 Ω for the body, and 5 kΩ for the feet–shoes. Thus, the largest current you might carry would be (120 V)/(8.2 kΩ), about 15 mA. From Fig. 6.34 we judge that there is a fair chance that you will be unable to release your grasp and that you might hence be unable to breathe. This is, therefore, a dangerous situation.

WHAT IF?
What if you had on rubber-soled shoes?[36]

---

[36] No danger according to the chart. Never willingly touch a live circuit.

Skin resistance. One factor that Table 6.2 does not contain concerns the breakdown voltage of skin resistance. At approximately 700 V (ac), the resistance of the skin drops to near zero: in effect, the current burns a hole in the skin. Thus, the resistance of the skin, which might save your life for a lower voltage, becomes ineffective at such high voltages. For this reason, high voltages are seldom used in industrial applications, and 240 V is the highest voltage used in residential wiring.

An interesting, and alarming, calculation the reader might wish to perform is the following: Making the most pessimistic assumptions about body resistance, resistance to ground, and loss of skin resistance (say, cuts or blisters on the hands and feet), calculate the least voltage that might prove fatal. Such a calculation would have you standing in water or on a metal floor without shoes. Although we are describing an unusual situation, we still urge you to make the calculation.[37]

Precautions. Of more importance are the factors that increase your safety when working around electrical power. Make sure that the power is off before working on any electrical wiring or electrical equipment. Wear gloves and rubber-soled shoes. Avoid standing on a wet surface or on moist ground. Avoid working alone around exposed electric power.

floating

Grounding. From our discussion of safety, you might notice that, from one point of view, the danger increases because the electrical power system is grounded. If the circuit were *floating*, that is, not grounded, the only way to get shocked would be to come into contact with both wires simultaneously, an unlikely event. How can we reconcile this viewpoint with the common idea that electrical circuits are grounded for safety? Actually, both ideas are valid, and there is no contradiction; there is more involved in the issue than we have discussed thus far.

If the power system were floating. Suppose that the electrical power system in your building were floating. Everything would function correctly and safely until something went wrong with the equipment. But if, say, the wiring of the transformer on the pole became defective and a connection developed between primary and secondary, in this event, the 120-V circuits could float at 12,000 V or even 24,000 V. This would endanger virtually every piece of equipment on the line. Also, if the voltage suddenly were floated at 12,000 V, the danger to you becomes much greater. For this reason, the secondary of the transformer is grounded for protection of life and property. For if the secondary is grounded and a fault develops in the transformer, a large current flows immediately, a fuse or circuit breaker opens the circuit, and the source of power becomes disconnected from the offending part of the circuit.

Why equipment is grounded. Given that the power system in the building should be grounded, the necessity for grounding the equipment with the three-wire system in Fig. 6.31 becomes apparent. If a fault develops between the hot side of the power and the metal chassis of a piece of equipment, such as a washing machine, a fuse or circuit breaker will respond to the large current flowing through the ground connection and will remove power from the circuit containing the faulty connection. Hence, the safest way to install the power system involves grounding the power system

---

[37] A 28-volt dc power supply can do it.

and the equipment with which you might come into contact. But avoid being grounded yourself.

Ground fault interrupters (GFIs). The ground fault interrupter senses the current in the hot and neutral wires and opens the circuit when the current differs by about 5 mA. Thus, in Fig. 6.35, the most current that could pass a potential victim would be 5 mA, below the danger level. Modern building codes require GFIs be installed in bathrooms, and increasingly GFIs are built into appliances such as hair dryers.

Check Your Understanding

1. An ideal transformer has no losses. True or False?
2. In a transformer, does a large number of turns on a winding go with high or low voltage for that winding? What about current?
3. Transformers can be used to change voltage levels, current levels, impedance levels, or all three?
4. If we want to make a load impedance look smaller, which winding (primary or secondary) should have more turns?
5. To reduce losses in an electrical power transmission system, should the voltage level be raised or lowered?
6. If a larger voltage needs to be transformed to a smaller voltage, which side of the transformer should have the larger wire size?
7. What range of current through the human body is most likely to result in death, $i < 1$ mA, 100 mA $< i <$ 1 A, or 10 A $< i$?
8. In most electrocutions, current passes from the hot wire through the victim to the neutral wire or to earth ground?
9. When working on electrical devices, your body should be grounded for safety. True or False?

Answers. (**1**) True; (**2**) more turns go with the larger voltage and the smaller current; (**3**) all three; (**4**) secondary; (**5**) raised; (**6**) the low-voltage side; (**7**) 100 mA $< i <$ 1 A; (**8**) to earth ground; (**9**) false.

CHAPTER SUMMARY

Understanding of energy processes requires insight and wise application in any physical system. This chapter focuses on both theoretical and practical aspects of energy processes in ac circuits, with an emphasis on electric power engineering.

Objective 1: To understand how to calculate the time-average and effective values of periodic waveforms. The process of time averaging is explained and applied to the calculation of the effective or root-mean-square value of a sinusoidal waveform. The effective value gives the power-producing ability of a waveform and is what is indicated by ac meters.

Objective 2: To understand the energy and power requirements of resistors, inductors, and capacitors in the sinusoidal steady state. The behavior of an interconnected system depends upon the behavior of its individual components. We show that energy shuttles in and out of inductors and capacitors, with energy needed in diffcrent times in the ac cycle. By contrast, energy flows one way into resistors, in which electrical energy is transformed into nonelectrical form, usually heat.

Objective 3: To understand how to calculate apparent, real, and reactive power flow in an electric network based on time-domain analysis. We show that the energy flow into a general ac circuit containing resistors, inductors, and capacitors consists of two terms, the real power carrying energy to the resistors, and the reactive power representing energy lent periodically to the inductors and capacitors of the circuit. However, the inductors and capacitors share energy and only the imbalance between electric and magnetic energy needs must be lent to the circuit periodically by the source.

Objective 4: To understand how to calculate apparent, real, and reactive power in the frequency domain. The energy processes described before appear in the frequency-domain description of the circuit. The various types of ac power are represented in a power triangle that pictures real, reactive, and apparent power, plus the angle of the power factor.

Objective 5: To understand how to correct the power factor of a load. This section presents a practical aspect of electric power utilization that falls to the user. Because power companies require that the customer's power factor be high, capacitors are often added locally to an installation to balance the energy needs of motors and other magnetic devices.

Objective 6: To understand how to use a Thévenin equivalent circuit to establish the load to withdraw maximum power from an ac source. This section describes a technique used in electronic systems to make optimal use of the small amounts of available energy. To effect maximum energy transfer, a local resonance is created between source and load, and resistive values of source and load are made equal.

Objectives 7 and 8: To understand the voltage-, current-, and impedance-transforming properties of a transformer, especially their role in the transmission of electric power. Transformers are used both in power and electronic circuits to change voltage, current, and impedance levels. This section introduces the ideal transformer as a circuit element and explores the use of transformers in effecting maximum energy transfer in electronic circuits and in improving efficiency in power distribution systems.

Objective 9: To understand safe practice around electrical equipment. Finally, we describe the type of circuit used in domestic power distribution. The dangers of electricity are discussed and safe practice is described. Electrical equipment should be grounded, but personnel should be isolated from electrical ground as much as possible.

Chapter 7 builds on these concepts to investigate the distribution of electric power in three-phase systems and its utilization in electric motors.

GLOSSARY

Apparent power, p. 284, the product of the effective values of the voltage and current.

Average, p. 272, a number that characterizes a body of information in one particular way, omitting all the rest of the information.

Complex power, p. 291, a complex number with the real power its real part and reactive power its imaginary part.

DC value or component, p. 274, the time-average value of a signal.

Effective value of a time-varying waveform, p. 275, the equivalent dc value that would heat a resistor as hot as the time-varying waveform heats it. Also called the root-mean-square (rms) value.

Floating, p. 317, a circuit in which neither wire is grounded.

Ideal transformer, p. 302, a lossless transformer that stores no energy. Voltage and current are transformed reciprocally.

In-phase current, p. 288, the component of the current working in concert with the voltage to convey real power to a load.

Matching impedances, p. 299, when the load has the same resistance as the output impedance of the source but the opposite reactance, required for maximum power transfer.

Out-of-phase current, p. 289, the current required for reactive power flow.

Polarized plug, p. 314, an electrical plug where the hot and neutral connections differ in size.

Power factor, p. 284, the real power divided by the apparent power. The power factor is the cosine of the phase angle between the voltage and current.

Power factor correction, p. 295, placing a bank of capacitors in parallel with an inductive load to store energy locally.

Power triangle, p. 293, a right triangle formed by the real and reactive power as legs and the apparent power as hypotenuse.

Primary, p. 302, the input transformer coil connected to the ac source.

Reactive power, p. 288, the reciprocating flow of electrical energy to the reactive elements in a load.

Real power, p. 288, the time average of the instantaneous power for a load in sinusoidal steady state.

Root-mean-square (rms) value of a time-varying waveform, p. 277, the equivalent dc value that would heat a resistor as hot as the time-varying waveform heats it. Also called the effective value.

Secondary, p. 302, the output transformer coil supplying electrical power to a load.

Time average, p. 274, the dc value or dc component of the signal.

Turns ratio, p. 305, the ratio of primary to secondary turns in a transformer, hence, ratios of voltage and current transformation.

PROBLEMS

Section 6.1: AC Power and Energy Storage: The Time-Domain Picture

6.1. Compute the time average and the rms value for the waveforms shown in Fig. P6.1.

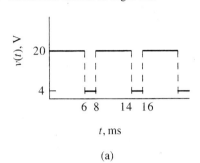

(a)

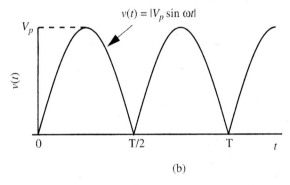

(b)

Figure P6.1

6.2. Show that ac and dc power are additive. That is, if $i(t) = I_{dc} + I_p \cos(\omega t)$, then the time average of $i^2(t)$ is $I_{dc}^2 + (I_p/\sqrt{2})^2$.

6.3. Figure P6.3 shows a periodic waveform.
  (a) Find the average value.
  (b) Find the effective value.

6.4. For the sawtooth waveform in Fig. P6.4, find the following:
  (a) The average value, V_{dc}.
  (b) The effective value, V_e.
  (c) What would be the average power if this were the voltage across a 10-Ω resistor?

6.5. (a) A current source having an rms value of 0.85 A is connected to a 12-Ω resistor. What is the power in the resistor?
  (b) If the current source in part (a) is sinusoidal with a period of 10 μs, what is the equation of the current as a function of time? (Assume the cosine form and phase = $+12°$.)

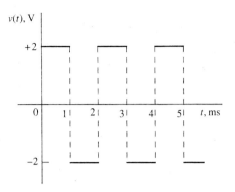

Figure P6.3

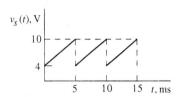

Figure P6.4

  (c) If the current in part (b) is put through an impedance of $12 \angle -60°$ Ω rather than a pure resistance, what is the time-average power now?

6.6. Figure P6.6 shows a periodic voltage waveform.
  (a) What is the time-average voltage?
  (b) If this voltage is applied to a 30-μF capacitor, what would be the effective value of the current in the capacitor?

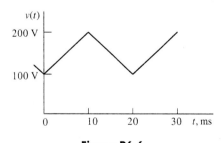

Figure P6.6

6.7. For the circuit in Fig. P6.7, find the following:
  (a) Minimum instantaneous power into R.
  (b) Time-average power into R.
  (c) Maximum instantaneous power into R.

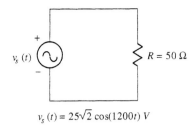

$v_s(t) = 25\sqrt{2}\cos(1200t)$ V

Figure P6.7

(d) Time between maxima in the instantaneous power into R.

6.8. For the circuit shown in Fig. P6.8, $v(t) = 240\cos(\omega t)$ V and $i(t) = 5.5\cos(\omega t)$ A. Find the following:
(a) Instantaneous power into the load at $t=0$.
(b) Time-average power into the load.
(c) Effective value of the voltage.
(d) Impedance of the load in polar form.

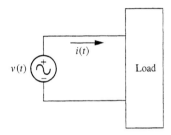

Figure P6.8

6.9. A circuit in sinusoidal steady state is shown in Fig. P6.9.
(a) Transform this circuit into the frequency domain.

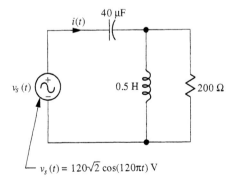

$v_s(t) = 120\sqrt{2}\cos(120\pi t)$ V

Figure P6.9

(b) What is the impedance, $\mathbf{Z}_{eq}$, seen by the voltage source?
(c) Find the time-domain current, $i(t)$.
(d) Find the time-average power given to the circuit by the voltage source.
(e) Find the peak electric energy stored in the capacitor.

6.10. A voltage source with $v_s(t) = 120\sqrt{2}\cos(250t)$ V is connected in series with a 100-Ω resistor, a 0.2-H inductor, and a 25-μF capacitor.
(a) Find the impedance of the circuit at the source frequency.
(b) Determine the sinusoidal steady state current, $i(t)$, in the series connection.
(c) Find the time-average power in the resistor.
(d) Determine the reactive power in VARs given to the circuit by the source.

6.11. A 120-V(rms), variable-frequency source is connected in series with a 100-Ω resistor and a 100-μF capacitor. Find the input power to the circuit from the source at the frequency where the power factor of the load is 0.75.

6.12. For the circuit shown in Fig. P6.12, $v(t) = 120\sqrt{2}\cos(120\pi t)$ V and $i(t) = 6.5\cos(120\pi t + 30°)$ A. Find the following:
(a) Effective current.
(b) Time-average power out of the source.
(c) Resistance, R.
(d) Peak stored energy in the capacitor.

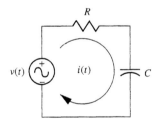

Figure P6.12

6.13. A 117-V rms, 60-Hz household circuit has two 75-W lights, $PF = 1$, and a fan using 500 VA at 0.78 PF (lagging).
(a) Draw the circuit, representing each load by its impedance. Include the switches for each load.
(b) Determine the total current required for all loads operating simultaneously.
(c) What capacitor in parallel with the loads will give unity power factor?

6.14. For the circuit shown in Fig. P6.14, find the following:
(a) How much time-average power flows one way between the source and load?
(b) How much energy is exchanged between the source and the load?
(c) How much energy is exchanged between the capacitor and the inductor?

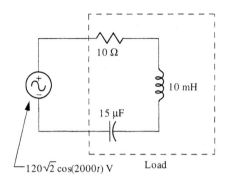

Figure P6.14

6.15. For the circuit shown in Fig. P6.15, find the following:
(a) How much time-average power flows one way between the source and load?
(b) How much energy is exchanged between the source and the load?
(c) How much energy is exchanged between the capacitor and the inductor?

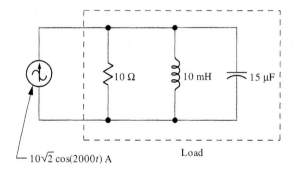

Figure P6.15

6.16. For the circuit in Fig. P6.16, find the following:
(a) Find $i(t)$. Use time- or frequency-domain techniques.
(b) What is the power factor for the load?

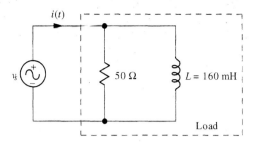

$v_s(t) = 100\sqrt{2}\cos(300\pi t)$ V

Figure P6.16

(c) Compute the time-average power into the load.
(d) Show that the time-average power into the load, P, is equal to the power dissipated in the resistor, P_R.
(e) Calculate the peak and time-average magnetic energy stored in L.

6.17. For the circuit shown in Fig. P6.17, find the following:
(a) Solve for $v(t)$. Let the phase of $i(t)$ be zero.
(b) Compute the time-average power into the circuit using the power factor.
(c) Show that the time-average power into the entire circuit, P, is equal to the time-average power dissipated in the resistor, P_R.
(d) Find the time-average electric stored energy.

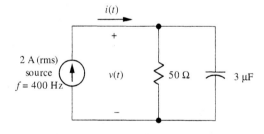

Figure P6.17

6.18. The circuit shown in Fig. P6.18 is in sinusoidal steady state.
(a) Find $v_a(t)$ using nodal analysis.
(b) What is the time-average power out of the left-hand source?

6.19. For the circuit shown in Fig. P6.19, find the time-average power out of the ac current source.

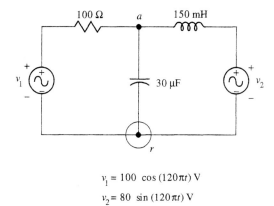

$v_1 = 100 \cos(120\pi t)$ V

$v_2 = 80 \sin(120\pi t)$ V

Figure P6.18

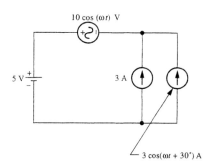

Figure P6.19

Section 6.2: Power and Energy in the Frequency Domain

6.20. For the circuit shown in Fig. P6.20,
$v(t) = 442\sqrt{2} \cos(120\pi t + 30°)$ V and
$i(t) = 21\sqrt{2} \cos(120\pi t + 60°)$ A.
 (a) What is the real power?
 (b) What is the reactive power?
 (c) What is the apparent power?
 (d) What is the impedance of the load in polar form?

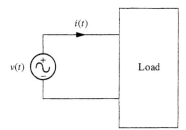

Figure P6.20

6.21. For the 60-Hz load shown in Fig. P6.21, the voltmeter measures 125 V, the ammeter measures 5.1 A, and the wattmeter measures 480 W. The load consists of a resistor and inductor. Find the following:
 (a) Power factor.
 (b) Leading or lagging?
 (c) Real power.
 (d) Apparent power.
 (e) Reactive power.
 (f) Average stored energy.
 (g) Draw a phasor diagram.

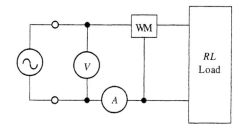

Figure P6.21

6.22. An impedance requires 80 V and 12 A as measured by standard ac meters. Figure P6.22 shows the phasor diagram. Find the following:
 (a) Apparent power (give units for all powers).
 (b) Power factor.
 (c) Real power.
 (d) Reactive power.
 (e) Impedance in rectangular form.

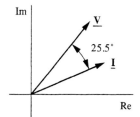

Figure P6.22

6.23. The complex power in a load is $1200 \angle 25°$ volt-amperes. Determine the following:

(a) Apparent power (give units).
(b) Real power (give units).
(c) Reactive power (give units).
(d) Power factor, leading or lagging.

6.24. For the circuit shown in Fig. P6.24, the voltage and current are $\mathbf{V} = 120\sqrt{2} \angle 0°$ and $\mathbf{I} = 10\sqrt{2} \angle +25°$.
  (a) Find the time-average power to the load.
  (b) Find the peak instantaneous power to the load.
  (c) Find the apparent power to the load.

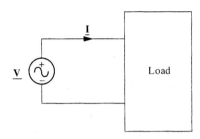

Figure P6.24

6.25. For the circuit shown in Fig. P6.25, find the following:
  (a) The complex power.
  (b) The real and reactive power, giving correct units.
  (c) The apparent power, giving correct units.
  (d) Draw a power triangle for this circuit.
  (e) Verify Eq. (6.44) for this circuit.

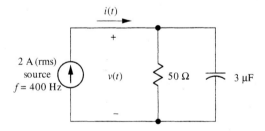

Figure P6.25

6.26. An electric motor is monitored with an ammeter, voltmeter, and wattmeter, which indicate 5.2 A, 120 V, and 480 W, respectively. Assume 60 Hz.
  (a) Draw a phasor diagram of the voltage and current, assuming the voltage at zero phase and lagging current.
  (b) What is the reactive power to the motor, including the units?

(c) To improve the power factor of the motor, a capacitor is hung directly across the motor terminals. What value of capacitance will give unity power factor?

6.27. A 460-V load draws 18 kVA at 0.82 PF, lagging. Find the kVAR of capacitance required to correct the power factor to 0.93, lagging.

6.28. A 60-Hz single-phase, $\frac{1}{2}$-hp motor in a washing machine (120 V) has an efficiency of 78% at full load (rated output power) and a power factor of 0.72, lagging. Find the line current, the reactive power, and the apparent power to the motor. Draw a phasor diagram.

6.29. For the circuit shown in Fig. P6.29, the wattmeter reads 2400 W, the ammeter reads 14.0 A, and the voltmeter reads 45 V.
  (a) What is R?
  (b) What would be the peak value of the source voltage, V_s?

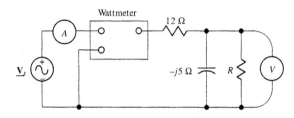

Figure P6.29

6.30. For the circuit shown in Fig. P6.30, the voltage and current are $v(t) = 110\sqrt{2} \cos(120\pi t + 32°)$ V and $i(t) = 4.8\sqrt{2} \cos(120\pi t - 21°)$ A.
  (a) Find the complex power into the load, $\frac{1}{2}\mathbf{V}\mathbf{I}^*$.
  (b) Draw a power triangle showing the numerical values of the real, apparent, and reactive powers.

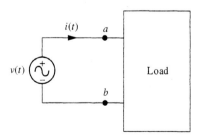

Figure P6.30

(c) Assuming that the load contains no electric energy storage, what is the peak value of the magnetic energy stored in the load?

(d) What value of capacitor connected between a and b will make the power factor as seen by the source to be 0.92, lagging?

6.31. For the circuit shown in Fig. P6.31, what should be the load impedance to draw maximum power from the circuit? Give component values, not just impedance values. Assume a series circuit for the load.

6.32. The output impedance of a typical 120-V appliance wall outlet might be $0.2 + j0.4\ \Omega$.
(a) What is the theoretical available power from the outlet?

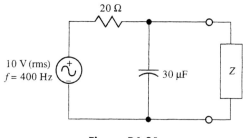

Figure P6.31

(b) Why would it be a bad idea to actually try to obtain that much power?

Section 6.3: Transformers

6.33. An ideal transformer has 120 turns on the primary and 240 turns on the secondary. Find the primary and secondary currents for the circuit shown in Fig. P6.33.

6.35. For the ideal transformer circuit shown in Fig. P6.35 a voltmeter measures 30 volts across the primary. What would the following measure?
(a) A voltmeter across the secondary.
(b) An ammeter in the primary.
(c) An ammeter in the secondary.
(d) A wattmeter in the primary.
(e) A wattmeter in the secondary.

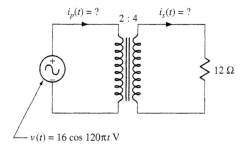

Figure P6.33

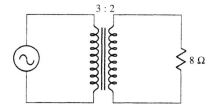

Figure P6.35

6.34. For the circuit in Fig. P6.34 containing an ideal transformer, find the following:
(a) Secondary voltage.
(b) Primary current.
(c) Secondary current.
(d) Power out of the source.

6.36. For the circuit in Fig. P6.36, containing an ideal transformer, find the following:
(a) Primary current.
(b) Load voltage.
(c) Power out of the source.

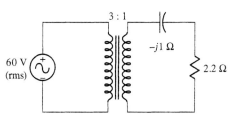

Figure P6.34

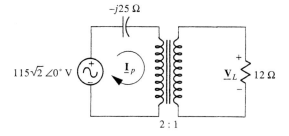

Figure P6.36

6.37. The circuit shown in Fig. P6.37 is already in the frequency domain.
(a) Find the rms value of the source voltage such that the power in the load is 1000 W.
(b) For this value of source voltage, what is the apparent power to the load?

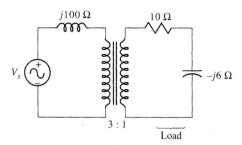

Figure P6.37

6.38. For the circuit shown in Fig. P6.38, find the following:
(a) The real power into the load, with units.
(b) The apparent power into the load, with units.
(c) If a capacitor were placed in series with the 10-Ω resistor, what value of capacitance would maximize the power to the load?

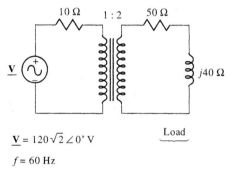

$\underline{V} = 120\sqrt{2}\angle 0°$ V

$f = 60$ Hz

Figure P6.38

6.39. The circuit shown in Fig. P6.39 is represented in the frequency domain. The source voltage is 120 V, rms.
(a) Find the phasor current in the primary of the transformer, considering the voltage source the phase reference.
(b) What is the effective voltage across the secondary of the transformer?

6.40. The circuit shown in Fig. P6.40 contains a step-up transformer, a transmission line with resistance of 0.5 Ω, a step-down transformer, and a load of 12 Ω.

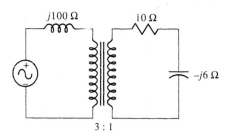

Figure P6.39

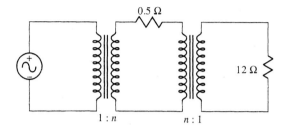

Figure P6.40

Find the value of n to make the efficiency of the transmission circuit 99%.

6.41. In the circuit in Fig. P6.41, the voltmeter measures 200 V. What would the following indicate?
(a) A voltmeter in the primary
(b) An ammeter in the secondary
(c) An ammeter in the primary
(d) A wattmeter in the secondary
(e) A wattmeter in the primary
(f) A dc ohmmeter connected between primary and secondary circuits

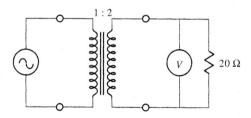

Figure P6.41

6.42. For the circuit shown in Fig. P6.42, find the turns ratio n_p/n_s such that the current in the secondary has a magnitude of 1 A (rms).

6.43. For the circuit shown in Fig. P6.43, find the turns ratio n_p/n_s such that the current in the secondary has a magnitude of 2 A (rms).

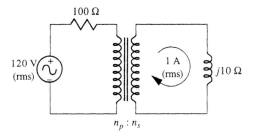

Figure P6.42

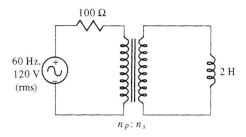

Figure P6.44

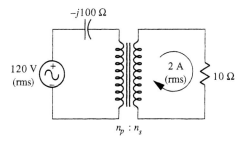

Figure P6.43

6.44. Figure P6.44 shows a circuit in which an inductor is coupled with a transformer. Find the number of secondary turns such that the power factor seen by the generator is 0.85, lagging. There are 1000 turns on the primary.

6.45. Power is generated at 480 V and utilized at 120 V. The generator is ideal, but the load is at some distance from the generator and hence line losses are appreciable, 1.2 Ω total for both wires. The equivalent load resistance would draw 15 kW at unity PF if 120 V were provided. Calculate the load voltage, the load power, and the efficiency of the transmission system under the following schemes:
(a) 4:1 transformer at the generator.
(b) 4:1 transformer at the load.
(c) 1:4 transformer at the generator and a 16:1 transformer at the load.

6.46. For the circuit shown in Fig. P6.46, the load impedance is fixed at $1000\,\Omega \parallel -j3000\,\Omega$, and the source output impedance is fixed at $10\,\Omega \parallel j10\,\Omega$. Maximum power transfer is to be achieved to the load with an ideal transformer with a turns ratio of $1:n$ and a "tuning capacitor" on the source side, represented by its reactance $+jX_C$ ($X_C < 0$). Find the values for n and X_C that achieve maximum power transfer.

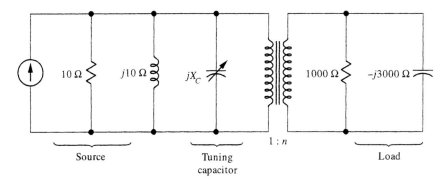

Figure P6.46

General Problems

6.47. With the circuit shown in Fig. P6.47, there is no value of the turns ratio, n, that will perfectly "match" the load to the source impedance, in the sense of eliminating all reactance and at the same time making the resistors match. However, there is still an optimum value of n that maximizes the power in the load. Find that value of n.

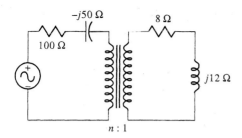

Figure P6.47

6.48. The circuit shown in Fig. P6.48 is the Thévenin equivalent circuit of a loop antenna operating at a frequency of 570 kHz. Assuming that the input circuit of the radio consists of a capacitor in parallel with a resistor, what values of R and C will extract maximum power from the antenna?

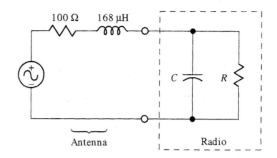

Figure P6.48

Answers to Odd-Numbered Problems

6.1. (a) 16.0, 17.4 V; (b) $2V_p/\pi$, $V_p^2/2$.
6.3. (a) 0 V; (b) 2 V.
6.5. (a) 8.67 W; (b) $0.85\sqrt{2}\cos(2\pi t/10^{-5} + 12°)$ A; (c) 4.34 W.
6.7. (a) 0 W; (b) 12.5 W; (c) 25.0 W; (d) 2.62 ms.
6.9. (a,b) 99.9 ∠ 19.6°; (c) $1.20\sqrt{2}\cos(120\pi t - 19.6°)$ V; (d) 136 W; (e) 0.254 J.
6.11. 81.0 W.
6.13. (a)

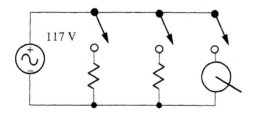

(b) $4.62 - j2.67$ A, rms; (c) 60.6 μF.
6.15. (a) 962 W; (b) 0.0962 J; (c) 0.144 J.
6.17. (a) $93.6\sqrt{2}\cos(800\pi t - 20.7°)$ V; (b) 175 W; (c) 175 W; (d) 0.0131 J.
6.19. −13.0 W.

6.21. (a) 0.753; (b) lagging; (c) 480 W; (d) 638 VA; (e) +420 VAR; (f) 0.557 J; (g) 41.2°.
6.23. (a) 1200 VA; (b) 1090 W; (c) +507 VAR; (d) 0.906, lagging.
6.25. (a) $175 - j66.0$ VA; (b) 175 W, −66.0 VAR; (c) 187 VA; (d) 0.0263 J electric energy ⇒ −66.0 VAR.
6.27. −4.47 kVAR ⇒ 56 μF.
6.29. (a) 42.2 Ω; (b) 268 V.
6.31. 6.11 Ω in series with 3.67 mH.
6.33. secondary $= 2.67\cos(120\pi t)$ A, primary $= 5.33\cos(120\pi t)$ A.
6.35. (a) 20 V; (b) 1.67 A; (c) 2.50 A; (d) 50.0 W; (e) same.
6.37. (a) 337 V; (b) 1170 VA.
6.39. (a) $1.19\sqrt{2}$ ∠ −27.1° A; (b) 41.5 V, rms.
6.41. (a) 100 V; (b) 10 A; (c) 20 A; (d) 2000 W; (e) 2000 W; (f) ∞ (no dc connection).
6.43. 5.74 and 1.74.
6.45. (a) 53.3 V, 2960 W, 44.4%; (b) 111 V, 12900 W, 92.8%; (c) 119.5 V, 14,900 W, 99.5%.
6.47. 2.78.

CHAPTER 7

Electric Power Systems

Three-Phase Power

Power Distribution Systems

Introduction to Electric Motors

Chapter Summary

Glossary

Problems

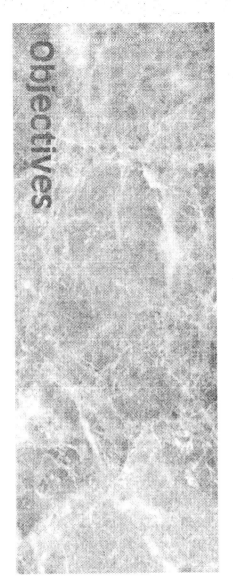

objectives

1. To understand the nature and advantages of three-phase power
2. To understand how to connect a generator in wye or delta
3. To understand how to calculate voltage, current, and power in balanced wye- and delta-connected loads
4. To understand how to derive and use the per-phase equivalent circuit of a balanced three-phase load
5. To understand how to connect single-phase transformers for three-phase transformation
6. To understand how transmission-line impedance affects power and voltage loss on the line
7. To understand how motor and load interact to establish steady-state operation
8. To understand how to analyze and interpret nameplate information of induction motors

This chapter applies the concepts of Chaps. 4 and 6 to power distribution systems and electric motors. Three-phase circuits are introduced, including three-phase transformers. The final section on electric motors examines motor/load interaction and shows how to analyze the nameplate information on the two most common types of electric motors.

7.1 THREE-PHASE POWER

Importance of Electric Power Systems

In Chaps. 4 and 6, we assumed that ac power is readily available as a sinusoidal voltage source with negligible output impedance. And this is what we assume in real life also. Our civilization is so dependent on reliable electric power that when the power fails, life seems to stop.

The generation and distribution of electric power is the business of large electric utility companies. Various fuels and sources are used, such as coal, natural gas, fuel oil, nuclear, and water power, not to mention some quaint and futuristic schemes such as wind power, tidal and wave power, solar power, and burning household garbage.

The normal way to generate[1] electrical power is to burn the primary fuel and produce high-pressure steam in a boiler. The steam drives a turbine that turns an electrical generator. The voltage from the generator is transformed to high voltage for transmission over long distances, as explained in Chap. 6. For reliability, all the generators in a geographic region are synchronized and are linked together with transmission lines to exchange real and reactive power.

Near the location of the industrial or residential customers, the voltage is lowered from transmission levels, typically, 120 – 500 kV, to distribution levels, typically 4.8 – 34 kV. For the residential consumer the voltage is lowered to 120/240 V, single-phase.

The power is generated, transmitted, and distributed in three-phase form. Only very near to the customer is the power changed from three-phase to single-phase power. This chapter begins with the introduction and study of three-phase electric systems. We then discuss three-phase transmission and distribution systems. Finally, we complete our study of electric power with an introduction to electric motors.

Introduction to Three-Phase Power Systems

Importance of three-phase systems. If you looked out your window at this moment, you would probably see some power lines. Count the wires and you will likely find there to be four. Go examine a pole closely and you will see that at each pole, one of the four wires is connected to a conductor that comes down the pole and enters the ground.

When you are driving cross country and see a large electrical transmission line, you will again see four wires. One of them, running along the top of the towers, will be noticeably smaller than the other three. If you look closely, you will again see that the small wire is grounded at every tower.

The three ungrounded wires in these transmission systems are driven by three ac generators. The grounded wire increases safety and protection from lightning. Power is conveyed by the three larger wires in the form of three-phase electric power. The overwhelming majority of the world's electric power is generated and distributed as three-phase power. For example, if you were to examine a catalog of industrial-grade motors, you will discover that all the larger electric motors, bigger than a few horsepower, would be three-phase motors.

---

[1] "Convert" would be a better word, because the energy is normally converted from chemical energy to heat, and finally to electrical energy.

LEARNING OBJECTIVE 1.

To understand the nature and advantages of three-phase power

IDEA 7. The Frequency Domain

What is three-phase power? Physically, there are three wires that carry the power, and often a fourth wire, called the *neutral*, which is grounded. In enclosed cables, the active wires are normally colored red, black, and blue; and the neutral, if present, is white or gray. The phases are traditionally designated A, B, and C, and the time-domain voltages between them are as shown in Fig. 7.1. The voltages are expressed mathematically as

$$v_{AB}(t) = V_p \cos(\omega t)$$
$$v_{BC}(t) = V_p \cos(\omega t - 120°) \quad (7.1)$$
$$v_{CA}(t) = V_p \cos(\omega t - 240°)$$

The frequency-domain picture for a three-phase system is shown in Fig. 7.2. We have used $\underline{V}_{AB}$ as our phase angle[2] reference and shown $\underline{V}_{BC}$ following by 120°, then $\underline{V}_{CA}$. This is known as an *ABC* phase sequence and corresponds to the time-domain representation in Fig. 7.1 and Eqs. (7.1).

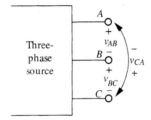

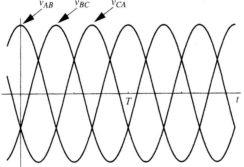

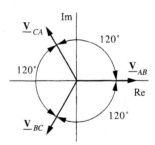

Figure 7.1 Voltages of a three-phase system in the time domain.

Figure 7.2 Voltages of a three-phase system in the frequency domain.

Advantages of three-phase systems. We have shown in Fig. 6.6 that single-phase power produces a pulsating flow of energy. A *smooth* flow of energy from source to load is achieved by a balanced three-phase system. If we have identical resistive (R)

---

[2] We apologize for using "phase" to mean two separate concepts in this section, but this vocabulary is standard. The three voltages have different phases for their sources, and from this property, the sources themselves came to be called "phases." So "phase" means the phase angle of a sinusoidal quantity as well as the source of that sinusoid. In this section, we use "phase angle" to refer to the phase of a sinusoid.

loads connected between the three phases, the instantaneous flow of power would be given by Eq. (7.2). We use the trigonometric identity $\cos^2 \alpha = [1 + \cos(2\alpha)]/2$ to derive the second and third forms.

$$\begin{aligned} p(t) &= \frac{v_{AB}^2(t)}{R} + \frac{v_{BC}^2(t)}{R} + \frac{v_{CA}^2(t)}{R} \\ &= \frac{V_p^2}{2R}[1 + \cos 2(\omega t) + 1 + \cos 2(\omega t - 120°) + 1 + \cos 2(\omega t - 240°)] \\ &= \frac{3V_p^2}{2R} + \frac{V_p^2}{2R} \underbrace{[\cos(2\omega t) + \cos(2\omega t - 240°) + \cos(2\omega t - 480°)]}_{\text{add to zero at all times}} \end{aligned} \qquad (7.2)$$

We see a constant term and a term that appears to be time-varying at twice the source frequency. Actually, *the second term adds to zero at all times*. This is easily shown by a phasor diagram; indeed, the phasors representing these terms give the same phasor diagram as Fig. 7.2.[3] Clearly, the phasor sum of the three symmetrical phasors is zero, and hence the fluctuating power term is also zero at all times. Thus, Eq. (7.2) reduces to

$$p(t) = \frac{3V_p^2}{2R} \quad \text{(a constant)} \qquad (7.3)$$

This constant flow of energy effects general smoothness of operation in three-phase electrical equipment. A rough analogy is suggested by comparing an engine having one cylinder with an engine having many cylinders—clearly, the multicylinder engine runs smoother.

Compared with a single-phase system, distribution losses are proportionally less for a three-phase system. Additionally, three-phase motors offer advantages over single-phase motors in both startup and run characteristics. In short, three-phase systems are supremely important for the generation, distribution, and use of electrical power, particularly in industrial settings.

Three-Phase Power Sources

LEARNING OBJECTIVE 2.

To understand how to connect a generator in wye or delta

Three single-phase sources. Three-phase generators produce three single-phase voltages with the required 120° phase-angle shifts, which are internally connected to produce a three-phase source. In this section, we pretend that the three single-phase voltages are brought out of the generator to a terminal board,[4] and our job is to connect the resulting six terminals together to produce a three-wire, three-phase source. The terminal board is shown in Fig. 7.3, and the phasor diagram of the available voltages is shown in Fig. 7.4.

terminal

Delta (Δ) and wye (Y) connections. The symmetry of the desired phasor voltages suggests that we require some sort of symmetrical connection for the three volt-

---

[3] A phase angle of −480° is the same as −120°.
[4] A *terminal* is the end of a wire from an electrical device. A *terminal board* is a place where terminals are made available for connection.

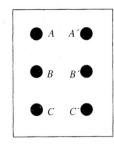

Figure 7.3 External terminals for the three separate phases. These must be connected to produce a three-phase system.

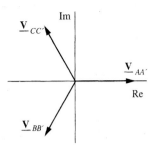

Figure 7.4 Phasor voltages in the three coils. These must be connected externally to make a true three-phase system.

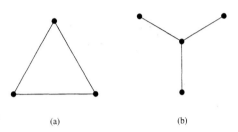

(a) (b)

Figure 7.5 The only two symmetrical configurations of three elements: (a) delta; (b) wye.

ages represented in Fig. 7.4. Only two symmetrical configurations exist for three elements connected end to end, and these are shown in Fig. 7.5. The closed ring is usually called a *delta* (for the Greek letter Δ), even when it is drawn upside down or on its side. The configuration with a common point is sometimes called a star configuration, but in three-phase terminology, it is more often called a *wye configuration* (a phonetic spelling of the letter Y), regardless of orientation.

These symmetric configurations suggest two solutions of the question posed earlier. We require three terminals for a source of three-phase power. The delta has three terminals, and hence the three-phase outputs connect these terminals. The wye, on the other hand, has four terminals, counting the common connection in the center. This gives us a place to connect the fourth wire mentioned earlier, the neutral wire that is grounded. We have to connect the terminals to have the geometric symmetry of the delta or wye configurations in Fig. 7.5, but we must also retain the electrical symmetry in Fig. 7.2.

delta connection

Delta connection. The delta ties the three generators in a closed ring. Figures 7.6(a) and 7.6(b) show one possible connection for the delta. We must be careful, however, when we close the ring, for the voltage must be small to avoid a large circulating current. Closing the delta is like jump starting a car having a weak battery, as shown in Fig. 7.6(c). The circuit can be closed if the polarities are correct. With the connection marked "Yes," there will be at most a small voltage across the gap and only small currents will flow through the batteries. But if the polarities are wrong, there will be approximately 24 V across the gap and a huge current will flow if the connection is made.

Similarly, if we are to close the ring of generators in Fig. 7.6(b), we require the voltage across the gap to be small. This requires that

$$\underline{V}_{AC'} = \underline{V}_{AA'} + \underline{V}_{BB'} + \underline{V}_{CC'} = 0 \qquad (7.4)$$

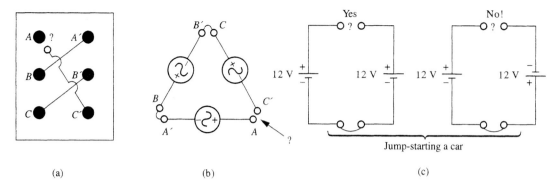

Figure 7.6 (a) Potential delta connection; (b) the voltage across A and C' must be zero if the ring is to be closed; (c) the polarity must correct before closing the circuit.

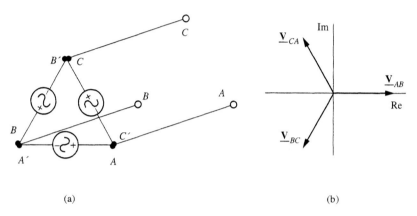

Figure 7.7 Delta connection with phasor diagram.

Notice that Eq. (7.4) follows from the rule for adding subscripted voltages given on page 18 because A' is connected to B and B' is connected to C. The general rule is

$$V_{AB} = V_{AX} + V_{XB} \tag{7.5}$$

We can extend this to $V_{AB} = V_{AX} + V_{YB}$ if X and Y are connected. The phasor diagram for the sum in Eq. (7.4) is easily derived from Fig. 7.4; clearly, the sum is small, ideally zero. Thus, it is safe to close the ring of generators and bring out the connected terminals as a three-phase source. Figure 7.7(b) shows the phasor diagram of the final connection in Fig. 7.7(a). Another possible delta connection would result with A connected to B', B connected to C', and C connected to A'. We leave the investigation of this possibility to the reader.

wye connection

Wye connection. A wye connection results from connecting A', B', and C', as shown in Fig. 7.8(a), with three wires for the three-phase power (A, B, C) and a common point ($A'B'C'$) for a neutral. With this connection, the magnitudes of $\underline{V}_{AC}$, $\underline{V}_{BA}$, and $\underline{V}_{CB}$ are $\sqrt{3}$ greater than those of the component voltages, $\underline{V}_{AA'}$, $\underline{V}_{BB'}$, and $\underline{V}_{CC'}$, and the phase angle of $\underline{V}_{AC}$ lies at $-30°$ relative to $\underline{V}_{AA'}$. These combinations are illus-

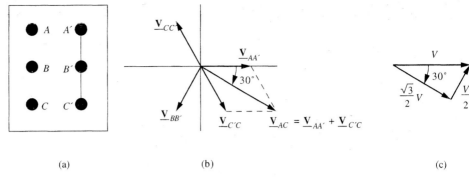

Figure 7.8 Possible wye connection.

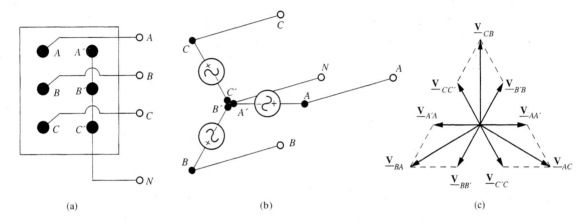

Figure 7.9 Wye connection. The line voltages are $\sqrt{3}$ larger than the phase voltages.

trated in Fig. 7.8(b) by the forming of $\underline{V}_{AC}$ through addition of $\underline{V}_{AA'}$ and $\underline{V}_{C'C}$, which is the negative of $\underline{V}_{CC'}$. The $\sqrt{3}$ comes from the 30° right triangle, as shown in Fig. 7.8(c). In Figs. 7.9(a) and 7.9(b), we label the three lines A, B, and C, and the neutral N. Figure 7.9(c) shows the final phasor diagram of the wye connection. The wye connection leads to the four-wire system that we described at the beginning of this section.

three-phase voltage, three-phase current, line voltage, line current

Three-phase voltage and current. The *three-phase voltage*, or *line voltage*, is the voltage between the lines carrying the power. The *three-phase currents*, or *line currents* are the currents in the three lines. For a balanced system, these quantities are simply the voltage and current measured by a voltmeter between any two lines and an ammeter in any of the lines. Hence, a 460-V three-phase system has 460 V (rms) between any two of the three lines carrying the power.

Line and phase voltage and current for delta and wye connections. A three-phase generator consists of three interconnected single-phase generators,[5] as

---

[5] Actually, there is only one generator, but it has three single-phase *windings* that are interconnected. Here we call these "generators" to indicate single-phase voltage sources.

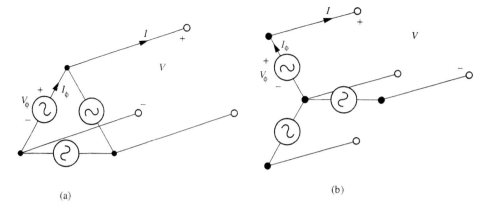

Figure 7.10 Three-phase (line) and phase voltage and current for a (a) delta- and (b) wye-connected three-phase system.

phase voltage, phase current

shown in Figs. 7.7 and 7.9. The *phase voltage*, V_ϕ, and *current*, I_ϕ, are the voltage across an individual generator and the current through that generator. For a delta connection, the phase voltage is the three-phase, or line voltage; but for a wye connection, the phase voltage is the three-phase voltage divided by $\sqrt{3}$, as shown in Fig. 7.9(c). The phase current is the same as the three-phase, or line current for a wye connection, but for the delta connection, the phase current is the three-phase current divided by $\sqrt{3}$. The phase and line voltage and current are shown in Fig. 7.10 and summarized in Table 7.1.

TABLE 7.1 Relationships between Phase and Three-Phase Voltages for Wye and Delta Connections

	Three-Phase Voltage, V	Three-Phase Current, I
Wye	$V_\phi = V/\sqrt{3}$	$I_\phi = I$
Delta	$V_\phi = V$	$I_\phi = I/\sqrt{3}$

EXAMPLE 7.1 Three-phase sources

Three single-phase sources are wye-connected as a three-phase source that measures 208 V for the three-phase voltage. What would be the three-phase voltage if the generators were delta-connected?

SOLUTION:
The phase, or line-to-neutral, voltage of the wye-connected system is $208/\sqrt{3} = 120$ V. Thus, the delta connection would give a three-phase voltage of 120 V.

> **WHAT IF?** What if one of the three phases in the delta is reversed, but the ring is not closed? What would be the voltages between the four sets of terminals?[6]

Phase rotation. The phase rotation came out *ABC* in both systems we developed, but *ACB* is also a possibility. The physical phase rotation is very important; for example, the rotational direction of a three-phase motor depends on the phase rotation of the input voltages. In practice, the three wires are arbitrarily labeled *A*, *B*, and *C*, and one of several techniques is used to determine whether the phase rotation is *ABC* or *ACB*.

Other possible connections. We have now shown the two ways for connecting the voltages in Fig. 7.3 to give a three-wire, symmetrical power system. Actually, we have shown one version of each way, for there is an alternative delta or wye. For example, we can make *A*, *B*, and *C* the neutral for a wye.

Three-Phase Loads

LEARNING OBJECTIVE 3.
To understand how to calculate voltage, current, and power in balanced wye- and delta-connected loads

Delta-connected resistors. Like three-phase generators, three-phase loads can be connected in delta or wye. Figure 7.11 shows a balanced three-phase resistive load connected in delta. The source of the three-phase power is not shown; we assume *ABC* phase rotation and a three-phase voltage $V = |\underline{V}_{AB}| = |\underline{V}_{BC}| = |\underline{V}_{CA}|$. We show no neutral because the load offers no place for connecting a neutral. In practice, however, a delta-connected load, such as a three-phase motor, is housed in a physical structure that is normally grounded directly to earth ground and through a neutral. However, the motor circuit would be floating.

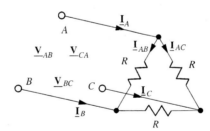

Figure 7.11 Delta-connected load.

Conservation of Charge

With a load connected in delta, we must distinguish between the line currents and the phase currents flowing in the resistors. Figure 7.12 shows the phase currents, $\underline{I}_{AB}$, $\underline{I}_{BC}$, and $\underline{I}_{CA}$. These currents have the same phase angle as their corresponding line voltages. Any line current, say, $\underline{I}_A$, can be determined by phasor addition of the phase currents. Kirchhoff's current law at the top node is

$$\underline{I}_A = \underline{I}_{AB} + \underline{I}_{AC} \tag{7.6}$$

---

[6] 120, 120, 120, and 240 V.

but

$$\mathbf{I}_{AC} = -\mathbf{I}_{CA} \qquad (7.7)$$

and hence the currents add as in Fig. 7.12. The other line currents could be determined similarly; indeed, the picture develops like that of the wye generator connection in Fig. 7.9(c). We see that the line currents are $\sqrt{3}$ greater than the phase currents. Thus, the second row of Table 7.1 is valid for delta-connected loads as well as sources. For the resistive load, the phase angle of the line current in A has the same phase angle as the average of the phase angles of $\mathbf{V}_{AB}$ and $\mathbf{V}_{AC}$. The phase-angle relationships for a resistive load are shown in Fig. 7.13.

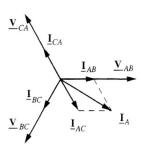

Figure 7.12 Phase current addition to yield line current.

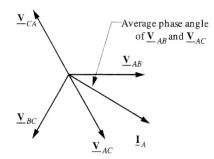

Figure 7.13 For a resistive load, the line current is in phase angle with the average phase angle of the voltages to the other two lines.

EXAMPLE 7.2 Delta-connected load

A delta-connected resistive load has a line voltage of 460 V and a line current of 32 A. Find the resistance values.

SOLUTION:

The phase voltage is $V_\phi = 460$ V and the phase current is $I_\phi = 32/\sqrt{3} = 18.5$ A. Thus, the phase impedance is

$$|\mathbf{Z}_\phi| = \frac{V_\phi}{I_\phi} = \frac{460}{18.5} = 24.9 \ \Omega \qquad (7.8)$$

WHAT IF? What if one of the resistors were missing? What would be the line currents?[7]

---

[7] 32, 18.5, and 18.5 A.

General loads in delta. With balanced loads that are not resistive, the phasor diagram shown in Fig. 7.12 changes only slightly. The magnitude of the phase currents is computed by dividing the line voltages, which are also the phase voltages, by the magnitude of the phase impedance. The phase currents will lead or lag the line voltages according to the angle of the phase impedance. Hence, the line currents will also be shifted in phase angle by the angle of the impedance. We give an example later.

Power in delta connections. The total power to a load, $P_{3\phi}$, is the sum of the powers delivered to the three phase impedances; and this would be

$$P_{3\phi} = 3\,P_\phi = 3\,V_\phi I_\phi \times PF \tag{7.9}$$

where PF is the power factor of the phase impedance. In Eq. (7.9), V_ϕ and I_ϕ represent the rms values of the phase voltage and current. We desire, however, to express the total power in terms of line voltage and current. The phase currents are often inaccessible for measurements, but the line voltage and current always can be measured. Consequently, we introduce the line voltage and current:

$$P = 3V \frac{I}{\sqrt{3}} \times PF = \sqrt{3}\,VI \times PF \tag{7.10}$$

where V and I are the rms line voltage and current, respectively. In the application of Eq. (7.10), the power factor is the cosine of the angle of the phase impedance and is *not* the phase angle between line current and line voltage. The power-factor angle is, however, the angle between the phase angle of the line current and the average phase angle of the voltages to the two other lines, as illustrated in Fig. 7.13.

EXAMPLE 7.3

Delta-connected RL impedances

A 230-V three-phase power system supplies 2000 W to a delta-connected balanced load with a power factor of 0.9, lagging. Determine the line currents, the phase currents, and the phase impedance. Draw a phasor diagram.

SOLUTION:
First, we calculate the magnitude of the line currents from Eq. (7.10).

$$P = \sqrt{3}\,VI \times PF \Rightarrow I = \frac{2000}{\sqrt{3}(230)(0.9)} = 5.58 \text{ A (rms)} \tag{7.11}$$

The phase currents are smaller by $\sqrt{3}$, so

$$I_\phi = \frac{I}{\sqrt{3}} = \frac{5.58}{\sqrt{3}} = 3.22 \text{ A} \tag{7.12}$$

This allows us to calculate the impedance in each phase of the delta. The angle of the impedance is implied by the power factor: $\theta = \cos^{-1}(0.9) = +25.8°$, + because the current is lagging (inductive).

$$Z_\phi = \frac{V_\phi}{I_\phi} \angle \cos^{-1}(PF) = \frac{230}{3.22} \angle \cos^{-1}(0.9) = 71.4 \angle +25.8° \ \Omega \tag{7.13}$$

We can now draw the phasor diagram, Fig. 7.14. We use $\underline{V}_{AB}$ for the phase-angle reference, with the other line voltages placed symmetrically in ABC sequence. The phase currents lag by 25.8°, as shown. The line currents can be computed by phasor addition of the phase currents, as we did in Fig. 7.12, but another approach is to use our earlier results to place the line currents behind the phase currents by 30° and greater by $\sqrt{3}$. Whichever way is chosen, only one line current need be determined, $\underline{I}_A$ for example, and the other two can be constructed by symmetry.

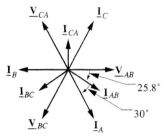

Figure 7.14 Line voltages, phase currents, and line currents. The phase currents are $\sqrt{3}$ smaller than the line currents.

WHAT IF? What if the phase impedances are $\underline{Z}_\phi = 60 \angle -20°$? Find the new line current and power.[8]

Wye-connected loads. Figure 7.15 shows a load connected in wye.[9] We have shown no connection to the neutral, labeled N, but there would often be a connection between the load neutral and the source neutral, if such existed, and the neutral is often grounded. For a perfectly balanced load, no current would flow in the neutral connection because the three line currents add to zero. We now calculate the line-to-neutral voltages, $\underline{V}_{AN}$, $\underline{V}_{BN}$, and $\underline{V}_{CN}$, and the line currents, $\underline{I}_A$, $\underline{I}_B$, and $\underline{I}_C$.

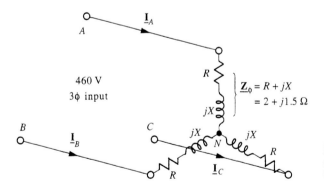

Figure 7.15 Wye-connected load. To find the line currents, we must determine the line-to-neutral voltages.

---

[8] 6.64 A and 2485 W.
[9] On the input voltage, "3ϕ" is a common abbreviation for "three phase."

Line-to-neutral voltages. The phase impedances are given, $\underline{Z}_\phi$; hence, it is clear that the line currents can be determined once the line-to-neutral voltages are known; for example,

$$\underline{I}_A = \frac{\underline{V}_{AN}}{\underline{Z}_\phi} \qquad (7.14)$$

The line-to-neutral voltages can be determined from consideration of the symmetry of the circuit. It is convenient to pretend that we know the line-to-neutral voltages (which we do not) and determine from them the line-to-line voltages. The relationship between these two sets of three-phase voltages then becomes known and we henceforth can deduce either set of voltages from the other. We assume the *ABC* phase sequence; hence, the line-to-neutral voltages must appear as in Fig. 7.16, assuming that we make $\underline{V}_{AN}$ the phase reference.

First, we determine $\underline{V}_{AB}$. We can express $\underline{V}_{AB}$ in terms of $\underline{V}_{AN}$ and $\underline{V}_{NB}$:

$$\underline{V}_{AB} = \underline{V}_{AN} + \underline{V}_{NB} \qquad (7.15)$$

Equation (7.15) becomes more useful when we reverse the subscripts on $\underline{V}_{NB}$ and change the sign:

$$\underline{V}_{BN} = -\underline{V}_{NB} \;\Rightarrow\; \underline{V}_{AB} = \underline{V}_{AN} - \underline{V}_{BN} \qquad (7.16)$$

Equation (7.16) is represented in Fig. 7.16, with the negative of $\underline{V}_{BN}$ drawn and added to $\underline{V}_{AN}$. We note that $\underline{V}_{AB}$ leads $\underline{V}_{AN}$ by 30° and is somewhat greater in magnitude. The phasor addition is identical to that shown in Fig. 7.8(b) and the magnitudes of phase and line voltages have the ratio $\sqrt{3}$, just as the currents do in the delta-connected load. The remaining line voltages, $\underline{V}_{BC}$ and $\underline{V}_{CA}$, may be determined by similar reasoning, or more directly by arranging them in *ABC* sequence, each 120° from $\underline{V}_{AB}$.

Figure 7.17 shows the results. The line-to-neutral voltages lag the corresponding line voltages by 30°, when we consider the two voltages with, say, *A* written first, like $\underline{V}_{AB}$ and $\underline{V}_{AN}$. But a better way to think about the phase angle is to realize that the phase angle of the corresponding line-to-neutral voltage lies between the phase angles of the two line-to-line voltages that connect to the same point. Thus, $\underline{V}_{AN}$ will lie halfway between $\underline{V}_{AB}$ and $\underline{V}_{AC}$. This phase-angle relation, together with the magnitude ra-

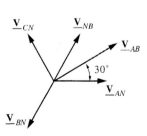

Figure 7.16 Determining line-to-line voltages from line-to-neutral voltages.

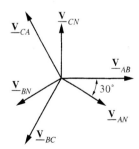

Figure 7.17 The line-to-neutral voltages are smaller by $\sqrt{3}$ and lag line-to-line voltages by 30°.

tio of $1/\sqrt{3}$, allows us to determine easily the line-to-neutral voltages from the set of line-to-line voltages, or vice versa, as shown in Fig. 7.17.

EXAMPLE 7.4 Wye-connected load

Find the line and phase current and line and phase voltage for the wye-connected load shown in Fig. 7.15.

SOLUTION:
We make $\underline{\mathbf{V}}_{AB}$ the phase-angle reference, that is, $\underline{\mathbf{V}}_{AB} = 460\sqrt{2} \angle 0°$ V. Hence, the line voltage is 460 V (rms). From Fig. 7.17, we see that $\underline{\mathbf{V}}_{AN}$ will lie at $-30°$ and be smaller by $\sqrt{3}$, or $\underline{\mathbf{V}}_{AN} = 266\sqrt{2} \angle -30°$ V; so the phase voltage is 266 V (rms). Hence, the current in line A is

$$\underline{\mathbf{I}}_A = \frac{\underline{\mathbf{V}}_{AN}}{\underline{\mathbf{Z}}_\phi} = \frac{266\sqrt{2} \angle -30°}{2 + j1.5} = \frac{266\sqrt{2} \angle -30°}{2.50 \angle 36.9°} = 106.2\sqrt{2} \angle -66.9° \text{ A} \qquad (7.17)$$

The other line currents can be determined similarly or by symmetry from $\underline{\mathbf{I}}_A$. Thus, the phase and line currents, which are the same for the wye-connected load, are 106.2 A.

Power in wye-connected loads. The total power to the wye-connected load is three times the power to each phase of the load, P_ϕ. Thus, we can compute the total power with

$$P_{3\phi} = 3 P_\phi = 3 V_\phi I_\phi \times PF \qquad (7.18)$$

where $P_{3\phi}$ is the power in the load, V_ϕ is the phase rms voltage, the line-to-neutral voltage in this instance, I_ϕ is the phase rms current, also the line current in this instance, and PF is the power factor of the phase impedance.

EXAMPLE 7.5 Power in wye-connected load

Find the total power to the three-phase load in the previous example.

SOLUTION:
The phase voltage and current were determined to be 266 V and 106.2 A, respectively. The power factor is the angle of the phase impedance, $2 + j1.5 = 2.50 \angle 36.9°$; so the power factor is $\cos 36.9°$, lagging. Using Eq. (7.18), we find the power in the load to be

$$P_{3\phi} = 3(266)(106.2)(\cos 36.9°) = 67.7 \text{ kW} \qquad (7.19)$$

WHAT IF? What if one phase impedance were missing? What would be the power?[10]

---

[10] 33.9 kW.

Using line voltage and current. The neutral of the wye-connected load might not be accessible for voltage measurement; hence, it is desirable to express the total power in terms of the line voltage and current. Using the $\sqrt{3}$ ratio between phase voltage and line voltage, we may convert Eq. (7.18) to

$$P_{3\phi} = \left(\frac{V}{\sqrt{3}}\right)(I) \times PF = \sqrt{3}VI \times PF \tag{7.20}$$

where V is the line voltage, I is the line current, and PF is the power factor of the load. In Eq. (7.20), the power factor is the cosine of the angle of the phase impedance.

Equation (7.20) is identical to Eq. (7.10), which was developed for the delta-connected load. Thus, the formula is general and applies to all balanced three-phase loads. Clearly, the two load configurations cannot be distinguished by external measurement but can be identified in practice only by examining the internal connections in the three-phase load.

Equivalent Circuits

Delta–wye conversions. This does not mean, however, that the *same* set of phase impedances are equivalent when connected first in delta and then in wye. Indeed, the appearance of the circuits suggests that the delta gives parallel paths, whereas the wye gives series paths. This appearance suggests that the line current for the delta would be larger than for the wye if the same phase impedance were used for each connection. It can be shown that the ratio is 3:1; that is, three identical impedances will draw three times the current (and three times the power) when connected in delta, as compared to when they are connected in wye.

Impedance Level

Thus, the delta and wye are equivalent if the phase impedances differ by a factor of 3, with the delta connection having the higher impedance level. This equivalence, shown in Fig. 7.18, is often useful in solving three-phase problems. We leave proof of this equivalence for a problem at the end of this chapter.

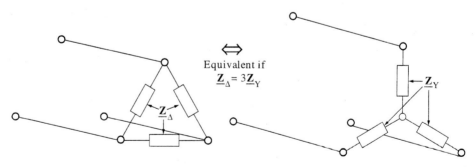

Figure 7.18 Delta and wye equivalence.

EXAMPLE 7.6 Delta load with line losses

Determine the power to the delta-connected load in Fig. 7.19.

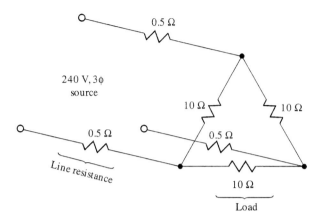

Figure 7.19 Find the power to the delta-connected load.

SOLUTION:

The 10-Ω resistors represent the load, but we must consider the resistance of the wires leading to the load, which is represented by the 0.5-Ω resistors. The presence of the wire resistance undermines our previous approach for solving delta-connected loads, but if we convert the delta to an equivalent wye load, we can solve the problem.

Figure 7.20 shows the circuit after conversion to wye. The wire resistance can now be combined with the load resistance to yield a phase resistance of 3.83 Ω, and the rms line-to-neutral voltage is $240/\sqrt{3} = 139$ V. The line current thus is $139/3.83 = 36.1$ A, and the total power to the wires plus load is $\sqrt{3}\,(240)\,(36.1) = 15.0$ kW. The wire losses are $3(36.1)^2(0.5) = 1960$ W, the rest of the power going to the delta-connected load.

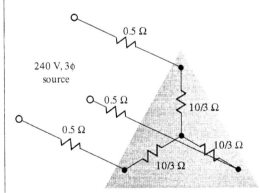

Figure 7.20 After converting the delta to a wye, we can determine the line currents.

WHAT IF?

What if you measured the voltage across a 10-Ω resistor in Fig. 7.19. What voltage would you measure?[11]

---

[11] 209 V.

Per-Phase Equivalent Circuits

LEARNING OBJECTIVE 4.
To understand how to derive and use the per-phase equivalent circuit of a balanced three-phase load

Equivalent Circuits

Need for a simpler model. Three-phase circuits are awkward to draw, and much of the drawing is unnecessary for a balanced circuit because each phase is identical. A single-phase model can represent the voltage, current, and power relationships in the three-phase circuit without this redundancy.

The *per-phase equivalent circuit* is a single-phase circuit that represents a balanced three-phase circuit. The per-phase model can represent any three-phase system, whether source or load is connected in delta or wye; thus, the system is that shown in Fig. 7.21. The power source establishes the voltage, V, and the load determines the current and the power factor. The voltage, current, and power factor on the line indicate the power delivered to the load.

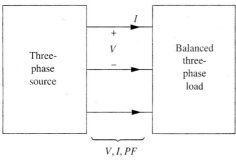

Figure 7.21 The per-phase circuit is based on the line voltage, current, and power factor and not on the internal connections of source and load.

Figure 7.22 The wye–wye circuit with neutral in place can be considered three single-phase circuits sharing a wire.

Wye–wye basis. The per-phase circuit is derived *as if* the three-phase source and load were wye-connected, as shown in Fig. 7.22. We have drawn the neutral connection dashed because it actually may be missing. Even if present, the current in the neutral will be zero for a balanced load, as indicated; thus, the neutrals of the source and the load are at the same voltage, as if connected by a short circuit. This equivalent short circuit connecting the neutrals divides the three-phase circuit into three, identical single-phase circuits sharing a common neutral. We may take the inner circuit, where we marked the voltage and current, as representing one-third of the three-phase system. The per-phase equivalent circuit, shown in Fig. 7.23, introduces the per-phase voltage, current, and impedance, $\underline{V}_{pp}$, $\underline{I}_{pp}$, and $\underline{Z}_{pp}$, respectively.

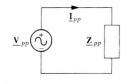

Figure 7.23 Per-phase equivalent circuit.

Relationship between three-phase and per-phase quantities. The relationship between the three-phase quantities and the per-phase quantities are shown in Table 7.2. The per-phase voltage is $V/\sqrt{3}$ because, as shown in Fig. 7.22, the single-phase circuit consists of one line of the three-phase system plus the neutral. The currents and the power factor are the same. The per-phase impedance is the same as the load phase impedance if the load is wye-connected, but is one-third the load phase impedance if delta-connected. The power in the per-phase circuit is one-third the full power in the three-phase circuit, as the name "per phase" suggests. The same would be true for apparent, reactive, and complex power.

TABLE 7.2 Relationships between Three-Phase and Per-Phase Quantities

	Three-Phase Circuit	Per-Phase Circuit
Voltage	V	$V_{pp} = V/\sqrt{3}$
Current	I	$I_{pp} = I$
Impedance	Z_Δ or Z_Y	$Z_{pp} = Z_Y$ or $Z_\Delta/3$
Power factor	PF	$PF =$ same
Power	$P_{3\phi} = \sqrt{3}\, VI \times PF$	$V_{pp} I_{pp} \times PF = P_{3\phi}/3$

EXAMPLE 7.7 Motor per-phase equivalent circuit

A three-phase induction motor has the following specifications: 3 hp, 3515 rpm, 230 V, 8.8 A, efficiency = 80.0%. Derive a per-phase equivalent circuit for the motor.

SOLUTION:

The per-phase voltage and current come directly from the three-phase voltage and current:

$$V_{pp} = \frac{230}{\sqrt{3}} = 133 \text{ V} \quad \text{and} \quad I_{pp} = 8.8 \text{ A} \tag{7.21}$$

and thus the magnitude of the per-phase impedance is established:

$$|Z_{pp}| = \frac{V_{pp}}{I_{pp}} = \frac{133}{8.8} = 15.1 \text{ }\Omega \tag{7.22}$$

To determine the angle of the impedance, we need the phase angle of the current, which is not given, or the power factor, which is not given either but may be determined from the power quantities. The output power of the motor is 3 hp $\times$ 746 W/hp = 2240 W, and the input power follows from this and the efficiency:

$$\text{Efficiency} = \eta = \frac{P_{out}}{P_{in}} \Rightarrow P_{in} = \frac{P_{out}}{\eta} = \frac{2240}{0.800} = 2800 \text{ W} \tag{7.23}$$

Thus, the power factor of both the three-phase and the per-phase circuits is

$$PF = \frac{P_{pp}}{V_{pp} I_{pp}} = \frac{2800/3}{133 \times 8.8} = 0.798 \tag{7.24}$$

and hence the angle of the per-phase impedance is $\cos^{-1}(0.798) = 37.1°$ and the per-phase impedance is $\mathbf{Z}_{pp} = 15.1 \angle 37.1° = 12.0 + j9.09 \text{ }\Omega$. *Note:* (1) we divided the total electrical input power by 3 to convert to power per phase, and (2) we assumed a positive angle for the impedance (inductive) because an induction motor draws lagging current. The per-phase equivalent circuit is that shown in Fig. 7.23 with $\mathbf{V}_{pp} = 133\sqrt{2} \angle 0° \text{ V}$, $\mathbf{Z}_{pp} = 15.1 \angle 37.1° \text{ }\Omega$, and $\mathbf{I}_{pp} = 8.8\sqrt{2} \angle -37.1° \text{ A}$.

> **WHAT IF?** What if the motor were connected for 460-V operation? In this case, the input current would be 4.4 A, but the power, speed, and efficiency would be unchanged. Find the new per-phase quantities.[12]

unbalanced three-phase systems

Unsymmetric loads. A per-phase equivalent circuit is possible only for a balanced system. A three-phase load becomes *unbalanced* when the phase impedances are not identical. This is an undesirable situation and is avoided in practice if possible. When unbalanced loads are connected in delta, calculation of the phase and line currents becomes tedious, though straightforward. All phase and line currents must be calculated individually because symmetry has been lost.

When unbalanced loads are connected in wye, the analysis is straightforward only when the neutral of the load is connected to the neutral of the three-phase source. With the neutral connected, the three loads operate in effect as single-phase loads that share the neutral connection. Current will flow in the neutral wire for an unbalanced load.

When there exists no neutral wire in the unbalanced wye connection, complications arise in the calculation of the line-to-neutral voltages and the line currents. Because the neutral of the load is no longer at the same voltage as the neutral of the source, the first step in solving the problem is to calculate the voltage of the neutral of the load. Then one can proceed to solve for the line currents. Such calculations are routinely performed with computers.

Check Your Understanding

1. A three-phase circuit measures 762 V (rms) between earth ground and line *A*. What would be the voltage between lines *B* and *C*?
2. For a delta-connected load, the phase and line currents are the same. True or false?
3. A three-phase load uses 50 kW at 480 V. The current is 64 A. Find the reactive power required by the inductive load.
4. Three 120-Ω resistors connected in wye are equivalent to what resistances connected in delta?

Answers. (1) 1320 V; (2) false; (3) +18.2 kVAR; (4) 360 Ω.

7.2 POWER DISTRIBUTION SYSTEMS

Introduction. The power that is indispensable to our civilization is generated in large central power plants, transformed to high voltages for transmission over long distances, lowered to moderate voltages for distribution throughout a geographic region such as a small city, and finally reduced to the voltage levels required by industrial and residential consumers. Only at the last stage is the power available in single-phase form; the generation, transmission, and distribution use three-phase.

---

[12] $V_{pp} = 266$ V, $I_{pp} = 4.4$ A, and $\mathbf{Z}_{pp} = 60.4 \angle 37.1°$ Ω.

In this section, we deal with the transmission and distribution aspects of the power system. In the next section, we deal with ac motors, which are a major consumer of electric power and have special importance to engineers.

Voltage levels. For reasons intrinsic to good design, large three-phase generators produce voltage in the range of 11–25 kV. As discussed in the previous chapter, the voltage is raised to much higher values, 120 – 500 kV, for transmission. The voltage is lowered, perhaps by degrees, to smaller values, 7.2–23 kV, for distribution over a geographic area. As near to the customer as possible, within the building for a commercial customer or on a pole outside for a residential customer, the voltage is transformed finally to the standard utilization voltages, 120 and 240 V for residential, perhaps higher voltages for industrial customers. All these voltage transformations, with the exception of the last one, are performed by three-phase transformers.

Three-Phase Transformers

> **LEARNING OBJECTIVE 5.**
> To understand how to connect single-phase transformers for three-phase transformation

Three-phase power may be transformed by three single-phase transformers or by a single three-phase transformer. Economics favors the latter for larger transmission transformers and the former for smaller distribution transformers. The principles are the same in both cases. We speak in the following as if we are using three single-phase transformers.

In three-phase transformation, primaries and secondaries can be connected in delta (Δ) or wye (Y). This gives four possible combinations: Y–Δ, Δ–Y, Δ–Δ, and Y–Y. We now analyze the Y–Δ connection and summarize the other three connections in a table.

The Y–Δ connection. Figure 7.24 shows a Y–Δ connection in two circuit representations. We assume three single-phase transformers with voltages of $V_p : V_s$ and currents

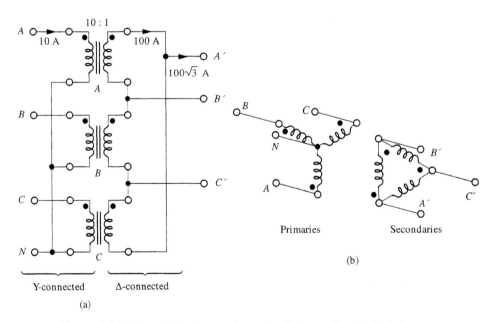

Figure 7.24 Three single-phase transformers in a Y-Δ connection: (a) circuit diagram; (b) schematic diagram.

of $I_p:I_s$ with apparent power $S = V_p I_p = V_s I_s$. Operated at rated voltage and current, the primary line voltage and current are $V = \sqrt{3}V_p$ and $I = I_p$, for an apparent power of $\sqrt{3}\,VI = 3V_p I_p = 3S$. Thus, the apparent power rating of the three single-phase transformers connected for three-phase transformation is three times the apparent power rating of each component transformer. The secondary line voltage is V_s and the rated current is $\sqrt{3}I_s$; hence, the apparent power rating again is $3S$. In both cases, the rating of the three-phase connection is three times the rating of the single-phase transformers. This is true for all four possible connections.

EXAMPLE 7.8 Y–Δ transformer connection

Three 2400/240-V, 24-kVA (each) single-phase transformers are connected with the 2400 side in wye and the 240 side in delta. Find the rated three-phase voltage, current, and apparent power on both sides of the transformer.

SOLUTION:
The allowed currents in the primary and secondary windings are 10 A and 100 A, respectively. Operated at rated voltage and current, the primary line voltage and current are $\sqrt{3} \times 2400 = 4160$ V and 10 A, respectively, for an apparent power of $\sqrt{3}\,VI = 72$ kVA. The secondary line voltage is 240 V and the rated current is $\sqrt{3} \times 100 = 173$ A. Hence, the apparent power in the secondary is also 72 kVA.

WHAT IF?
What if the transformers were connected Δ–Δ?[13]

The phasor diagram for the Y–Δ connection is shown in Fig. 7.25. The secondary voltages are one-tenth the primary line-to-neutral voltage and 30° out of phase angle. This suggests that care should be used in designing and installing such systems because if multiple paths exist, and they do in most power systems, phase angles must be correct.

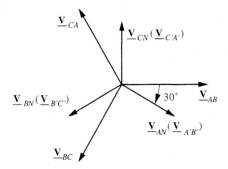

Figure 7.25 Phasor diagram for a Y–Δ transformer connection. Voltage magnitudes are not to scale.

---

[13] Primary: 2400 V, 17.3 A, 72 kVA; secondary: 240 V, 173 A, 72 kVA.

TABLE 7.3 Voltage and Currents Resulting from Three-Phase Transformer Connections*

Connections	Primary V	Primary I	Secondary V	Secondary I
Y–Y	$\sqrt{3}V_p$	I_p	$\sqrt{3}V_s$	I_s
Y–Δ	$\sqrt{3}V_p$	I_p	V_s	$\sqrt{3}I_s$
Δ–Y	V_p	$\sqrt{3}I_p$	$\sqrt{3}V_s$	I_s
Δ–Δ	V_p	$\sqrt{3}I_p$	V_s	$\sqrt{3}I_s$

*The winding (single-phase) voltage and current are V_p and I_p on the primary and V_s and I_s on the secondary. The three-phase voltage and current are V and I, respectively.

Applications of transformer connections. The four possible transformer connections are presented in Table 7.3. The winding (single-phase) voltage and current are V_p and I_p on the primary and V_s and I_s on the secondary, and the three-phase voltage and current are given in the table. Each of the connections find use:

- **Wye–wye.** This connection is occasionally used in high-voltage transmission lines. In addition to providing a neutral, the wye connection has the virtue of having lower voltage across the windings and is desirable for high-voltage connections. When used, Y–Y transformers often have a third set of windings, a *tertiary winding,* that is connected in delta to reduce harmonics in the power system.

- **Delta–wye.** This connection is used to step up the voltage from generation levels to the high voltages required for long-distance transmission. Also can be used to reduce from distribution levels to consumer voltage levels.

- **Wye–delta.** This connection is used to step down the voltage from transmission levels to distribution levels.

open-delta connection

- **Delta–delta.** This connection is used at distribution and consumer levels. This connection has the unique property that one leg of the delta can be omitted in the so-called *V* or open-delta connection.[14] This configuration is useful because it can operate at 58% of the rating that the full delta would have and hence the open-delta connection is often installed in a growing system for later expansion.

Per-phase equivalent circuits for transmission and distribution systems. All the connections just listed, and indeed the entire transmission and distribution system, can be represented by an equivalent per-phase (single-phase) circuit. After presenting per-unit calculations, we proceed to describe the properties of power transmission and distribution systems.

---

[14] The open-delta connection is drawn in the figure for Problem 7.21.

Per-Unit Calculations

per-unit calculations

Introduction. The *per-unit* system of calculation is a method of *normalizing* electrical circuit calculations so that voltage transformations in transformers become inconsequential. Other benefits of the per-unit system are (1) many of the $\sqrt{3}$'s inherent in three-phase calculations are eliminated, and (2) on a per-unit basis all generators, transmission lines, transformers, motors, etc., look more or less alike.

Base values. In power system calculations, we are concerned with calculating voltages, currents, impedances, and the three types of power, apparent power (S), real power (P), and reactive power (Q). We need base values of these quantities to normalize actual circuit quantities; but voltage, current, impedance, and power are interrelated such that only two may be chosen as base values and the other two base values may be calculated. Normally, voltage (V or kV) and apparent power (VA, kVA, and MVA) are used as primary base values, and current and impedance base values are secondary.

Converting to per unit (pu). To convert a circuit quantity to per unit (pu), divide by the base value. Thus

$$V_{pu} = \frac{V}{V_{base}} \quad \text{and} \quad S_{pu} = \frac{S}{S_{base}} \tag{7.25}$$

where the unsubscripted quantities are the circuit values. For example, we have a 240/120-V, 12-kVA transformer that is operated at 10 kVA with 220 V on the primary. If we use the rated values as base, the per-unit voltage and power are

$$V_{pu} = \frac{220\,\text{V}}{240\,\text{V}} = 0.917 \quad \text{and} \quad S_{pu} = \frac{10\,\text{kVA}}{12\,\text{kVA}} = 0.833 \tag{7.26}$$

The per-unit quantities are unitless. We can normalize current and impedance also, but we must first establish consistent bases from the base voltage and apparent power.

Base values for current and impedance. Because the apparent power is voltage times current, we can derive the current base as

$$I_{base} = \frac{S_{base}}{V_{base}} = \frac{12\,\text{kVA}}{240\,\text{V}} = 50\,\text{A} \tag{7.27}$$

and similarly for impedance

$$Z_{base} = \frac{V_{base}}{I_{base}} = \frac{V_{base}^2}{S_{base}} = \frac{(240\,\text{V})^2}{12\,\text{kVA}} = 4.8\,\Omega \tag{7.28}$$

EXAMPLE 7.9 Per-unit circuit calculations

A 240/120-V, 12-kVA transformer that is operated at 220 on the primary has a 10-Ω load on the secondary. Derive a per-unit circuit and calculate the per-unit current in both primary and secondary and the per-unit power. Figure 7.26 shows the circuit.

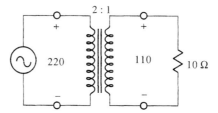

Figure 7.26 The transformer will disappear in the per-unit circuit.

SOLUTION:

We have the base values for the primary in Eq. (7.26). We choose the rated voltage and the kVA for the secondary base values, which makes the transformer disappear, as we shall see. Thus, the base values for the secondary are, from Eqs. (7.27) and (7.28)

$$I_{base} = \frac{12\,\text{kVA}}{120\,\text{V}} = 100\,\text{A} \quad \text{and} \quad Z_{base} = \frac{(120\,\text{V})^2}{12\,\text{kVA}} = 1.2\,\Omega \quad (7.29)$$

The per-unit voltage on both primary and secondary is 0.917, and the per-unit load impedance is

$$Z_{pu} = \frac{10\,\Omega}{1.2\,\Omega} = 8.33; \quad \text{hence} \quad I_{pu} = \frac{V_{pu}}{Z_{pu}} = \frac{0.917}{8.33} = 0.110 \quad (7.30)$$

Thus, the current in the secondary is $0.110 \times 100\,\text{A} = 11\,\text{A}$. The per-unit current in the primary is the same, 0.110, because both actual and base currents are reduced by the turns ratio. Thus, the current in the primary is $0.110 \times 50\,\text{A} = 5.5\,\text{A}$. Note the per-unit voltage and current are the same on both sides of the transformer; it becomes in effect a 1:1 transformer, as shown in Fig. 7.27. Hence, the per-unit power is the same on both sides of the ideal transformer:

$$P_{pu} = V_{pu} I_{pu} = 0.917 \times 0.110 = 0.101 \quad (7.31)$$

so the actual power is $0.101 \times S_{base} = 0.101 \times 12{,}000\,\text{VA} = 1210\,\text{W}$.

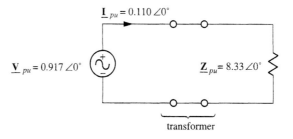

Figure 7.27 The transformer is gone.

> **WHAT IF?** What if we used 10 kVA as the base apparent power? What would be the new bases? Would the per-unit quantities change? Would the actual quantities change?[15]

Transmission Properties

LEARNING OBJECTIVE 6.
To understand how transmission-line impedance affects power and voltage loss on the line

Introduction. In this section, we investigate how the performance of a transmission system depends upon the real and reactive power passing through that system. Figure 7.28 shows a simple per-phase circuit model of a transmission system composed of a generator, a line represented by resistance and inductance, and a load. The transformers of the system are not shown because we can eliminate them either through reflecting all impedances to the transmission line part of the circuit or by using per-unit calculations. The resistance of the system will be ignored except when power loss calculations are made, and the reactance of the system consists of the inductive reactance of the transmission line and the transformers. The generator is represented by a voltage magnitude, $\underline{V}_g = V_g \sqrt{2} \angle \delta$, at an angle, δ, with respect to the load voltage.

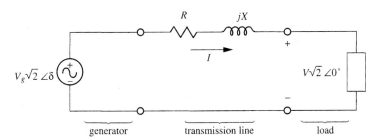

Figure 7.28 Per-phase model of power transmission system.

Analysis of the system. In the following analysis, we assume that the load voltage, $\underline{V} = V\sqrt{2} \angle 0°$, is fixed both in magnitude and angle. Thus, the current, and consequently the real and reactive power in the system, depend only on the load. We analyze the circuit to show how the rms voltage loss in the line, $\Delta V = V_g - V$, the phase angle shift, δ, and the losses depend on the power flow in the system. Of course, we could perform an exact analysis, including the resistance, but we will gain adequate insight by ignoring the resistance.

Figure 7.29 shows the phasor diagram of the system with all quantities on a per-phase basis. We have assumed the voltage across the impedance of the transmission system, $jX\underline{I}$, to be small relative to the voltages of source and load because this is typical. If we ignore the small difference between $\underline{V}_g$ and its projection on the real axis, we have

---

[15] On the primary, the new base current would be 41.7 A and base impedance 5.76 Ω. On the secondary, the new base current would be 83.3 A and base impedance 1.44 Ω. The per-unit quantities change to $I_{pu} = 0.132$ and $S_{pu} = 0.121$, but the actual currents and power are unchanged.

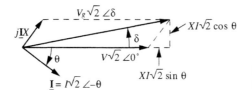

Figure 7.29 Phasor diagram for the transmission system.

$$V_g = V + XI \sin\theta = V + \frac{XQ}{V} \tag{7.32}$$

because $Q = VI \sin\theta$. Thus, the voltage loss in the line, for fixed system reactance and load voltage, depends approximately on the reactive power flow:

$$\Delta V = \frac{X}{V} \times Q \tag{7.33}$$

angle of transmission

The *angle of transmission*, δ, can be derived geometrically:

$$\delta = \sin^{-1}\left(\frac{XI \cos\theta}{V_g}\right) \approx \frac{XI \cos\theta}{V} = \frac{X}{V} \times P \tag{7.34}$$

because $P = VI \cos\theta$. Thus, the angle of transmission depends on the real power on the line. To find the line losses, we must include the line resistance, R:

$$P_{loss} = I^2 R = \frac{S^2}{V^2} \times R = \frac{R}{V^2} \times P^2 + \frac{R}{V^2} \times Q^2 \tag{7.35}$$

Thus, both real and reactive power flow contribute to the loss in the transmission system.

Example 7.10 Transmission line

A 345-kV (receiving end voltage) transmission line is 160 km long. The impedance/km is $0.034 + j0.32$ Ω/km per phase. The load requires 500 MW and 100 MVAR. At each end are 1000-MVA transformers with 0.1 per-unit reactance per phase based on the transformer voltage and apparent power rating. Find the voltage loss, ΔV, the angle of transmission, and the line losses on a per-unit basis.

Solution:

We use the line voltage and the transformer apparent power rating as bases. We need to calculate the base impedance to normalize the transmission line impedance.

$$Z_{base} = \frac{V_{base}^2}{S_{base}} = \frac{(345\,\text{kV})^2}{1000\,\text{MVA}} = 119\,\Omega \tag{7.36}$$

so the per-unit line impedance is

$$\underline{Z}_{pu} = \frac{\underline{Z}_{line}}{Z_{base}} = \frac{(0.034 + j0.32) \times 160}{119} = 0.0457 + j0.430 \tag{7.37}$$

and hence $R_{pu} = 0.0457$ and $X_{pu} = 0.430 + 0.2$ after we add the reactance of the transformers. The per-unit powers are $P_{pu} = 500$ MW/1000 MVA $= 0.5$, and similarly $Q_{pu} = 0.1$. Thus, Eq. (7.33) gives the voltage drop to be

$$\Delta V_{pu} = \frac{X_{pu}}{V_{pu}} \times Q_{pu} = \frac{0.630}{1} \times 0.1 = 0.063 \tag{7.38}$$

Hence, there is a 6.3% voltage drop over the transmission line. The angle of transmission is given by Eq. (7.34):

$$\delta \approx \frac{X_{pu}}{V_{pu}} \times P_{pu} = \frac{0.630}{1} \times 0.5 = 0.315 \text{ radians} = 18.1° \tag{7.39}$$

Thus, there exists an 18.1° phase-angle shift between generator and load. Finally, the power loss is given by Eq. (7.35):

$$P_{loss(pu)} = \frac{R_{pu}}{V_{pu}^2} \times P_{pu}^2 + \frac{R_{pu}}{V_{pu}^2} \times Q_{pu}^2 \tag{7.40}$$

$$= \frac{0.0457}{1^2} \times (0.5)^2 + \frac{0.0457}{1^2} \times (0.1)^2 = 0.0119$$

so the per-unit line loss is 1.19%, which on a base of 1000 MVA would be 11.9 MW. Thus the transmission efficiency is $(500 - 11.9)/500 = 97.6\%$.

WHAT IF? What if the load were 800 MVA at a lagging power factor of 0.9? Find voltage and power loss in percent.[16]

Power-factor correction in a transmission system. Because the voltage and power loss in the transmission system depend on the reactive power flow, the power company has an interest in minimizing Q in the system. This it does by forcing large consumers to control their power factor and by using capacitors in transmission and distribution systems. Figure 7.30(a) models a source, load, and distribution line, which includes resistance and inductance, and Fig. 7.30(b) shows capacitors added to the distribution system by the power company to improve performance. With leading current, the load voltage can be equal to or greater than the source voltage. Thus, adding capacitors to the system can reduce line power and voltage losses.

Figure 7.31(a) shows the phasor diagram for the system with 0.9 power factor, lagging current. The lagging current, combined with the inductive impedance of the line, results in a large voltage loss in the line. Figure 7.31(b) shows the phasor diagram with

---

[16] $\Delta V = 22.0\%$ and loss $= 4.06\%$.

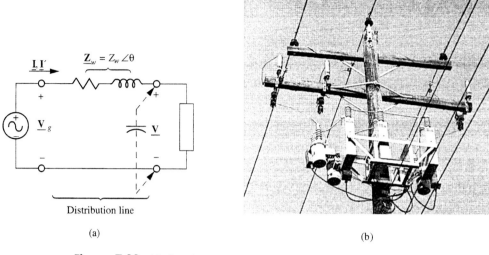

Figure 7.30 (a) Capacitors placed near the load improve efficiency and voltage regulation because the transmission system is inductive; (b) a 12.5-kV three-phase distribution power line with six capacitors between the lines and neutral.

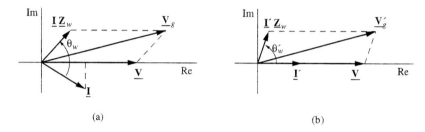

Figure 7.31 (a) Phasor diagram for the system with 0.9 power factor and no capacitors; (b) phasor diagram for the systems with the same load and unity power factor (with capacitors). Note that line current is smaller and $\underline{V}$ and $\underline{V}_g$ are more nearly equal in (b).

unity power factor and the same load. Here line current is smaller, and load and source voltages are more nearly equal.

Check Your Understanding

1. If a line voltage is 0.9 per unit, what is the per-phase voltage (line-to-neutral) in per unit?

2. The base voltage is 13,200 V and the base power is 100 kVA. Find the base current and impedance.

3. If the *PF* is 0.9 and the line loss is 1000 W, what would be the line loss if the *PF* were corrected to unity?

Answers. (1) 0.9, the $\sqrt{3}$'s go away; (2) 7.58 A, 1742 Ω; (3) 810 W.

7.3 INTRODUCTION TO ELECTRIC MOTORS

Introduction. A large part of the load on an electric power system consists of electric motors. At home (air conditioners, hair dryer, blender), in the shop (tools, fans), in the office (printers, cooling fans for electronic devices), and of course in factories, electric motors are everywhere. Although there are many types of electric motors, at least 90% are induction motors: single-phase induction motors for small jobs and three-phase for big.

In this section, we give information relevant to all motors and discuss the steady-state and dynamic responses of motors. We give detailed analysis of nameplate information on three-phase and single-phase induction motors. Much practical information about a motor can be deduced from its nameplate by applying basic knowledge of ac circuits and simple mechanics.

Terminology

rotor, stator

Rotor and stator. Figure 7.32 suggests an electrical motor. It has an electrical input of voltage and current, a mechanical output in the form of torque and rotation, and losses represented as heat. The motor consists mechanically of a *stator*, which does not rotate, a *rotor*, which can rotate, and an air gap to permit motion.

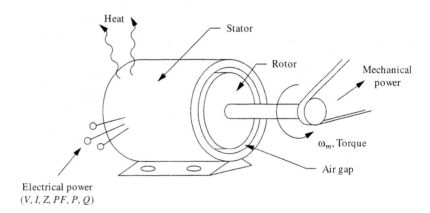

Figure 7.32 A basic motor.

armature, field

Armature and field. Electrical force (torque) is produced by currents interacting with magnetic flux. We distinguish electrically between the *field* circuit, which produces a magnetic flux, and the *armature* circuit, which carries a current. The field usually has many turns of wire carrying relatively small currents, and the armature normally has few turns of larger wire carrying relatively large currents. Depending on the type of motor, the field might be on the rotor or stator, and the armature is always the opposite. For induction motors, the field is on the stator and the armature on the rotor.

Types of electric motors. Although there are many types of electric motors, we deal here with two types: the three-phase induction motor and the single-phase induction motor. After discussing certain general principles that relate to all motors, we discuss and analyze the nameplate information for both these types of motors.

Motor Characterization in Steady-State Operation

Electrical input. We consider first steady-state operation, such as a fan motor turning at constant speed. The electrical quantities of interest in steady state are the input voltage, current, and the power factor. In many cases, analysis of the motor suggests an equivalent circuit that accounts for losses, energy storage, and the conversion of electrical power into mechanical power. The electrical operation depends in part on the mechanical load, as will be discussed presently.

Mechanical output. The output characteristic of a motor is the output torque, $T_M(\omega_m)$, as a function of rotation speed, ω_m. The operating speed of the motor is jointly determined by the output torque characteristic of the motor and the torque requirement of the load.

Basic equations for three-phase motor. The input electrical power is

$$P_{in} = \sqrt{3} VI \times PF \tag{7.41}$$

and the output mechanical power is

$$P_{out} = \omega_m \times T_M \tag{7.42}$$

where ω_m is the mechanical speed in radians/second and T_M is the output torque in newton-meters. Two other quantities of interest are the losses, P_{loss}, and the efficiency, η:

$$P_{loss} = P_{in} - P_{out} \text{ and } \eta = \frac{P_{out}}{P_{in}} \tag{7.43}$$

Basic equations for single-phase motors. Equations (7.41) through (7.43) are also valid for single-phase motors except that the input power equation lacks the $\sqrt{3}$ for single phase.

EXAMPLE 7.11 Single-phase motor

Consider a 60-Hz, single-phase induction motor with the following nameplate information:[17] 1 hp, 1725 rpm, 115 V, 14.4 A, efficiency = 68%. Find the output torque, the power factor, and the cost of operating the motor if electric power costs 9.6 cents/kWh.

SOLUTION:
The output power is 1 hp × 746 W/hp = 746 W, so the input power is, from Eq. (7.43)

$$P_{in} = \frac{P_{out}}{\eta} = \frac{746}{0.68} = 1097 \text{ watts} \tag{7.44}$$

---

[17] We will discuss nameplate information presently. For now, assume that the actual operation of the motor is given by the nameplate information.

The power factor follows from Eq. (7.41) without the $\sqrt{3}$:

$$PF = \frac{P_{in}}{VI} = \frac{1097}{115 \times 14.4} = 0.662 \qquad (7.45)$$

and the current is lagging for this type of motor. The output torque is

$$T_{out} = \frac{P_{out}}{\omega_m} = \frac{746}{1725(2\pi/60)} = 4.13 \text{ N-m} \qquad (7.46)$$

where the mechanical speed in rpm has been converted to radians/second. For continuous operation and nameplate conditions, the cost would be

$$\frac{1097 \text{ W}}{1000 \text{ W/kW}} \times \frac{24 \text{ hours}}{1 \text{ day}} \times \frac{9.6 \text{ cents}}{1 \text{ kWh}} \times \frac{1 \text{ dollar}}{100 \text{ cents}} = \$ \, 2.53/\text{day} \qquad (7.47)$$

WHAT IF? What if the motor is running without a load?[18]

The Motor with a Load

LEARNING OBJECTIVE 7.
To understand how motor and load interact to establish steady-state operation

Typical motor and load characteristics. Figure 7.33 shows the diverse torque characteristics that can be achieved for several types of electrical motors. Even within a specific motor type, the designer can tailor the motor characteristics within broad limits. Figure 7.34 shows representative torque requirements for various mechanical loads.

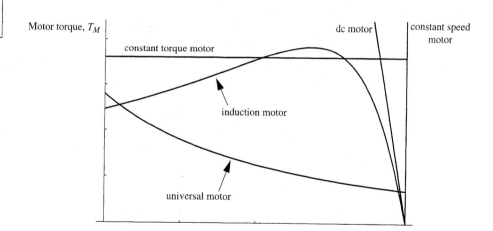

Figure 7.33 Motor torque characteristics.

---

[18] All we know in that case is that $P_{out} = 0$ and $T_M = 0$. But the cost would not be zero due to losses of the motor.

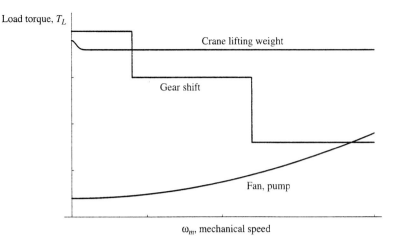

Figure 7.34 Load characteristics.

blocked-rotor torque, breakover torque

Motor–load interaction. System operation is jointly determined by load requirements, $T_L(\omega_m)$, and motor characteristics, $T_M(\omega_m)$. Consider, for example, connecting a three-phase induction motor to a fan starting from rest, as shown in Fig. 7.35. We have identified the motor-starting torque, often called the *blocked-* or *locked-rotor torque*, the maximum torque, often called the *breakover torque*, and the no-load speed. For rotation to occur, the starting torque of the motor must exceed the starting-torque requirement of the load. The excess torque, $\Delta T(\omega_m) = T_M - T_L$, will accelerate the system to the speed where the two characteristics cross, which would be the steady-state speed of the motor–load system. This condition, $T_M = T_L$, determines the steady-state speed of the system. Thus, the speed, output torque, and output power of the motor depend in part on the load requirements. If, for example, the load required less torque, the motor would run slightly faster and supply less power.

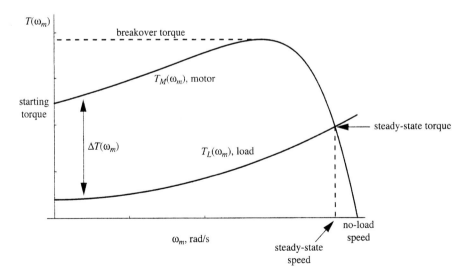

Figure 7.35 Motor and load characteristics. The load is accelerated by the excess torque, ΔT.

Dynamic Operation

run-up time

In addition to steady-state operation, we are also interested in transient, or dynamic, operation. For example, we may need to predict the starting current and the time required to reach the steady-state speed, which is called the *run-up time*. In dynamic operation, the rotor and load moments of inertia become factors, and energy processes are more complicated than for steady-state operation.

Run-up time. We can calculate the time required to reach steady-state speed by integrating the equation of motion. In general, for a rotational system,

$$J\frac{d\omega_m}{dt} = T_M - T_L = \Delta T(\omega_m) \tag{7.48}$$

where ω_m represents mechanical rotation speed in radians/second, J the combined moment of inertia of motor and load, T_M the motor output torque, and T_L the load torque requirement. We may integrate Eq. (7.48) from zero time, when the motor is stopped, to the run-up time, t_{ru}, when the motor reaches the steady-state speed

$$t_{ru} = \int_0^{\omega_{ss}} \frac{J d\omega_m}{\Delta T(\omega_m)} \approx \sum \frac{J \Delta \omega_m}{\Delta T(\omega_i)} \tag{7.49}$$

where ω_{ss} is the steady-state speed. We may approximate the run-up time as indicated by the second form of Eq. (7.49) and illustrated in Fig. 7.36.

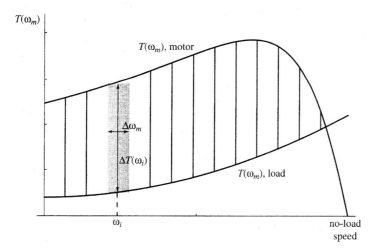

Figure 7.36 Run-up time calculation.

EXAMPLE 7.12 Run-up calculation

Consider a motor with a constant output torque of 5 N-m, driving a load requiring a torque proportional to speed. The steady-state speed is 800 rpm, and the combined moment of inertia of motor and load is 0.02 kg-m². Find the time required to reach equilibrium speed, starting from standstill.

SOLUTION:
The load requirement would be $T = C\omega_m$, where C is a constant to be determined. The equilibrium speed would be $\omega_{ss} = 800 \times 2\pi/60 = 83.8$ rad/s; hence, the constant is

$$T_M - T_L = 0 \Rightarrow 5 = C \times 83.8 \Rightarrow C = 0.0597 \text{ N-m/(rad/s)} \quad (7.50)$$

By Eq. (7.49), the run-up time is

$$t_{ru} = 0.02 \int_0^{83.8} \frac{d\omega_m}{5 - 0.0597\omega_m}$$

$$= \frac{0.02}{-0.0597} \ln(5 - 0.0597\omega_m)\Big|_0^{83.8} \quad (7.51)$$

Equation (7.51) leads to an infinite result at the upper limit. The difficulty is that the system speed approaches steady state asymptotically, so theoretically it never reaches equilibrium.[19] One way to approximate the run-up speed is to assume a "final" speed slightly below the steady-state speed. For Eq. (7.51), changing the upper limit to 98% of the steady-state speed gives a run-up time of 1.311 s.

WHAT IF? What if we define the final speed as 99.5% of final speed? What would be the run-up time?[20]

Summary. In this section, we discussed some fundamentals that apply to all motors. We presented equations relating to the input electrical variables and the output mechanical variables. We discussed how the output-torque characteristic of the motor interacts with the torque requirement of the load to establish the steady-state speed, torque, and power in the motor–load system.

In the next sections, we present and interpret the nameplate information for a three-phase induction motor and a single-phase induction motor, with most of the emphasis on the former. These two types of motors account for the vast majority of motor applications. Our purpose is to give sufficient information to analyze nameplate information.

LEARNING OBJECTIVE 8.
To understand how to analyze and interpret nameplate information of induction motors

Three-Phase Induction Motor Nameplate Interpretation

Induction motor principle. In an induction motor, the field windings are on the stator. The currents in the field windings set up a rotating magnetic flux, which in turn induces voltage in the rotor. The voltage in the rotor drives currents that interact with the flux to produce torque. Because the relative motion between the rotating flux and the rotor conductors produces the voltage, the rotor will always turn slower than the flux rotates. The induction motor works like a fluid clutch or automatic transmission. Instead of fluid, the working medium is magnetic flux.

---
[19] Of course, a real motor does reach equilibrium. Our linear *model* leads here to a mathematical problem.
[20] 1.78 seconds.

Nameplate information. Three-phase induction motors are available from one-third horsepower to thousands of horsepower. The nameplate of a specific 60-Hz three-phase induction motor includes the following information:

- 50 horsepower
- Three phases
- 1765 rpm
- NEMA 326T frame
- 208–230/460 V
- 140–122/61 A
- Time rating: continuous
- Insulation class: F
- 1.15 service factor
- Maximum ambient temperature: 40°C
- NEMA code: G
- NEMA design: B
- NEMA nominal efficiency: 92.4%
- NEMA minimum efficiency: 91.0%

The motor nameplate also might give the motor type: drip-proof, totally enclosed fan cooled (TEFC), explosion-proof, etc.

service factor

Power rating. The *power rating*[21] of this motor is 50 hp. This is a nominal rating and does not mean that the motor will put out 50 hp on all, or possibly on any, occasions. The actual power out of the motor depends on the load demands, as explained earlier—nor is 50 hp the maximum power of the motor. The maximum power that the motor can put out on a continuous basis is the nameplate power times the service factor, 50 hp × 1.15 = 57.5 hp in this case. Thus, the **service factor** is something of a "safety factor"; the 50-hp rating is, as we said before, the nominal power rating of the motor.

The nameplate power is significant in the following sense: *If* the motor is supplied with rated voltage, 460 V, and *if* the load demands exactly 50 hp, *then* the motor speed will be 1765 rpm *and* the input current will be 61.0 A.

Motor speed. The nameplate *speed* is 1765 rpm, and the significance of this speed is given in the preceding paragraph. With a power frequency of 60 Hz, induction motors run slightly below the standard speeds of 3600, 1800, and 1200 rpm, for the reason given above. If this motor is unloaded, the motor speed will be approximately 1800 rpm.

Motor output torque. The nameplate torque, T_{NP}, is not given but may be deduced from the output power and speed. Equation (7.42) yields

$$T_{NP} = \frac{P_{out}}{\omega_m} = \frac{50 \times 746}{1765 \times (2\pi/60)} = 202 \text{ N-m} \qquad (7.52)$$

where we changed horsepower to watts and rpm to radians/second to make the units consistent. Again we stress that this is the nameplate torque; we do not know the actual torque unless we know what the load demands.

Torque versus speed. In the normal operating range, the output torque of a three-phase induction motor is approximately a straight line between zero torque at the no-load speed, ω_{NL}, and nameplate torque at nameplate speed, as shown in Fig. 7.37. Thus, the torque equation is of the form

$$T_M(\omega_m) = \text{slope} \times (\omega_m - \omega_{NL}) \qquad (7.53)$$

---

[21] The nameplate terms will be in bold italics in the following discussion, for easy reference.

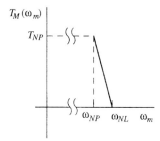

Figure 7.37 The torque characteristic is a straight line between nameplate and no-load conditions.

where the slope is $T_{NP}/(\omega_{NP} - \omega_{NL})$. Thus, the torque as a function of speed in the normal operating region is

$$T_M(\omega_m \text{ or } n) = T_{NP}\left(\frac{\omega_{NL} - \omega_m}{\omega_{NL} - \omega_{NP}}\right) = 202\left(\frac{1800 - n}{1800 - 1765}\right) \quad (7.54)$$

where we expressed speed in both rpm and rad/s.

EXAMPLE 7.13 Pump load

The 50-hp motor with the torque given by Eq. (7.54) drives a pump with a torque requirement of

$$T_L = 1.2 + 100\left(\frac{n}{1800}\right)^2 \text{ Nm} \quad (7.55)$$

Find the speed at which the motor–pump operates and the power required by the pump.

SOLUTION:
The motor–pump will operate where the motor output matches the pump demand

$$1.2 + 100\left(\frac{n}{1800}\right)^2 = 202\left(\frac{1800 - n}{1800 - 1765}\right) \quad (7.56)$$

which is a quadratic equation yielding $n = 1783$ rpm and $-188{,}598$ rpm. Clearly, the second answer represents the analytic extension of both motor and pump characteristics into regions where their models are invalid; the correct answer is 1783 rpm. The power is the speed in rad/s times the torque, which we can get from either motor or pump:

$$P_{out} = \omega_m T_M \text{ (or } \omega_m T_L\text{)}$$
$$= 1783(2\pi/60)\left[1.2 + 100\left(\frac{1783}{1800}\right)^2\right] = 18{,}540 \text{ watts} \quad (7.57)$$

which is slightly less than 25 hp.

WHAT IF? What if the pump torque required $T_L = 100 + n/18$ N-m?[22]

---

[22] $n = 1766$ rpm and power = 49 hp.

NEMA frame 326T. NEMA is an acronym for the National Electrical Manufacturers Association, an industry association that has standardized many aspects of electrical power equipment. In this case, the mechanical dimensions of motors are standardized; for example, the motor output shaft diameter is $2\frac{1}{8}$ inches and is centered 8 inches above the base of all motors built on a *NEMA 326T frame*.

Voltage and current ratings. The motor *voltage rating* is 208–230/460 V and the *current rating* is 140–122/61 A. We set aside the 208-V/140-A rating for a moment and concentrate on the 230/460-V rating. This reveals that each phase of the motor has multiple windings that can be connected in series or parallel. Figure 7.38(a) shows one phase of the high-voltage (460 V)/low-current (61 A) series connection, and Fig. 7.38(b) shows the low-voltage (230 V)/high-current (122 A) parallel connection. As you can see, the individual windings get the same voltage and current for both connections and the motor performance would be unchanged if indeed the voltage were exactly 230 or 460 volts. In practice, the series connection is preferred if 460-V three-phase power is available because the lower current requires smaller wire to supply the motor. Put another way, the impedance level of the motor is four times higher and the output impedance requirements of the power system are less demanding with the high-voltage/low-current connection.

Impedance Level

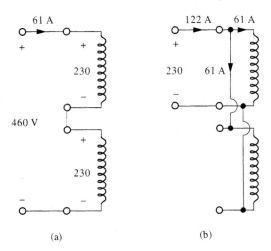

Figure 7.38 (a) A series connection for high voltage and low current; (b) a parallel connection for low voltage and high current.

The 208-V rating. The 208-V rating tells us the lowest voltage the motor should be supplied, and the current for 50 hp would be 140 A. First of all, $208 = 120\sqrt{3}$, so 208-V systems exist as 120-V three-phase connected in wye. But more common is low voltage due to older standards or long connecting wire, say, with pumps scattered around an oil field. Low voltage is dangerous to the motor because motors tend to overheat when the voltage is low, as explained in what follows, and heat shortens the motor lifetime.

Constant impedance or constant power? Up to this point, we have spoken of impedance as if it were constant, and this is true for many circuit elements. But a motor tends to take constant power from the electrical system and hence does not offer constant impedance. To see the difference, examine the contrast in Eq. (7.58):

$$Z = \frac{V}{I} = \frac{(\downarrow)}{(\downarrow)} \text{ but } P = \sqrt{3}VI \times PF = \sqrt{3}(\downarrow)(\uparrow) \times PF \qquad (7.58)$$

For constant impedance, the current goes down when the voltage goes down. This is the characteristic of resistors, inductors, capacitors, and many other devices as well. But for constant power, the current goes *up* when the voltage goes down provided the power factor does not change. This is how a motor behaves because the power of the motor depends largely on load demands.

Impedance Level

Thus, low voltage forces a high current, even if the voltage source is ideal. In practice, the high current lowers the voltage further due to the output impedance of the power system. Thus, when you use a portable electric tool such as a chain saw or lawn mower, the manufacturer will warn you to use only 100 feet of extension cord: Too long a cord produces low voltage, which produces higher current, which produces even lower voltage, etc. Because the resistive losses of the motor increase with the square of the current, the motor can rapidly overheat if the extension cord is too long.

EXAMPLE 7.14 Long extension cord

A saws-all electric saw is rated at 120 V, 8 A single phase. The saw is used with 100 ft of 16/3[23] extension cord. How much do the motor losses increase as a result of the extension cord?

SOLUTION:
The resistance for No. 16 wire is 4.02 Ω/1000 ft, and we have 200 feet of wire considering both conductors. Thus, the nominal 8 A of current causes a voltage drop of

$$\Delta V \approx \frac{4.02 \; \Omega}{1000 \; \text{ft}} \times 200 \; \text{ft} \times 8 \; \text{A} = 6.4 \; \text{V} \tag{7.59}$$

and hence the voltage at the saw is about 113.6 V, assuming 120 V for the supply. Because the saw requires constant power, this loss in voltage causes a further increase in current to

$$\Delta I \approx 8 \; \text{A} \times \frac{120}{113.6} = 8.45 \; \text{A} \tag{7.60}$$

which will cause a further decrease in voltage, but we will ignore further effects. The increase in motor loss will be

$$\Delta \text{loss} \approx \left(\frac{8.45}{8}\right)^2 = 1.116, \text{ or a } 11.6\% \text{ increase} \tag{7.61}$$

WHAT IF? What if 150 ft of extension cord is used?[24]

Thermal specifications. The *time rating* is continuous operation. The *insulation class* of F tells us that the maximum temperature of the insulation should not exceed 155°C. The importance of the *ambient temperature* of 40°C is self-evident. These

---
[23] 16/3 = No. 16 power wires with a ground.
[24] The losses increase by 18.3%.

specifications remind us again that the motor lifetime depends on the thermal environment affecting the temperature of its windings.

NEMA code G. The *NEMA code* tells us the starting-current requirements of the motor. NEMA classifies motors according to locked rotor (starting) kVA, and code G means 5.60 to 6.29 kVA/hp. Hence, for this motor, the starting current lies between 609 and 684 A with the 460-V connection, and twice these values for the 230-V connection. Because of these large currents, large induction motors are frequently started with reduced voltage and the voltage is increased as the motor speed increases.

NEMA design B. All three-phase induction motors have similar characteristics, but motors may be tailored somewhat for specific properties:

- Design A has high run efficiency and high breakover torque, with moderate starting torque and high starting current.
- Design B has moderate run efficiency, moderate starting torque with low starting current, and moderate breakover torque.
- Design C is similar to design B but has higher starting torque and is more expensive.
- Design D has relatively low run efficiency, extremely high starting torque which is also the breakover torque, and low starting current.

Our motor is *design B*, which is the general-purpose motor.

Efficiency. The nameplate gives a nominal (92.4%) and minimum (91.0%) *efficiency*. The nominal is what one would expect on an average, and we assume this value to calculate the nameplate input power from Eq. (7.43):

$$P_{in} = \frac{P_{out}}{\eta} = \frac{50 \times 746}{0.924} = 40{,}370 \text{ W} \tag{7.62}$$

and the input electrical power allows us to calculate the nameplate power factor from Eq. (7.41):

$$PF = \frac{P_{in}}{S} = \frac{40{,}370 \text{ W}}{48{,}600 \text{ VA}} = 0.831 \tag{7.63}$$

where $S = \sqrt{3} \, VI = \sqrt{3} \times 460 \times 61 = 48{,}600$ VA is the nameplate VA rating of the machine. We also calculate the nameplate reactive power required by the motor:

$$Q_{in} = \pm\sqrt{S^2 - P_{in}^2} = +\sqrt{(48.6)^2 - (40.4)^2} = +27.0 \text{ kVAR} \tag{7.64}$$

where the positive sign is used because an induction motor has lagging current.

The motor is designed to have maximum efficiency at nameplate conditions and hence the efficiency will be fairly constant in the vicinity of nameplate operation. However, the efficiency falls as the motor is lightly loaded.

Summary. We discussed the nameplate information for a three-phase induction motor. From the nameplate information, we deduced nameplate torque, the torque–speed characteristic, power factor, and reactive power. We stress again that the actual motor conditions depend on the load demands.

Single-Phase Induction Motor

Nameplate information. Single-phase induction motors come in many varieties and sizes; we now discuss the type of single-phase motor used to power farm machinery, stationary shop tools, compressors, and the like.[25] Such induction motors are made in sizes from one-fourth horsepower to 10 horsepower and are used where three-phase power is unavailable. The nameplate information on one motor is as follows:

- Phase 1
- Frame L56
- Hz 60
- Volts 115
- 40°C
- SF 1.35

- Type CS
- HP 1/3
- RPM 1725
- Amps 5.8
- Insulation class A

We now discuss the nameplate information that is not self-evident.

Type CS. Single-phase induction motors come in three types, CS, SP, and CR:

- CS: A capacitor-start/induction-run motor has high starting torque and moderate run efficiency.
- SP: A split-phase motor has low starting torque and moderate run efficiency.
- CR: A capacitor-start/capacitor-run motor has high starting torque and high run efficiency, and is more expensive.

RPM 1725. This is self-evident. However, we should mention that the no-load speed of this motor would not be 1800, as for a three-phase motor, but would be somewhere around 1780–1790 rpm. Thus, we cannot derive the torque–speed characteristics without measuring the no-load speed. Once we know the no-load speed, we can derive the torque–speed characteristic as for a three-phase motor.

Other specifications. The remainder of the specifications are similar to those for three-phase motors. Torque can be calculated from power and speed. Efficiency is not normally given, so losses and power factor cannot be calculated from the nameplate information.

Check Your Understanding

1. In a motor the field current is usually larger than the armature current. True or false?
2. The output torque of a motor in operation depends in part on the load requirements. True or false?
3. The product of the power factor and the efficiency of an ac motor is always equal to the output mechanical power divided by the input electrical apparent power. True or false?

Answers: (1) False; (2) true; (3) true.

---

[25] Small fans, hand-held tools, and many household appliances use other types of single-phase motors.

CHAPTER SUMMARY

Chapter 7 introduces electric power distribution and utilization in electrical motors. Three-phase circuits are investigated. The per-phase model is developed and per-unit normalization is defined and illustrated. System aspects of electric motors are discussed, and the analysis of the nameplate information for three- and single-phase electric motors is illustrated.

Objective 1: To understand the nature and advantages of three-phase power. The three-phase system is described in the time and frequency domains. The advantages include efficiency of distribution, smoothness of power flow, and characteristics in starting and operating induction motors.

Objectives 2 and 3: To understand how to analyze a generator or load in wye or delta connection. We derive the phase and amplitude relationships between line voltage, line current, and power and phase voltage, phase current, and power for both wye and delta connections. We show that three-phase power depends only on line voltage, current, and power factor and not on the load or source connections.

Objective 4: To understand how to derive and use the per-phase equivalent circuit of a balanced three-phase load. The per-phase equivalent circuit is a single-phase circuit that models the state of a balanced three-phase circuit. The per-phase current and power factor are the same as the three-phase line current and power factor, but the per-phase voltage is equal to the line-to-neutral voltage of the three-phase system. The per-phase power is one-third the three-phase power. The per-phase equivalent circuit is independent of the actual connection of the three-phase load or source.

Objective 5: To understand how to connect single-phase transformers for three-phase transformation. Three-phase power may be transformed either by one three-phase transformer or by three single-phase transformers. Connecting primaries and secondaries in wye or delta gives four possible connections, each having different voltage and current ratios and different advantages and disadvantages.

Objective 6: To understand how transmission-line impedance affects power and voltage loss on the line. Transmission of electrical power over large distances involves significant effects of line impedance. An approximate analysis shows the effect of line inductance on power angle and power magnitude. We introduce the definitions and advantages of the per-unit system of normalization.

Objective 7: To understand how motor and load interact to establish steady-state operation. We examine dynamic and steady-state interactions of motors with their loads. Motor power and speed depend on load demands.

Objective 8: To understand how to analyze and interpret nameplate information of induction motors. Nameplate information gives a benchmark for comparing motors. The nameplate directly or indirectly gives power, speed, torque, voltage, current, efficiency, and other information. The dangers of operating a motor at low voltage are illustrated.

GLOSSARY

Angle of transmission, p. 356, the difference in phase angle between the voltage at sending and receiving ends of a power transmission system; depends on real power flow.

Base values, p. 353, voltage, current, and power quantities to normalize circuit voltage, current, and power in per-unit calculations.

Delta (Δ) connection, p. 334, a three-phase connection that ties three generator windings or loads in a closed ring, with no neutral connection.

Open-delta connection or V connection, p. 376, a three-phase transformer connection using two single-phase transformers.

Per-phase equivalent circuit, p. 347, a single-phase circuit that represents a balanced three-phase circuit.

Per-unit calculations, p. 353, a method of normalizing electrical circuit calculations.

Phase current, p. 338, the current in a generator winding supplying a three-phase system or in one leg of a three-phase load.

Phase voltage, p. 338, the voltage across a generator winding supplying a three-phase system, or the voltage across one leg of a three-phase load.

Run-up time, p. 363, the time required for a motor/load to reach steady-state speed.

Service factor, p. 365, the ratio of maximum power to nameplate power on a motor, something of a safety factor.

Terminal, p. 334, the end of a wire connected to an electrical device.

Terminal board, p. 334, a place where electrical terminals are made available for connection.

Three-phase current, p. 338, the current in the lines carrying the power in a three-phase system.

Three-phase voltage, p. 338, the voltage between the lines carrying the power in a three-phase system.

Unbalanced load, p. 349, when phase impedances are not identical, resulting in unequal line currents.

Wye (Y) connection, p. 336, a three-phase connection that ties three generator windings or three loads to a common point, which is the neutral connection.

PROBLEMS

Section 7.1: Three-Phase Power

7.1. Pick two values of ωt in Eq. (7.2) and show by direct calculation that the time-varying terms cancel.

7.2. For Fig. 7.6(a), develop a delta connection by first connecting A to B'. Draw a phasor diagram of the resulting system. Is the phase rotation ABC or ACB?

7.3. Three 230-V (rms) generators are connected in a three-phase wye configuration to generate three-phase power. The load consists of three balanced impedances, $\mathbf{Z}_L = 2.6 + j1.8\ \Omega$, connected in delta.
(a) Find the line current an ammeter would measure.
(b) Find the apparent power.
(c) Find the real power to the load.
(d) What is the phase angle between $\mathbf{I}_A$ and $\mathbf{V}_{AB}$, assuming ABC rotation?

7.4. For the three-phase power system shown in Fig. P7.4, a voltmeter measures 146 V between line A and the neutral N. Find the following:
  (a) Line voltage.
  (b) Line current.
  (c) Load power factor.
  (d) Apparent power.
  (e) Real power.

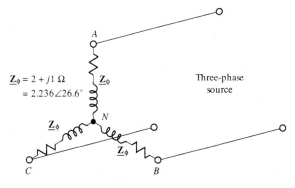

Figure P7.4

7.5. In Fig. P7.5, the voltage between A and N is 120 V as measured by a standard meter. Let this voltage be the phase reference. The phase impedance is $\mathbf{Z}_\phi = 6.2 + j2.7 = 6.76 \angle 23.5° \, \Omega$.
  (a) What is $\mathbf{V}_{AB}$ as a phasor?
  (b) What would an ammeter measure as the line current?
  (c) What is the apparent power?
  (d) What is the real power?

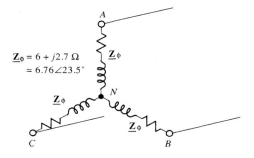

Figure P7.5

7.6. Figure P7.6 shows a three-phase source and load.
  (a) Draw appropriate connections.
  (b) Find the phase voltage.
  (c) Find the line current.
  (d) Determine the power factor.

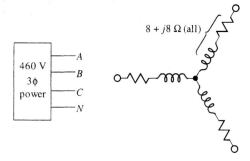

Figure P7.6

  (e) What is the power in the load?
  (f) Assuming $\mathbf{V}_{AB} = 460\sqrt{2} \angle 0°$, find $\mathbf{I}_A$ as a phasor.

7.7. For the three-phase delta-connected load in Fig. P7.7, we have the line voltage and current to be $\mathbf{V}_{AB} = 480\sqrt{2} \angle 0°$ V and $\mathbf{I}_A = 10\sqrt{2} \angle -30$ A.
  (a) What is $\mathbf{V}_{CA}$?
  (b) What is the phase current in the load, rms value?
  (c) What is the time-average power into the load?
  (d) What is the phase impedance?

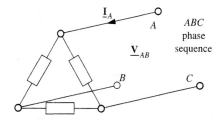

Figure P7.7

7.8. Three 230-V rms generators are connected in delta to form a three-phase source. A balanced wye-connected load has phase impedances of $12 + j7 \, \Omega$. Find the rms values of the following:
  (a) The phase voltage of the source.
  (b) The line voltage of the system.
  (c) The phase voltage of the load.
  (d) The phase current of the load.
  (e) The line current of the system.
  (f) The phase current of the source.
  (g) What is the power factor of the load?
  (h) What is the real power to the load?

7.9. In the three-phase circuit shown in Fig. P7.9, find the following:
  (a) The line current that would be measured by an ammeter.

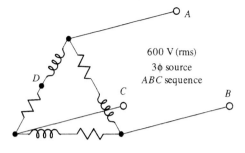

All $\underline{Z}$'s = $20 + j11$ Ω

Figure P7.9

  (b) The power factor of the three-phase load.
  (c) The voltage that would be measured between B and D by a voltmeter.

7.10. In the three-phase circuit shown in Fig. P7.10, find the following:
  (a) The line current that would be measured by an ammeter.
  (b) The power factor of the three-phase load.
  (c) The voltage that would be measured between B and D by a voltmeter.

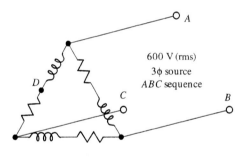

All $\underline{Z}$'s = $15 + j8.5$ Ω

Figure P7.10

7.11. For the three-phase circuit connected in delta shown in Fig. P7.11, find the following:
  (a) The load power factor. Assume lagging.
  (b) The line current, rms.
  (c) The magnitude of the phase impedance.
  (d) The reactive power to each phase impedance.

7.12. For the three-phase circuit connected in delta shown in Fig. P7.12, find the following:
  (a) The load power factor. Assume lagging.
  (b) The line current, rms.
  (c) The magnitude of the phase impedance.
  (d) The reactive power to each phase impedance.

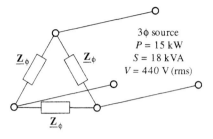

Figure P7.11

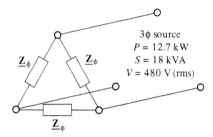

Figure P7.12

7.13. For the three-phase system shown in Fig. P7.13, find the following:
  (a) Phase voltage.
  (b) Line voltage.
  (c) Phase current.
  (d) Line current.
  (e) Phase impedance.
  (f) Apparent power.
  (g) Power factor.
  (h) Real power to the load.

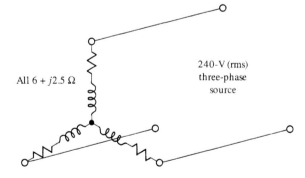

Figure P7.13

7.14. A balanced, wye-connected three-phase load is shown in Fig. P7.14. The current in line A is $\underline{I}_A = 8\sqrt{2} \angle 0°$ A. The voltage from B to the neutral point is $\underline{V}_{BN} = 120\sqrt{2} \angle -90°$ V.

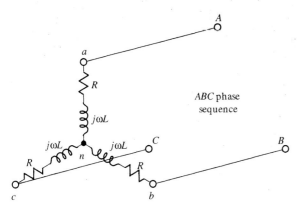

Figure P7.14

(a) Find the three-phase voltage that a voltmeter would read.
(b) Find the real and reactive power into the entire three-phase load.
(c) Determine R and L, assuming 60-Hz operation.

7.15. Demonstrate the equivalence of the delta and wye circuits in Fig. 7.18. Assume an impedance $Z_\Delta \angle \theta$ for the delta and an impedance $Z_Y \angle \theta$ for the wye and compute the complex power for each. Equate these powers and confirm the 3:1 ratio shown in Fig. 7.18.

7.16. Three identical resistors are placed in a wye configuration and draw a total of 150 W from a three-phase source. What power would the same resistors draw if placed in delta?

7.17. For the three-phase circuit shown in Fig. P7.17, the 0.1-Ω resistors represent the resistance of the distribution system. Find the following:
(a) Total power out of the source, including line and load.
(b) Line losses.
(c) Distribution system efficiency.

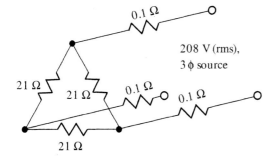

Figure P7.17

7.18. Using delta–wye transformations, determine the total power given to the delta and wye loads in Fig. P7.18, not counting the losses in the 0.1-Ω resistors that represent losses in the connecting wires. *Hint:* The neutrals of two balanced wye loads have the same voltage and hence may be considered as connected.

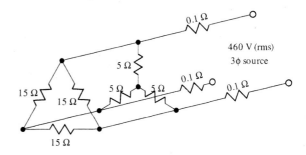

Figure P7.18

7.19. The city of Austin, Texas, distributes power with a three-phase system with 12.5 kV between the power-carrying wires. But each group of houses is served from one phase and ground, and transformed to 240/120 V by a pole transformer, as shown in Fig. P7.19.
(a) What is the turns ratio (primary/secondary turns), of the pole transformer to give 240 V, center-tapped?
(b) When a 1500-W hair dryer is turned on, how much does the current increase in the high-voltage wire? Assume the power factor is unity and the transformer is 100% efficient.

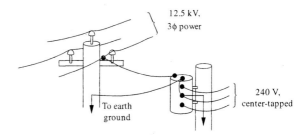

Figure P7.19

7.20. A workman finds a three-phase cable with four wires. He labels the wires 1, 2, 3, and 4 and measures the following voltages: $V_{12} = 150$ V and $V_{23} = 260$ V. What are V_{13} and V_{34}?

In the text, we showed how to transform three-phase power with the use of three single-phase transformers. There are two ways to transform three-phase power with *two* single-phase transformers. The next two problems investigate these methods. In them, we will transform 460 V three phase to 230 V three phase; hence, the transformers have a turns ratio of 2:1. *Hint*: In both figures, the geometric orientation hints of the phasor relationships.

7.21. The configuration shown in Fig. P7.21 is called the "open-delta" or V connection, for obvious reasons. Identical 2:1 transformers are used.
  (a) Show that if *ABC* is 460-V balanced three-phase, *abc* is 230-V balanced three-phase. Consider the *ABC* voltages to be a three-phase set and prove the *abc* set is three-phase.
  (b) If the load is 30 kVA, find the required kVA rating of the transformers to avoid overload. [You can solve this independent of part (a).]

7.22. The circuit shown in Fig. P7.22 is called the T connection. For this connection, the 2:1 transformers are not identical but have different voltage and kVA ratings. The bottom transformer is center-tapped so as to have equal, in-phase voltages for each half.
  (a) Find the voltages V_1 and V_2 to make this transform 460-V to 230-V three-phase.
  (b) If the load is 30 kVA, find the required rating of each transformer to avoid overload.

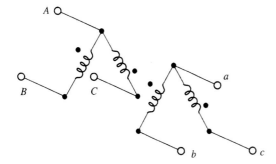

Figure P7.21 The V or open-delta transformer connection.

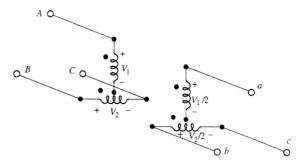

Figure P7.22 The T transformer connection.

Section 7.2: Power Distribution Systems

7.23. A wye-connected, balanced, three-phase load has a three-phase voltage of 480 V and requires a kVA of 20 kV at a *PF* of 0.94, lagging.
  (a) What is the per-phase, per-unit equivalent of the load, using the three-phase voltage and 42 kVA for bases?
  (b) What if the load is in delta?

7.24. A 500-MVA, 22-kV ac generator is represented on a per-phase equivalent circuit as an ideal ac voltage source in series with an inductive reactance. The internal reactance is Y-connected, 1.1 per unit. Find the actual reactances that are connected in Y.

7.25. A 12/138-kV, 50-MVA three-phase transformer has a per-phase inductive reactance of $j0.005\ \Omega$, referred to the primary side. Find the per-unit reactance for the transformer and give a per-unit, per-phase equivalent circuit for the transformer.

7.26. A transmission line has an impedance of $100 + j500\ \Omega$, including both wires. The 3ϕ voltage at the generator output is 22 kV. The voltage is stepped up to 345 kV in a Δ–Y transformer connection, sent over a transmission line, and then stepped down to a nominal 23 kV with a Y–Δ transformer connection. The load is 30 MW at a lagging *PF* of 0.92.
  (a) Give a per-phase equivalent circuit showing the per-phase voltage and current on the line.
  (b) Find the approximate per-unit voltage loss. Ignore the transformer inductance.
  (c) Find the approximate per-unit power loss.

7.27. The circuit shown in Fig. P7.27 represents a 60-Hz power distribution system. The distribution voltage is 8 kV and the load is 20 houses, each requiring on the average 12 kW at 0.95 *PF*, lagging. The line impedance is shown. Assume an ideal transformer.
  (a) What would be the magnitude of the current in the primary of the transformer?
  (b) What are the line power losses?
  (c) What value of capacitor across the transformer primary minimizes line losses? Consider that the load voltage is constant.

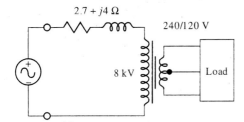

Figure P7.27

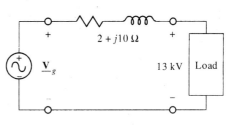

Figure P7.28

(d) What are the line power losses with the corrected *PF*?

7.28. The circuit in Fig. P7.28 shows a load that requires 500 kVA at a lagging *PF* of 0.84, a voltage of 13 kV (maintained constant), and frequency of 60 Hz. The $2 + j10\ \Omega$ represents the impedance of the distribution line. Making no approximations, you are to calculate the line losses and voltage difference between load and source voltage under two conditions:
(a) The circuit as shown.

(b) The circuit with a capacitor at the load to correct the power factor to unity with the same real power. Also determine the value of the required capacitor.

7.29. A 138-kV overhead line has a series resistance and inductive reactance of 0.16 and 0.41 Ω/km, respectively. Find the magnitude of the voltage required at the sending end of the line for 138 kV at the receiving end if the line is 80 km long. The apparent power on the line is 12.5 MVA at 0.95 *PF*, lagging.

Section 7.3: Introduction to Electric Motors

7.30. A motor has the output torque characteristic shown in Fig. P7.30. The load torque characteristic is

$$T_L(n) = 10\left(\frac{n}{1200}\right)^2 \text{ N-m}$$

(a) Determine the operating speed.
(b) Determine the output power.
(c) What is the maximum possible output power from the motor?

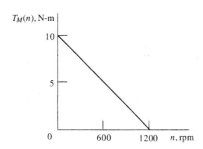

Figure P7.30

7.31. A motor has the parabolic output torque characteristic shown in Fig. P7.31. The load torque characteristic is

$$T_L(n) = 3 + 4\left(\frac{n}{1800}\right) \text{ N-m}$$

(a) Determine the operating speed.

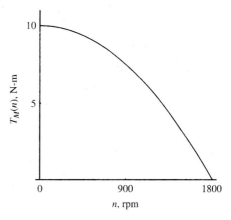

Figure P7.31

(b) Determine the output power.
(c) What is the maximum possible output power from the motor?

7.32. A motor has an output power given by the formula

$$P(\omega_m) = 12\omega_m - \frac{10\omega_m^2}{120\pi} \text{ watts}$$

(a) What is the starting torque of the motor?
(b) Find the no-load speed in rpm.
(c) Find the higher speed for 1-hp output power.
(d) Find the maximum output power.

7.33. A motor has an output torque characteristic given by the equation

$$T_M(\omega_m) = 50 - \frac{(\omega_m - 40)^2}{80} \text{ N-m}$$

where ω_m is the mechanical speed of the motor in radians/second.
(a) Find the no-load speed of the motor.
(b) What is the maximum torque of the motor?
(c) What is the maximum power of the motor?

7.34. The torque characteristics of two motors are shown in Fig. P7.34. The load requires a constant torque of 3 N-m.
(a) Which motor would have the longer run-up time if starting the load from standstill?
(b) Determine the power required for the load in steady state if driven by motor #2.

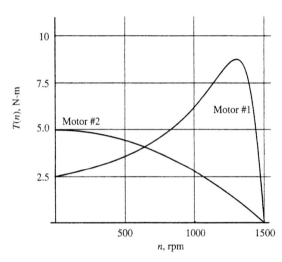

Figure P7.34

7.35. An electric motor has a torque–speed characteristic given by the equation

$$T_M(\omega_m) = \frac{100}{(5 + K\omega_m)^2} \text{ N-m}$$

where K is a constant. The motor puts out 1 hp at 12,000 rpm.
(a) Find K.
(b) What is the starting torque?
(c) What is the maximum power out of the motor?

7.36. The output torque of a motor is given in Fig. P7.36.
(a) Find the blocked-rotor torque.
(b) Find the no-load speed in rpm.
(c) At what speed does the motor put out 12 hp?

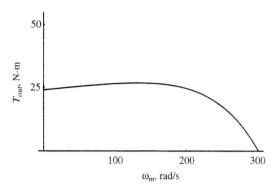

Figure P7.36

(d) If a load requires 20 N-m, what speed would the motor–load run in rpm?

7.37. A motor has the torque given by the equation

$$T_M(\omega_m) = 50 + 0.1\omega_m - \frac{(\omega_m - 40)^2}{80} \text{ N-m}$$

(a) Find the starting torque.
(b) Find the no-load speed of the motor.
(c) Find the maximum torque of the motor.
(d) Find the maximum power of the motor.

7.38. A motor has an output torque characteristic that is well approximated by the function

$$T(n) = 12 \cos\left(\frac{\pi}{2} \times \frac{n}{1200}\right) \text{ N-m}$$

where $0 < n < 1200$ is the speed in rpm.
(a) Find the starting torque.
(b) Find the no-load speed.
(c) Find the maximum power out of the motor in hp.

7.39. A motor with a moment of inertia $J = 0.5$ kg-m² is turning a load at a constant speed of 1160 rpm. When the load is suddenly disconnected, the instantaneous angular acceleration is $+12$ rad/s². Find the load torque at 1160 rpm.

7.40. Write the differential equation for a motor–load system where $T_M(\omega_m) = T_M$, a constant, and $T_L(\omega_m) = C\omega_m$, similar to the example on page 363. Use the notation $J =$ moment of inertia, $K =$ torque constant, and $\omega_{ss} =$ steady-state speed. This system fits the conditions of Chap. 3 transients; hence the initial value, final value, and time constant establish the response. Determine the time constant. Calculate the run-up time for 98% of equilibrium speed based on the differential equation solution in terms of T_M, J, and ω_{ss}.

7.41. A fan requires a driving torque of the form $T_L(\omega_m) = K\omega_m^2$. The fan requires $\frac{1}{2}$ hp of drive power on the shaft to turn 1800 rpm. The fan is driven by an electrical motor with the output characteristic given in Fig. P7.41.
  (a) Find the constant K, with torque and speed expressed in mks units, N-m and rad/s.
  (b) Find the speed in rpm at which the fan will operate.
  (c) Find the approximate time it takes the fan to reach its final speed. The moment of inertia of the motor–load is $J = 0.06$ kg-m².

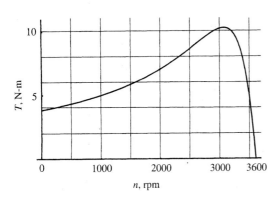

Figure P7.41

7.42. Calculate the steady-state speed and run-up time for the motor and load used in the example on page 363 if the load torque requirement is changed to
$T_L(\omega_m) = 0.001\omega_m^2$ N-m.

7.43. A motor–load system has a run-up characteristic given by
$n(t) = 1180(1 - e^{-t/0.6s})$ rpm.
The motor–load moment is $J = 0.2$ kg-m², and the load torque requirement is constant at 8 N-m. Find the following:
  (a) Motor-starting torque.
  (b) Motor torque at 1180 rpm.
  (c) Time required to reach 96% of final speed.

7.44. A motor–load has a combined moment of inertia of 2 kg-m². The output torque of the motor is
$T_M(n) = 10\left[1 - \left(\dfrac{n}{1800}\right)\right]$ N-m
where n is the speed in rpm. The required load torque is a constant 6 Nm.
  (a) What is the equilibrium speed in rpm?
  (b) How long would it take the system to reach 99% of the final speed starting from a standstill?

7.45. Figure P7.45 shows the input current, efficiency, and power factor of a 230-V three-phase motor from no load to full load.
  (a) What is the rated output power of the motor in hp?
  (b) What is the reactive power used by the motor at 50% load? The current is lagging.
  (c) What are the losses of the motor at no load?

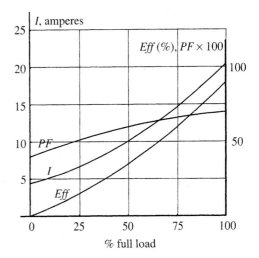

Figure P7.45

7.46. A 208-V three-phase motor runs 1720 rpm at a load requiring 9.0 N-m torque. The motor draws 5.5 A of current and the power factor is 0.92, lagging.
  (a) Find the input power.
  (b) Find the losses of the motor.
  (c) Find the efficiency of the motor.

7.47. A 60-Hz, three-phase induction motor has the following nameplate information: 25 hp, 1755 rpm, 91.7% eff., 230/460 V, 61.8/30.9 A, service factor 1.15.
  (a) Find the torque output at nameplate conditions.
  (b) What is the apparent power into the machine at nameplate conditions?
  (c) Find the power factor at nameplate conditions.
  (d) Determine the reactive power into the machine at nameplate conditions, assuming lagging current.

7.48. An induction motor has the following nameplate information: 60 Hz, three phase, 7.5 hp, 1750 rpm, 230/460 V, 22.0/11.0 A, 86.0% eff., NEMA frame

213T, 1.15 SF, $376 list, $250.79 wholesale, 129.0 lb shipping weight. For this type of motor, the blocked rotor torque is 135% and the breakover torque is 185% of the nameplate torque.
(a) Find the input power factor at nameplate conditions.
(b) Find the nameplate starting torque of the motor.
(c) Find the losses of the motor at nameplate conditions.
(d) What range of power in hp can the motor sustain constantly without failure under normal thermal conditions?

7.49. An engineer is comparing two 60-Hz, 1.5-hp, three-phase motors. Motor A is a high-efficiency motor with the following nameplate information: 1.5 hp, 230/460 V, 1740 rpm, 85.5% eff., 4.4/2.2 A, 1.15 SF, $146.41. Motor B has the following nameplate information: 1.5 hp, 230/460 V, 1740 rpm, 82.5% eff, 5.2/2.6 A, 1.15 SF, $130.07. Based upon 2000 hours of operation a year at nameplate rating and 8¢/kWh, how much money would motor A save per year, not including the motor investment cost?

7.50. An induction motor has the following nameplate information: single-phase, $\frac{1}{2}$ hp, 50 Hz, 1425 rpm, 110/220 V, 9.2/4.6 A, 1.25 SF, Class B insulation, 40°C.
(a) Find the output torque under nameplate conditions.
(b) What is the maximum output power that can be produced by the motor in sustained operation without overheating? (Assume 40°C ambient temperature.)
(c) Assuming that the motor power factor and efficiency are numerically equal, what are the motor losses at nameplate condition?

7.51. A three-phase, 60-Hz induction motor has the following nameplate information: 7.5 hp, 1755 rpm, 230/460 V, 19.4/9.7 A, 88.5% eff., 1.15 SF. We may assume the no-load speed of the motor to be 1800 rpm.
(a) Find the power factor at nameplate conditions.
(b) Find the output torque at nameplate conditions.
(c) Find the speed at which the motor and load will operate if the load requires a torque of
$$T_L(n) = 10 + 12.8(n/1800)^2 \text{ N-m}.$$

7.52. A three-phase, 60-Hz induction motor has the following nameplate information: 30 hp, 39 A, 1760 rpm, 460 V, 87.5% eff, 1.15 SF. The no-load speed of the motor is 1799.4 rpm. The load torque is given by the equation
$$T_L(n) = 10 + 12\left(\frac{n}{1800}\right) + 15\left(\frac{n}{1800}\right)^2 \text{ N-m}$$
where n is the speed in rpm. Find the operating speed of the motor–load system.

7.53. A three-phase, 60 Hz induction motor has the following nameplate information: 25 hp, 1755 rpm, 230/460 V, 64.0/32.0 A, 91.0% efficiency, 1.15 SF.
(a) Find the reactive power requirement at nameplate conditions.
(b) Find the motor losses at nameplate conditions.
(c) Find the nameplate output torque.
(d) At what speed will the motor produce 10 hp?

7.54. The torque–speed characteristic shown in Fig. P7.54 is the normal operating region (no load to nameplate load) for a three-phase induction motor. At full load the efficiency is 67.7% and the current and voltage are 1.1 A and 230 V, respectively.
(a) What is the hp rating of the motor?
(b) Find the motor losses at nameplate conditions.
(c) If the voltage dropped to 220 V at nameplate power, what would be the current, approximately?

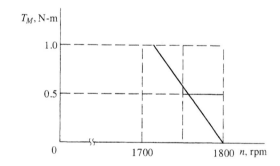

Figure P7.54

7.55. A 60-Hz, single-phase induction motor has the following nameplate information: 1 hp, 1740 rpm, 115/230 V, 13.6/6.8 A, service factor 1.0, power factor 0.73, starting current 50 A at 115 V, starting torque 9.0 lb-ft.
(a) What is the output torque under nameplate conditions?
(b) Determine the efficiency at nameplate conditions.
(c) Determine the reactive power into the machine under nameplate conditions, assuming lagging current.

7.56. A motor catalog lists the following motor: 5 hp, 230 V, 12.8 A. 1740 rpm, 60 Hz, 85.5% efficiency, 1.15 SF. Assume operation at nameplate conditions and assume losses are equally divided between rotor and stator.
(a) Is this a three-phase or single-phase motor? Explain how you know.
(b) What is the power factor of the motor at nameplate conditions?
(c) Find the motor losses at nameplate conditions.

7.57. A 3-hp, 120-V, 1160-rpm, 60-Hz, single-phase motor has input current of 30 A and a power factor of 0.76. At nameplate conditions, find the motor losses and the output torque.

7.58. A single-phase induction motor has the following nameplate information: $\frac{1}{2}$ hp, 1140 rpm, 115/230 V, 10.0/5.0 A, SF 1.25. Assume the power factor at nameplate is 0.78, lagging. The torque–speed curve is shown in Fig. P7.58.
(a) Find the efficiency of the motor at nameplate conditions.
(b) Find the ratio of the starting torque to full-load torque.
(c) Determine the torque as a function of speed in rpm in the region from zero to nameplate power.

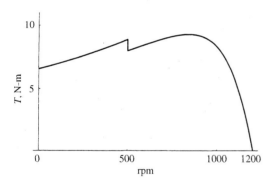

Figure P7.58

General Problems

7.62. Two pump jacks in the oil field require three-phase induction motors to drive the pumps. The wells are 325 yards apart and are connected by three No. 10 wires (1 Ω/1000 ft). One of the wells is shallow and requires a 10-hp motor; the other well is deep and requires a 50-hp motor. For the 10-hp motor the nameplate information is: 60 Hz, 1750 rpm, 230/460 V, 26.2/13.1 A, 88.5% eff., 1.35 SF. For the 50-hp motor, the nameplate information is: 60 Hz,

7.59. A motor catalog gives the following nameplate information for a 60-Hz, single-phase induction motor: 3 hp, 1740 rpm, 115/230 V, 38.0/19.0 A, SF 1.15. Losses are known to be 17% of the output power at nameplate condition.
(a) Find the output torque of the motor at nameplate conditions.
(b) Find the power factor of the motor at nameplate conditions.
(c) Find the maximum power that the motor can sustain for a long period.
(d) What is the maximum power that the motor can produce for a short period without stalling?

7.60. A motor catalog lists the following information about 60-Hz, 1-hp motors: **Motor A:** single-phase, 1725 rpm, 115/230 V, 15.0/7.5 A, no efficiency given, 30 lb; **Motor B:** three-phase, 1750 rpm, 230/460 V, 3.5/1.75 A, 77.2% efficiency, 32 lb.
(a) Calculate the efficiency of the single-phase motor, assuming that the motors have the same power factor.
(b) Which motor has the greater output torque at nameplate conditions and what is it?
(c) Which motor has the greater apparent power requirement at nameplate conditions and what is it?

7.61. Two single-phase induction motors are being considered for an application: **Motor A:** 1 hp, 1725 rpm, 115/230 V, 9.2/4.6 A, 1.25 SF, $225.27, 37 lb. **Motor B:** 1 hp, 1725 rpm, 115/230 V, 14.8/7.4 A, 1.0 SF, $182.99, 30 lb.
(a) Assume that for each motor, the power factor and the efficiency are equal. Find the efficiencies of both motors.
(b) If electric energy cost 3.6 ¢/kWh and the motors were run continuously at nameplate conditions, how soon (days) would Motor A justify its increased cost through saving of electric energy?

1775 rpm, 230/460 V, 90.2% eff., 121.4/60.7 A, 1.35 SF. The motors are connected for 460-V operation, and must be operated between 440 and 460 V for the warranties to be valid. Assume the motors are operated at their nameplate power. Power is brought to the motors from a 13-kV feeder line, and three single-phase transformers, connected in Y on the primaries and Δ on the secondaries, supply 460-V power to the line that serves the motors.

(a) Determine a suitable location for the transformers. That is, at what point between the motors should the transformers be located in order to minimize losses and to provide the acceptable voltages to the motors?

(b) Give the voltage and current in primary and secondary windings of the single-phase transformers.

(c) Find the overall efficiency of the system with both motors running at nameplate power.

7.63. A junior engineer needs 12 hp to drive a load, so he takes a 10-hp motor and a 2-hp motor, both three-phase, and puts them on the same shaft. The motor nameplates read: 10 hp, 1750 rpm, 60 Hz, 230/460 V, 26.2/13.1 A, 88.5% eff., 1.1 service factor; 2 hp, 1725 rpm, 60 Hz, 230/460 V, 6.6/3.3 A, 80.0% eff., 1.35 service factor. Assume operation in the small-slip region and negligible rotational loss.

(a) What speed does the system run, assuming exactly 12 hp for the load?

(b) Is either motor overloaded? To answer, you must calculate the motor power for both motors and explain your conclusion.

Answers to odd-numbered problems

7.1. $\cos(0°) + \cos(-240°) + \cos(-480°) = 0$ and $\cos(20°) + \cos(-220°) + \cos(-460°) = 0$.

7.3. (a) $218\sqrt{2} \angle -34.7°$ A; (b) 151 kVA; (c) 124 kW; (d) $-64.7°$.

7.5. (a) $120\sqrt{3}\sqrt{2} \angle +30°$ V; (b) 17.7 A; (c) 6390 W; (d) 5860 W.

7.7. (a) $480\sqrt{2} \angle -240°$ V; (b) 5.77 A rms; (c) 8.31 kW; (d) $83.1\sqrt{2} \angle 0°$ Ω.

7.9. (a) 45.5 A rms; (b) 0.876; (c) 311 V, rms.

7.11. (a) 0.833; (b) 23.6 A rms; (c) 32.3 Ω; (d) 3.32 kVAR.

7.13. (a) 139 V rms; (b) 240 V rms; (c) 21.3 A rms; (d) 21.3 A rms; (e) $6 + j2.5$ Ω; (f) 8860 W; (g) 0.923; (h) 8180 W.

7.15. Proof.

7.17. (a) 6090 W; (b) 85.8 W; (c) 98.6 %.

7.19. (a) 30.1:1; (b) 0.208 A, rms.

7.21. (a) Calculate V_{bc} from the other two and it comes out right; (b) 34.6 kVA for the two transformers.

7.23. (a) $2.10 \angle -19.9°$; (b) same.

7.25. (a) 0.00174 pu; (b) transformer is 1:1 with $j0.00174$ in primary or secondary.

7.27. (a) 31.6 A, rms; (b) 2690 W; (c) 3.27 µF; (d) 2430 W.

7.29. $140,050 \sqrt{2} \angle 1.01°$ V.

7.31. (a) 1188 rpm; (b) 702 W; (c) 726 W at 1039 rpm.

7.33. (a) 103.2 rad/s; (b) 50 N-m; (c) 2740 W at 65.5 rad/s.

7.35. (a) 6.35×10^{-3}; (b) 4 N-m; (c) 787 W at 7520 rpm.

7.37. (a) 30 N-m; (b) 110 rad/s; (c) 54.2 N-m at 44.0 rad/s; (d) 3200 W at 70.1 rad/s.

7.39. 6 N-m at 1160 rpm.

7.41. (a) 2.23×10^{-4}; (b) about 3500 rpm; (c) 3.6 s.

7.43. (a) 49.2 N-m; (b) 8 N-m; (c) 1.93 s.

7.45. (a) 7.1 hp; (b) 3030 VAR; (c) 771 W.

7.47. (a) 101 N-m; (b) 24.6 kVA; (c) 0.826; (d) +13.9 kVAR.

7.49. $7.61/year.

7.51. (a) 0.818; (b) 30.4 N-m; (c) 1767 rpm.

7.53. (a) 15.2 kVAR; (b) 1845 W; (c) 101.5 N-m; (d) 1782 rpm.

7.55. (a) 4.09 N-m; (b) 65.3%; (c) 1070 VAR.

7.57. 498 W and 18.4 N-m.

7.59. (a) 12.3 N-m; (b) 0.599; (c) 3.45 hp; (d) no way to determine.

7.61. (a) 84.0% for A and 66.2% for B; (b) 205 days.

7.63. (a) 183.0 rad/s; (b) 10.6 hp for 10-hp motor and 1.4 hp for 2-hp motor. Neither is overloaded.

Index

A

admittance, 198
alternating current (ac), 164, 168, 223
    power, 275
    residential, 312
ammeter, 74
ampere, unit, 7
Ampere's force law, 5
angle of transmission, 356
antenna, 68, 298
apparent power, 276, 284
    autotransformer, 307
    frequency domain, 292
    per-unit, 345
    starting, 361
    three phase, 333
    transformer, 296
    units, 286
arc welders, 302
armature, 359
asymptotic Bode plot, 245
    high-pass filter, 249
    low-pass filter, 246
autotransformer, 307
available power, 71, 291
average, 257, 272
    sinusoidal function, 274

B

bandpass filter, 250
base values, per-unit calculations, 353
battery, 26, 33
battery chargers, 302
battery-switch, Thévenin equivalent circuit, 144
blocked-rotor torque, 362
Bode, H. W., 245
Bode plot, 242, 245
    high-pass filter, 248
    low-pass filter, 245
branch currents, 91
breakover torque, 362
bridge circuit, 87

C

calculator, 31, 36
capacitor, capacitance, 118, 123
    circuit symbol, 123
    final values in transient, 138
    generalized impedance, 222
    impedance, 189, 239
    integrating current, 125
    KCL, 123
    mechanical analog, 124
    parallel, 127
    power factor correction, 295
    power, 281
    series, 127
    stored energy, 127
capacitor-start/capacitor-run motor, 370
capacitor-start/induction-run motor, 370
cascaded systems, 244
    dB gain, 250
center-tapped transformer, 313
charge, 5
circuit analysis, 6, 7
    methods, 92
circuit elements, parallel, 35
circuit symbol, circuit symbol, 24
coil of wire, model, 122
common, voltmeter, 84
complex conjugate, 291
complex frequency, 216
    ac, 218
    dc, 217
    frequency domain, 220
    time domain, 220
complex number, 170
    absolute value, 173
    complex conjugate, 172
    division, 172
    exponential form, 174
    magnitude, 173
    notation, 179
    polar form, 173
    real part, 177
    rectangular form, 173
    square root, 171
complex plane, 170
    rotating point, 177
complex power, 291
conductance, 24, 37, 198
conduction electrons, 7
conservation,
    charge, 10
    electric energy, 4, 14, 22
    energy in ideal transformer, 304
constrained node, 77, 81
coulomb, 5
critical damping, 233, 235, 241
critical frequencies, 247
current, 5, 7
    base value, 353
    current divider, 39
    dividers, 37
    physical, 8, 33
    physiological effects, 314
    rating, 367
    sign convention, 10
    transformation, 303
current source, 27, 60
    Norton, 70
    turned OFF, 59
cutoff frequency, 241
    high pass filter, 248
    low pass filter, 241
cycles per second, hertz, Hz., 164

D

damped oscillation, 233
damping constant, 233
dB, 242
    loss, 243
    power, 243
dB gain, cascaded circuits, 250
dBm, 243
dBw, 243
dc, direct current, 26, 217
    component, 274
    transformer, 304
    zero frequency, 193, 241
decade, 245
delta connection, 335
    power, 341
    three phase transformers, 350
    unbalanced, 349
delta-delta transformer, 352
delta-wye transformer, 352
delta-wye conversion, 345
differential equation
    first order, 129, 255
    particular integral, 129

differential equation (*cont.*)
    second order, 148
differentiation, frequency domain, 182, 222
distribution system
    per-phase circuit, 352
    power, 332
    voltages, 350

E

Edison, Thomas, 164
effective value, 277
efficiency
    motor, 360, 369
    transmission system, 309
eigenfunction, 220
electric energy, 118, 282
    conservation, 21
    electronics, 4
    information, 4
    mechanical analogy, 118
    storage, capacitor, 127
electric motors, 332
electrical analogs, mechanical analogs, 127
electrical engineering, 4
electron, 5
electrostatic forces, 5
energy, 5, 19
    exchanged, 21
    inductor, 144
    initial conditions, 130
equivalent, resistance, 30
equivalent circuit, 4, 31, 68, 345
    Norton, 70
    parallel resistors, 30, 36
    resistance, 30
    series resistors, 32, 34
    Thévenin, 66
Euler, 174
event frequency, hertz, Hz., 164

F

farad, 123
Faraday, Michael, 123
feedback, 4
    systems, 253
ferromagnetic materials, 122
field, 359
filter, 247
    frequency domain, 238
    high pass filter, 248
    low pass filter, 239
    radio, 200
    RLC, 251
final values, 132, 138
    capacitors, 138
    inductors, 138
first order transient, 128, 132
floating, 317, 339
forced response, 129, 222
forces, electrostatic, 5
    magnetic, 5, 367
frame, NEMA, 367
frequency, 164
    complex, 216
    familiar, 167

radio, 167
real, 219
frequency domain, 4, 182, 238
    apparent power, 292
    circuit representation, 188
    complex frequency, 220
    filters, 238
    Kirchhoff's laws, 188
    reactive power, 289
    real power, 280
    summary, 203
fuels, electric power generation, 332

G

generalized impedance, 221
    capacitors, 222
    inductor, 221
    resistors, 222
generation, power, 332
gigahertz (GHZ), 167
ground, 84, 314, 317, 315, 332
Ground Fault Interrupters (GFI), 318

H

half-power frequency, 242, 243
harmonics, 238
henry, 121
Hertz, Heinrich, 165
    unit, 165
high-pass filter, 248
    asymptotic Bode plot, 249
    cutoff frequency, 248
homogenous solution, 130
horsepower, 294
hot wire, 314
Hz., unit, 165

I

ideal current source, 27
ideal switch, 25
ideal transformer, 302
ideal voltage source, 26
ignition coil, 302
imaginary number, 170
impedance, 189
    antenna, 298
    base value, 353
    benefits, 189
    capacitor, 189, 239
    generalized, 221
    inductor, 186
    mechanical analogy, 193
    power, 283
    resistor, 188
    termination, 255
    transformer, 304
impedance level, 4, 73, 345
impedance match, 68, 299
in-phase current, 288
independent energy storage elements, 144
independent loop, 85
independent node, 78, 82
induction motor, single-phase, 370
    three-phase, 364

inductor, inductance, 118
    circuit symbol, 119
    energy, 144
    final values, 138
    generalized impedance, 221
    impedance, 186
    magnetic energy, 120
    mechanical analogy, 119
    parallel, 122
    power, 279
    reactance, 194
    series, 122
inertia, 363
initial condition, 130, 132, 138, 146, 148
    energy, 130
    RLC, 234
integrating current, capacitor, 125
iron, 122

K

Kirchhoff's Current Law (KCL), 4, 10, 78, 91
    capacitor, 123
Kirchhoff's laws, frequency domain, 188
Kirchhoff's Voltage Law (KVL), 4, 14, 18, 85
kVA, 295
kVAR, 295

L

ladder network, 111
lagging current, 193
Laplace transform, 223, 237
LC circuit, mechanical analog, 149
leading current, 193
lighting fixture, 314
lightning, 332
line current, 337
line voltage, 337
linear equations, 76
linear systems, 63, 252
load, 63
    motor, 361
load set, 24
loading of circuit, 72
loads, wye-connected, 342
loop, 15, 18
    feedback, 254
    gain, 354
loop current, 85, 90
    current sources, 88
    independent, 90
loss, dB, 243
    transmission line, 356
losses, electrical limits, 292
    motor, 360
low voltage and motor current, 367
low-pass filter, 239
    Bode plot, 245
    filter function, 239

M

magnetic forces, 5, 6
magnetic energy, 118, 280
    inductor, 120
    mechanical analogy, 118

maximum power transfer, 68, 298, 301
mechanical analogs
   capacitance, 124
   charge, 8, 21
   current, 8, 21
   power, 21
   resistance, 128
   summary, 128
   voltage, 13
mesh currents, 90
meter, input impedance, 72
microwave oven, 159
mobile charges, 7
motor, 351
   ambient temperature, 368
   dynamic operation, 363
   efficiency, 360, 369
   insulation class, 368
   load, 361
   losses, 360
   power factor, 369
   power rating, 365
   speed, 366
   steady state, 360
   time rating, 368
   torque, 366
music, 238
MVA, 295
MVAR, 295

N

nameplate interpretation, motor, 364
natural frequencies, 224, 226
   open circuit, 226
   RLC circuit, 232
   short circuit, 226
   zeros and poles, 228
natural response, 130, 222, 224
NEMA, 367
   code, 369
   design class, 369
neutral, 314, 332, 336, 349
no-load speed, induction motor, 366
   single-phase motor, 370
nodal analysis, 76, 145, 229
   subscripts, 76
   voltage sources, 80
node, 10
   counting, 82
   ground, 84
   reference, 84
node voltage, 76
   mechanical analogy, 84
nonlinear devices, superposition, 63
Norton, E. L., 70, 111
Norton equivalent circuit, 70, 74, 297
notation
   subscript, 3
   system, 252

O

ohm, unit, 23
Ohm's law, 23
open circuit, 25, 59
   infinite resistance, 59

   power, 25
   voltage, 64
open delta connection, 352
optimum, 69
oscillations, 233
out-of-phase current, 289
outlet, polarized, 314
output impedance, 64, 67, 73, 137
overdamped response, 233, 236

P

parallel, 30, 35, 37
   capacitors, 127
   inductors, 123
   notation, 37
   resistance, 37
parallel resonance, 202
   spectrum, 238
per-unit quantities, 353
period, 164
periodic function, average, 273
phase current, 338
phase rotation, 339
phase voltage, 338
phasor, 181
   Kirchhoff's laws, 188, 190
   nodal analysis, 190
   real power, 288
   sine form, 184
   voltage divider, 191
physical current, 8, 12, 33
   loop current, 86
physical inductor, model, 122
physical voltage, 14
pole, 228
   and zeros, 247
   filter, 248
   switches, 25
potential, 12, 84, 228
power, 19, 25, 63
   ac circuits, 275
   capacitance, 282
   dB, 243
   delta connections, 341
   electronics, 68
   generation, 332
   impedance, 283
   inductance, 279
   maximum, 73
   output impedance, 73
   real, 284
   reference direction, 19
   resistance, 275, 279
   sign, 19
   superposition, 63
   time average, 284
   transmission systems, 356
   wye connection, 344
power factor, 284
   measured, 293
   motor, 369
power factor correction, 285, 295
   transmission system, 357
power supplies, 302
power system, 301
   transformers, 302, 308

power triangle, 293
proton, 5
pulse circuits, 142
   Thévenin equivalent circuit, 144

R

radio, 251, 298
   filter, 200
reactance, 194
reactive part of impedance, 194
reactive power, 285
   electronics, 297
   frequency domain, 288
   importance, 295
   transmission systems, 356
real numbers, 170
real power, 284
   phasors, 288
reference directions, 8
   supplying, 40
   voltage, 14
reference node, 76
   choosing, 78
resistance, resistor, 23, 60
   energy flow, 276
   equivalent, 30
   generalized impedance, 222
   impedance, 188
   mechanical analog, 128
   power rating, 275
   series, 33
resistive part of impedance, 194
resonant charging, 235
RC circuit, parallel, 198
   series, 199
RL circuit, series, 192
   parallel, 195
RLC circuit, 147, 232
   parallel, 202
   poles, 232
   series, 200
   transients, 234
root-mean-square, rms, 277, 287
   sinusoid, 277
rotor, 359
rpm, 167
run-up time, 363

S

s-plane, 219
safety, 84, 314, 332
sampler, 254
satellite, 167
secondary of transformer, 302
sensor, 257
series connection, 32, 33
   capacitors, 127
   inductors, 122
   resistances, 33
series resonance, 200
service factor, 365
short circuit, 25, 59
   power, 25
   zero resistance, 59
siemens, 199

sign convention
   current, 10
   power, 19
   voltage, 15
single-phase induction motor, 370
sinusoid, 164
   average value, 274
   rms value, 277
   sinusoidal steady-state, 181, 247
skin resistance, 317
source set, 19
source transformation, 74, 82
spectra, 238
   transforms, 238
speed, torque, 366
split-phase induction motor, 370
stator, 359
steady state
   ac, 129
   dc, 168
   motor and load, 360
subscript, 14
   voltages, 19
summer, 252
supernode, 10, 81
superposition, 58, 67, 157
   limitations, 63
   nonlinear, 63
   principle, 59
susceptance, 199
switch, 23
   ideal, 25
system, 129, 216
   feedback, 254
   linear, 169, 252
   notation, 252

T

terminal, 334
termination, impedance, 255
Tesla, Nicola, 164
thermostat, 302
Thévenin, 63, 111
Thévenin equivalent circuit, 66, 74, 136, 297
   battery-switch, 144
three phase, 332
   advantages, 333
   current, 337
   frequency domain, 333

induction motor, 364
loads, 339
transformers, 350
voltage, 337
throws, switches, 25
time average, 273
   power, 276, 284
time constant, 130, 132, 146, 168, 217, 224, 257
   RC circuit, 135
   RL circuit, 130
time domain, 182, 239
   complex frequency, 220
   summary, 203
torque, 360
   motor, 366
   speed, 366
transfer function, 229
   RLC circuit, 232
transformer, 301
   apparent power, 304
   circuit analysis, 306
   connections, 352
   conservation of power, 304
   dc, 304
   ideal, 302
   impedance, 304
   power distribution, 308
   primary and secondary, 302
   three phase, 350
transforms, spectra, 238
transient, 128, 130
   ac, 181
   first order, 128, 132
   higher order, 144, 232
   response, 217, 222
   RLC, 234
transmission systems, 349
   per-phase circuit, 353
   power-factor correction, 357
   voltages, 350
turned-OFF sources, 59, 67
turns ratio, 305

U

unbalanced three-phase systems, 349
undamped response, 232
underdamped response, 233
units
   ac power, 294

ampere, 7
coulomb, 5
farad, 123
henry, 121
hertz, 165
mho, 25
ohm, 23
siemens, 25
voltage, 12

V

VAR, 286
VAR, 295
velocity, 8
Volt-Ampere (VA), 292, 295
Volt-Ampere Reactive (VAR), 286
Volta, Count Alessandro, 12
voltage, 12
   active, 20
   line-to-neutral, 343
   mechanical analog, 13
   reactive, 20
   reference direction, 14
   sign conventions, 15
   subscripts, 18
voltage divider, 33, 38, 61, 229
voltage rating, 367
voltage sources, 26
   turned OFF, 59
voltmeter, 72, 84, 111, 278

W

watts, 295
wire loss, 346, 354
wires, 24
work, 6
wye, three-phase transformers, 350
wye connection, 336
   unbalanced, 349
   power, 344
wye delta, 352
wye-wye connection, 347, 352

Z

zero, 228